DICTIONARY OF ROCKS

DICTIONARY OF ROCKS

Richard Scott Mitchell

VNR VAN NOSTRAND REINHOLD COMPANY

New York

Copyright © 1985 by **Van Nostrand Reinhold Company Inc.**
Library of Congress Catalog Card Number: 84-22062
ISBN: 0-442-26328-7

Manufactured in the United States of America.

Published by Van Nostrand Reinhold Company Inc.
135 West 50th Street
New York, New York 10020

Van Nostrand Reinhold Company Limited
Molly Millars Lane
Wokingham, Berkshire RG11 2PY, England

Van Nostrand Reinhold
480 Latrobe Street
Melbourne, Victoria 3000, Australia

Macmillan of Canada
Division of Gage Publishing Limited
164 Commander Boulevard
Agincourt, Ontario MIS 3C7, Canada

15   14   13   12   11   10   9   8   7   6   5   4   3   2   1

Library of Congress Cataloging in Publication Data
Mitchell, Richard Scott, 1929–
  Dictionary of rocks.

  Bibliography: p.
  1. Petrology—Dictionaries.   I. Title.
QE423.M58   1985   552′.003′21   84-22062
ISBN 0-442-26328-7

In remembrance of the former
Mineralogical Laboratory
University of Michigan
and
my mentors there:
E. H. Kraus
W. F. Hunt
L. S. Ramsdell
C. B. Slawson
E. W. Heinrich

The Earth from her deep foundations unites with the celestial orbs
that roll through boundless space, to declare the glory and show forth
the praise of their common Author and Preserver; and the voice of
Natural Religion accords harmoniously with the testimonies of
Revelation, in ascribing the origin of the Universe to the will of One
eternal, and dominant Intelligence, the Almighty Lord.—William
Buckland, *Bridgewater Treatise VI, 1836.*

CONTENTS

Preface   ix

Introduction   1

    How Rocks and Minerals Differ   /   1
    Major Rock Classes   /   2
    Rock Nomenclature   /   3
    Derivation of Rock Names   /   4
    Bibliography   /   6

Abbreviations and Symbols   9

Sample Entry   10

Definitions   11

Glossary   219

PREFACE

This is the first dictionary in the English language devoted exclusively to the names of rocks. Although the terms included here are scattered throughout earlier dictionaries, glossaries, lists of nomenclature, and indexes, this work has focused on bringing together these terms to form a truly comprehensive vocabulary of rock names. In addition to a concise definition for each term, an attempt also has been made to indicate the derivation of the name, to credit the person who first used or defined it, and to give the date of its introduction. In compiling this book, the author has had a broad spectrum of readers in mind. He has written it for lay persons, for amateur rock collectors, and for students and professionals in petrology. Because some petrologic terms unfamiliar to the layman are used to define the rocks, a glossary is included. Recent igneous rock definitions published by the International Union of Geological Sciences (IUGS) are included, especially for the professionals in the science. The presence of a name in this book does not necessarily give approval to the name; many obscure, local, and obsolete terms are included solely for their informative and historical value.

When compiling the vocabulary of this work, the initial emphasis was on distinct igneous, sedimentary, and metamorphic rocks. However, it became clear that other materials should likewise be considered. Migmatites, tektites, impactites, and major meteorite types are included. The vocabulary also comprises natural organic resins, ambers, bitumens, and major coal varieties. Also included are cave rocks and formations. A relatively large number of gem-rock terms has been incorporated in this compilation. Because some of these are single-mineral aggregates, they traditionally have been considered mineral varieties but are included here as rocks. Numerous local and trade names for gem and decorative rocks are defined in this study, including various terms applied to jasper, agate, and other forms of quartz, as well as opal. Non-English terms commonly used in English-language publications are included.

Mineral names are not defined in this compilation, and minerals are not included except in those instances where a rock is composed of a single mineral and has not been given a separate rock name. Structural and textural terms commonly used in petrology and microscopic petrography are not included in the vocabulary of this book, but many of them appear in the glossary. Likewise, coal constituents (macerals) and other terms related to coal petrography are not emphasized. No native crystalline organic compounds are considered here, since they relate specifically to mineralogical nomenclature. Although some terms related to soil and its classification are included, there has been no attempt to cover the nomenclature of soil science. With only a few exceptions, terms related to rocks formed in ore deposits are covered only if the material is used as a gem or decorative stone.

Rock names consisting of a single word were easily placed in the alphabetical arrangement of this dictionary. On the other hand, many rock names are comprised of two or more words, usually a root name modified by descriptive adjectives. Generally the rock is listed under the root name. For example, terms like carbonaceous limestone, clastic limestone, lithographic limestone, and pellet limestone will be found under limestone (the root name). Furthermore, some common terms are not considered root names here, for example, the terms rock and stone. Examples such as howlite rock, prehnite rock, dumortierite rock, and rhodonite rock, or Eilat stone, Petoskey stone, and stomach stone are listed alphabetically according to the mineral or other modifying term and not under the terms rock or stone.

An effort has been made to determine the position of each rock within the threefold classification scheme of igneous, sedimentary, and metamorphic, and for igneous rocks additional data regarding the geological occurrence of the rock are usually given. Rocks not clearly fitting into the common classification scheme—for example, migmatites, tektites, impactites, and meteorites—or that have uncertainties about their origin are indicated.

As mentioned earlier, a serious attempt was made to determine the origin of each rock name. In those instances where the origin is not certain, but where a strong likelihood for the origin exists, I have prefaced the assumption by using the word *possibly*. An effort has also been made to determine the person or persons who first used a specific name and the date the name was introduced. In some instances the name was given by one person and first published by another.

Although rock names derived from personal names are relatively few, an effort has been made to include brief biographical data for each person when such is the case.

Earlier dictionaries that focused on petrology, including a combination of rock names and related petrologic terms, are *Petrographisches Lexikon* by F. Loewinson-Lessing (1893) and *The Nomenclature of Petrology* by Arthur Holmes (1920). These were especially useful in selecting and defining some of the older terms considered in the present work. The four-volume work, *A Descriptive Petrography of the Igneous Rocks* by Albert Johannsen (1931-1938), and *Spezielle Petrographie der Eruptivgesteine (mit 1. Nachtrag)* by W. E. Tröger (1969) have been of special help in treating igneous rocks. The *Glossary of Geology*, second edition, edited by Robert L. Bates and Julia A. Jackson (1980), has been especially helpful with newer definitions of sedimentary and other rock types. *Microscopic Petrography* by E. William Heinrich (1956) is a useful guide for defining metamorphic and numerous other rocks. Many other publications, especially those listed in the bibliography at the end of the Introduction, have also been used.

I wish to acknowledge the late John Reese Henley for his generous assistance in various aspects of this project, especially for his help in copying rock names from various indexes and lists during the initial compilation of the rock vocabulary. Special appreciation is extended to Kenneth W. Lawless, who prepared all the black-and-white and a majority of the color photographs used in this book. The specimens illustrated are in the Lewis Brooks Collection at the University of Virginia. Other color photographs, where credit is specifically acknowledged, were supplied by the Carolina Biological Supply Company of Burlington, North Carolina. To those whose efforts made the use of these illustrations possible, I wish to express my sincere thanks. I am also very grateful to the numerous persons who replied to my various inquiries about rock definitions and name derivations. Finally, I wish to express my gratitude to the many persons who have preceeded me in the compiling of glossaries, dictionaries, and lists of rock terms, for without their earlier efforts a compilation of this magnitude would be virtually impossible.

RICHARD S. MITCHELL

INTRODUCTION

How Rocks and Minerals Differ

For those not versed in the earth sciences there is usually some confusion regarding the difference between a rock and a mineral. Because this work is principally concerned with rocks rather than minerals, their differences are considered here.

A mineral is a substance formed in nature without the direct influence of organisms, it has a characteristic chemical formula and usually an ordered atomic arrangement (crystal structure), which is often expressed in external geometrical forms (crystals). Thus, each mineral is a unique chemical substance and is not a mixture of two or more substances. Mineral-like compounds manufactured by man or made in nature by organisms are not considered to be minerals even if they have compositions that can be expressed by chemical formulas and have crystal structures, for example, synthetic diamond, synthetic spinel, and synthetic sapphire; nor are seashells, bones, teeth, gallstones, pearls. Some of these materials, especially those formed by organisms, are called mineraloids.

In general terms, a mineraloid is a naturally occurring substance with distinctive properties that does not fully meet the requirements for a mineral. Many are formed directly by organisms or have compositions that cannot be expressed by a single formula, and many are amorphous, that is, they lack a crystal structure. Examples include shells, bones, amber, coals, bitumens, natural glass (obsidian, tektite), and others. Substances that fulfill all the requirements for a mineral except for having a crystal structure, for example, some varieties of opal or liquids like water or native mercury, are sometimes included with mineraloids, but in reality most opals and other so-called amorphous minerals exhibit ordered atomic arrangements either upon close study (X-ray, electron diffraction) or in controlled environments (lowered temperatures for water and mercury).

A rock is defined as a naturally formed aggregate composed of one or more minerals or mineraloids or both, including loose incoherent masses as well as firm solid masses, and constituting an essential or appreciable part of the Earth or other celestial body. To be specific, rocks are formed by natural processes, and thus man-made materials such as concrete, plaster, terra cotta, and various ceramics are excluded. Rocks are also aggregates of discrete particles (or, in natural glass or resins, of randomly aggregated ions or molecules) between which there is no crystallographic continuity. The constituent particles may consist of one mineral species or one mineraloid, or of a mixture of two or more minerals, mineraloids, or both. Rocks containing one essential mineral are often termed monomineralic rocks, for example, dolomite marble, dunite, and anorthosite. Unconsolidated aggregates are also usually included in the term rock, for example, beach and dune sand, earths, till, and others. Most rocks, however, are solid, well-indurated materials. Some petrologists suggest that, to qualify as a rock, the mass should constitute a sizable body of material. The minimum size has not been defined. Some have suggested it should be large enough to be mappable on a geologic map, but in practice much smaller bodies are referred to as rocks, for example, concretions, nodules, spherulites, geodes, and the like, which may be only an inch or less across. Finally, although the rocks most familar to man are from the Earth's crust, they also include aggregates from the mantle, from the Moon, and from outer space (e.g., meteorites).

The components of a specific rock usually can be classified into three types depending upon their relative importance to that rock. The essential minerals (and mineraloids) are those whose presence is necessary before the root name of the rock can be assigned. For example, granite must have quartz and potassic feldspar, syenite must have potassic feldspar, limestone must have calcite, sandstone must have quartz, and so on. Characterizing accessory minerals (and mineraloids) are those that do not determine the root name but are important enough to modify the name, for example, biotite granite, hornblende syenite, bituminous limestone, siliceous (quartz) dolomite, glauconitic sandstone. Other minerals, often only observed during studies under the microscope, are usually present in small amounts; these generally have no bearing on the rock name and are termed minor accessory minerals. For example, most granites have magnetite, zircon, apatite, ilmenite, titanite, and many others.

Major Rock Classes

Nearly all rocks can be classified according to their origin into three major groups. These divisions include igneous rocks, sedimentary rocks, and metamorphic rocks. A few rocks are hybrid in nature or of other or uncertain modes of origin and will be mentioned subsequently.

Igneous [L. *ignis,* fire] rocks are those that have formed from the solidification (usually crystallization) of a magma, a high-temperature, silicate-rich, fluid rock material. Magma that appears at the surface is known as lava, but large volumes of magma also crystallize at depth to form a wide variety of coarse to fine-grained igneous rocks. Many of these rocks are now exposed because erosion has removed the overlying rocks that originally covered them. Apparently some rocks that resemble igneous rocks in almost every detail have not been formed from magmas but by a replacement process (called metasomatism) whereby earlier solid rocks are changed into igneous-like rocks by the systematic migration and reconstitution of chemical components essentially in a solid environment. Granitization, where nongranite rocks are converted to rocks of granitic composition and texture by this process, is a concept accepted by many petrologists. Rocks formed by these metasomatic replacement processes (related to metamorphism) are often classified with igneous rocks.

Sedimentary [L. *sedimentum,* settling] rocks consist of chemical and mineralogical materials derived from the destruction of older rocks (and more rarely, organisms). Usually these materials are transported from their sources and deposited by the action of water, wind, or glacial ice. In composition they can consist of one or more of the following: particles of primary minerals from older rocks, particles of undissolved decomposition products of older rocks (e.g., clays derived from feldspar), fragments of organically derived materials, all originating outside the area of deposition; and of minerals and mineraloids originating by chemical precipitation within the area of deposition. Residual materials, formed in situ by advanced decomposition and not having undergone appreciable transportation, for example, saprolite and coals, are also classified as sedimentary. It is obvious, therefore, that a large variety of rocks is included under sedimentary. Furthermore, unconsolidated aggregates like sand, gravel, and earths are also included here.

Metamorphic [Gk. *meta,* after *morphe,*

form] rocks are derived from pre-existing rocks by changes in texture or mineralogy in an essentially solid environment, usually through processes involving increased temperatures, increased pressures, or chemically active fluids. Because these changes occur below the near-surface zones of weathering and cementation, sedimentary rocks such as saprolite and soils or those rocks in which diagenetic changes are important are not considered to be metamorphic. The parent materials (the pre-existing rocks) of metamorphic rocks can be igneous, sedimentary, and earlier metamorphic rocks and therefore involve a very wide variety of mineralogical and chemical compositions. Environments in which metamorphic changes occur vary widely and may be a result of the following: contact with high-temperature igneous magmas; invasion of old rocks by chemically active fluids derived from high-temperature igneous magmas or from fluids released from old rocks when they are deeply buried or heated by magmas; increased temperatures and pressures (hydrostatic) caused by deep burial; or pressures (stress) caused by the buckling or folding of crustal rocks during mountain-building processes. Some metamorphic rocks are formed by extreme milling and pulverization along fault zones (e.g., mylonite). Although many of the well-known metamorphic rocks have foliated structures (parallel structures or arrangements of components) like gneiss, schist, phyllite, and slate, many others are granoblastic (equigranular grains oriented at random).

Although most rocks can clearly be classified into the three major categories outlined here, some are not so easily placed. Migmatites, although often classed as metamorphic, have a hybrid or mixed origin. They consist of two different parts: the older country rock, often more or less metamorphosed, and material of magmatic (igneous) origin that has been injected as dikes and/or sills. Another hybrid rock is tuffite, composed of igneous pyroclasts mixed with ordinary sediments such as detrital quartz, calcite, and clay minerals.

Rocks as well as minerals and ore deposits, usually named hydrothermal, likewise are a bit difficult to classify. Some of these, especially hydrothermal ore deposits, are believed to owe their origin to residual fluids derived from igneous magmas; others were formed from water and associated components derived from older rocks and minerals through remobilization processes related to metamorphism.

Although many meteorites are clearly related to typical igneous rocks, some of them show significant differences. Tektites, whose origin is still uncertain, do not fit into the common threefold classification. Once believed to be of extraterrestrial origin, like meteorites, a current alternative view is that they are the product of large hypervelocity meteorite impacts on terrestrial rocks. These and other impactite glasses therefore have a very specialized origin, which approaches metamorphic.

Rock Nomenclature

As the science of rocks (petrology) developed, many rock names were introduced into the scientific literature. Often a different name was assigned to every rock that showed a slight difference from any known rock. Consequently, the nomenclature of petrology is vast and contains many terms that, in light of our present knowledge of rocks, are unnecessary. Many of the names were assigned to varieties of common or universal rock types. Modern petrologists are in favor of defining a relatively small number of rock types, assigning to these root names, and then expressing variations within these types by the use of appropriate modifiers. For example, root names like syenite, shale, sandstone, slate, and schist can be modified to describe rocks like corundum syenite, phosphatic shale, feldspathic sandstone, chiastolite slate, and magnetite-chlorite schist.

Over the years different attempts have been made to reclassify, redefine, and simplify rock nomenclature. The International Union of Geological Sciences in recent years has established a Commission on Petrology to study rock classification and rock nomenclature.

Results of their deliberations on igneous rocks are included in this book for those rocks that have been defined.

Derivation of Rock Names

From ancient times rocks and minerals have been assigned specific names. The work *On Stones* by Theophrastus contains names such as achates (agate), asphaltos, Lydian stone, marmaros (marble), ochra (ocher), and sardion (sard). Pliny the Elder, in *Natural History,* used the names obsidianus, pumice, silex, syenites, tephrias, and many more. These terms, as well as others from different cultures and time periods have accumulated over the centuries to form our present rich vocabulary of rock names.

The ending *-ite* has been added to rock (and mineral) names since antiquity. Generally this suffix signified a quality, constituent, use, or locality of the stone. Because it appeared desirable to James D. Dana (*System of Mineralogy,* 5th ed., New York, 1868, p. *XXXIV*) that names of rocks should have some difference from names of minerals, he suggested using the suffix *-yte* for rocks. Thus names like diorite, tonalite, and rhyolite would be dioryte, tonalyte, and rhyolyte, but this innovation was never accepted. Other important endings for rock names are *-lite, -lith,* and *-lithite* [Gk. *lithos,* stone]. A casual glance at the names in this dictionary will show other common endings such as *-oid* [Gk. *-oeides,* shape], *-phyre* [Gk. *porphyra,* porphyry], *-ide* [L. having the quality of], and *-fels* [G. *Fels,* rock].

Most names without these endings were not invented by petrologists but have come into the nomenclature from various languages through common usage, for example, earth, limestone, mud, sandstone, shale, and slate.

The derivations of rock names are manifold. The greatest number of names (over 30%) were derived from geographical localities. This is especially true for igneous rocks, but it applies to other rock types as well. Names of oceans, countries, states, counties, provinces, territories, cities, towns, mountains, valleys, islands, rivers, lakes have been included in rock names. A few examples include andesite (Andes Mountains), atlantite (Atlantic ocean), corsite (Corsica), dunite (Dun Mountain, New Zealand), helsinkite (Helsinki, Finland), italite (Italy), missourite (Missouri River), monzonite (Monzoni, Tyrolean Alps), nelsonite (Nelson County, Virginia), nevadite, norite (Norway), pacificite, tinguaite (Tinguá Mountains, Brazil), vesuvite (Mt. Vesuvius, Italy), wyomingite, and yosemitite (Yosemite, California).

Names derived from Greek are second in abundance (over 15%). Generally these terms relate to a mode of formation or to a property of the rock. For example, mylonite [Gk. *mule,* mill, alluding to the milling of the rock]; pelagite [Gk. *pelagos,* sea]; perknite [Gk. *perknos,* dark]; phyllite [Gk. *phullon,* leaf]; schist [Gk. *schistos,* cleft]; scoria [Gk. *skoria,* that which is thrown off, dross]; trachyte [Gk. *trachus,* rough].

Third in abundance (over 11%) are those rock names derived from older rock names by adding appropriate prefixes or suffixes or by modifying the name in some other way. Therefore, to cite a few, there are apogranite [Gk. *apo-,* from], calcibreccia [*calci-,* calcium], kaligranite [L., *kalium,* potassium], leucogabbro [Gk. *leukos,* white, light color], melagabbro [Gk. *melas,* black], metabentonite [Gk. *meta-,* with, after], and paragneiss [Gk. *para-,* beside]. Rock names have also been altered so they can be applied to different rocks: from syenite are syenitite, syenitoid, syenoid, syenodiorite, syenogabbro. Many other examples can be cited, especially in compound rock names like granogabbro (granite-gabbro), monzogabbro (monzonite-gabbro), rhyodacite (rhyolite-dacite), and trachyandesite (trachyte-andesite).

Many of the rocks included in this compilation (nearly 10%) have names common in English. Very familiar are names such as brownstone, graystone, greenstone, limestone, mill rock, mud stone, oil rock, puddingstone, rottenstone, sandstone, soapstone, and tile stone. There are also terms like pastelite, vistaite, and liverite. The origins of many English words can be traced to Old English: chalk, clay, coal, cobble, loam, ooze, quernstone, shale, tar, and whetstone; or to Middle English: peat, rammell, silt, slate. Rarely, some of the terms have come from Scottish (blaes), as well as Cornish (gossan, growan) and other Celtic dialects (elvanite).

Some rock names (8%) have been formed from standard mineral names. Terms like anhydrite rock, garnet rock, gypsum rock, halite rock, howlite rock, olivine rock, prehnite rock, and sulfur rock are commonly used. In other instances the mineral's name becomes a rock name when modified by an appropriate suffix: biotitite, dolomitite, magnetitite, melilitite, oligoclasite, quartzite, spinellite; halilith, hauynolith, nephelinolith; dololithite and others.

Names derived from Latin are sixth in abundance (less than 7%). Usually these terms are derived from the appearance or from a mode of formation of the rock. Some examples include concretion [L. *concretus,* grown together], laterite [L. *later,* brick], nodule [L. *nodus,* knob, knot], novaculite [L. *novacula,* razor], orbiculite [L. *orbis,* circle], pinolite [L. *pinus,* pine cone], septarium [L. *septa,* partitions], and variolite [L. *varius,* spotted]. Although rare, and not preferred, there are some names derived from Latin and Greek combinations: sideroferrite [Gk. *sideros,* iron + L. *ferrum,* iron].

Although the origins of the names considered so far account for nearly three-fourths of the rocks cataloged in this book, there are many additional origins of names. None of the remaining derivations accounts for as much as 5% of the vocabulary, but many of the names are important in petrology and have interesting sources. They are considered here in their order of decreasing numbers.

In contrast with the modern practice in mineralogy to name minerals for persons, rocks named for individuals are relatively few (fewer than 100). Personal names have been given to igneous, metamorphic, and sedimentary rocks, but many have been applied to organic substances initially classified with minerals. A few examples of rocks named for persons include: beekite (Henry Beeke), buchite (Christian L. von Buch), catlinite (George Catlin), charnockite (Job Charnock), dolo-

mite (Déodat G. S. T. G. de Dolomieu), muckite (Fritz Muck), obsidian (Obsidius, a Roman consul), pallasite (Peter S. Pallas), pencatite (Giuseppe Marzari-Pencati), saussurite (Horace B. de Saussure), and ulrichite (George H. F. Ulrich).

A considerable number of rock names used in English have come from German: gneiss, greisen, kieselguhr, knotenschiefer, loess, meerschaum, schalstein, sinter, wacke.

Chemistry has influenced rock names: acidite, basite, ferrolite, manganolite, oxidite, phosphorite, radiumite, silicalite.

Use also has been made of portmanteau words, where a new term is created by combining parts of existing words: alkremite [aluminum + Russ. *kremei,* silicon], apaneite [apatite + nepheline], auganite [augite + andesite], coralgal [coral + algae], criquina [crinoid + coquina], dismicrite [disturbed microcrystalline], grospydite [grossular + pyroxene + disthene (kyanite)], kamafugite [katungite + mafurite + ugandite], marundite [margarite + corundum], rhyoandesite [rhyolite + andesite], sporbo [smooth + polished + round + black, blue, brown + object], and usamerite [United States of America].

In addition to terms derived from Greek, English, Latin, and German, many other languages have contributed to petrologic nomenclature. From Italian are breccia, lava, travertine, tufa, to cite only a few. From French are griotte, minette, névé, peridotite, and vase. Spanish terms are caliche, coquina, jade, ricolite, and sillar. From Swedish are gyttja, trap, and varve. Names have been introduced from Afrikaans, Arabic, Cariban, Chinese, Dutch, Finnish, Hindi, Icelandic, Indonesian, Japanese, Polish, Portuguese, and Russian. Languages and dialects of the following places and peoples also are represented: Brazil, Ecuador, Hawaii, Madagascar, Mexico, Peru; Armoricans, Australian aborigines, Aztecs, Bantus, Mayas, Mongols, New Zealand Maoris, North American Indians, Tartars.

Although uncommon, some rock names have been derived from biological terms: algarite (alga), diatomite (diatom), encrinite (genus *Encrinus*), foraminite (foraminiferan), radiolarite (radiolarian), spongolite (sponge), as well as terms like Bayard egg, chrysanthemum stone, desert rose, poppy jasper, rosette rock. Names referring to crystallographic features have been applied to some meteorite types: hexahedrite (for cube), octahedrite. Names derived from the geologic formation in which the rock was initially found are: bentonite (Upper Cretaceous, Benton shale), cherokite (Mississippian, Cherokee limestone), and taconite (obsolete Taconic system of Emmons, 1842). Although not formation names, some terms were introduced to indicate the age of the rocks: propylite (Gk. *propylon,* before the gate) alluding to its existence at the beginning of the Tertiary. Other names have their roots in mythology: vulcanite [Vulcan, Roman god of fire], plutonite (Pluto and the lower world), and Pele's hair (Hawaiian fire goddess). Finally, several rocks have been named in honor of tribal peoples: bojite (Boii, Celtic tribe in Germany), gondite (Gonds, central India), onkilonite (Onkilones, Siberian tribe), ossipite (Ossipee Indians, New Hampshire), shoshonite (Shoshone Indians, Wyoming), suevite (Suevians of medieval Schwaben), taurite (Tauroi, ancient peoples, southern Crimea), vulsinite (Vulsinii, Etruscan tribe, Italy).

Bibliography

American Federation of Mineralogical Societies. 1978. *Approved Reference List of Lapidary Material Names.* 4th ed.

Bailey, Dorothy, and Kenneth C. Bailey. 1929. *An Etymological Dictionary of Chemistry and Mineralogy.* London: E. Arnold.

Bates, Robert L., and Julia A. Jackson, eds. 1980. *Glossary of Geology.* 2nd ed. Falls Church, Va.: American Geological Institute.

Carozzi, Albert V. 1960. *Microscopic Sedimentary Petrography.* New York: John Wiley & Sons.

Cotta, Bernhard von. 1862. *Die Gesteinslehre.* Freiberg: J. G. Engelhardt.

English, George L. 1939. *Descriptive List of the New Minerals, 1892-1938.* New York: McGraw-Hill.

Fay, Albert H. 1920. Glossary of the mining and mineral industry. *U.S. Bureau of Mines Bulletin 95.*

Folk, R. L. 1959. Practical petrographic classification of limestones. *Am. Assoc. Petroleum Geologists Bull.* **43:**1-38.

Folk, R. L. 1980. *Petrology of Sedimentary Rocks.* Austin, Texas: Hemphill Publishing.

Grabau, Amadeus W. 1904. On the classification of sedimentary rocks. *Am. Geologist* **33:**228-247.

Grout, Frank F., ed. 1927. *Kemp's Handbook of Rocks.* 6th ed. New York: D. Van Nostrand.

Grout, Frank F. 1932. *Petrography and Petrology.* New York: McGraw-Hill.

Gümbel, Karl W. 1886. *Grundzüge der Geologie.* Kassel: T. Fischer.

Harker, Alfred. 1939. *Metamorphism.* 2nd ed. New York: E. P. Dutton.

Harker, Alfred. 1919. *Petrology for Students.* 5th ed. Cambridge: University Press.

Hatch, Frederick H. 1949-1950. *Textbook of Petrology.* 2 vols. London: Murby.

Heinrich, E. William. 1966. *The Geology of Carbonatites.* Chicago: Rand McNally.

Heinrich, E. William. 1956. *Microscopic Petrography.* New York: McGraw-Hill.

Hill, Carol A. 1976. *Cave Minerals.* Huntsville, Ala.: National Speleological Society.

Holmes, Arthur. 1920. *The Nomenclature of Petrology.* London: Murby.

Humble, William. 1860. *Dictionary of Geology and Mineralogy.* 3rd ed. London: R. Griffon.

Johannsen, Albert. 1931-1938. *A Descriptive Petrography of the Igneous Rocks,* 4 vols. Chicago: University of Chicago Press.

Kalkowsky, Ernst. 1886. *Elemente der Lithologie.* Heidelberg: C. Winter.

Kemp, James F. 1908. *A Handbook of Rocks; with Glossary of the Names of Rocks.* 4th ed. New York: D. Van Nostrand.

Kinahan, George H. 1873. *A Handy Book of Rock Names.* London: R. Hardwicke.

Lasaulx, Arnold C. P. F. von. 1875. *Elemente der Petrographie.* Bonn: E. Strauss.

Lawrence, Philip H. 1878. *Rocks Classified and Described.* (Trans. of work by B. von Cotta.) London: Longmans, Green.

Leonhard, Karl C. von. 1823-1824. *Charakteristik der Felsarten.* 3 vols. Heidelberg: J. Engelmann.

Loewinson-Lessing, Franz J. 1893. *Petrographisches Lexikon.* Jurjew: C. Mattiesen.

Loewinson-Lessing, Franz J. 1898. *Petrographisches Lexikon, Supplement.* Jurjew: C. Mattiesen.

Mason, Brian. 1962. *Meteorites.* New York: John Wiley & Sons.

Mehnert, Karl R. 1968. *Migmatites and the Origin of Granitic Rocks.* Amsterdam: Elsevier.

Naumann, Karl F. 1850-1854. *Lehrbuch der Geognosie.* 2 vols. Leipzig: W. Engelmann.

Oldham, Thomas. 1879. *Geological Glossary.* London: E. Stanford.

Page, David. 1865. *Handbook of Geological Terms, Geology and Physical Geography.* 2nd ed. London: W. Blackwood & Sons.

Pettijohn, Francis J. 1975. *Sedimentary Rocks.* 3rd ed. New York: Harper & Row.

Rice, Clara M. 1955. *Dictionary of Geological Terms.* Ann Arbor: Edward Brothers.

Roberts, George. 1839. *An Etymological and Explanatory Dictionary of the Terms and Language of Geology.* London: Longman, Orme, Brown, Green, & Longmans.

Rosenbusch, Harry. 1887. *Mikroskopische Physiographie der Mineralien und Gesteine.* 2 vols. Stuttgart: E. Schweizerbart.

Rosenbusch, Harry. 1910. *Elemente der Gesteinslehre.* 3rd ed. Stuttgart: E. Schweizerbart.

Schaller, Waldemar T. 1920. Gems and precious stones: gem names. *U.S. Geological Survey: Mineral Resources of the United States for 1917* 145-164.

Schmid, R. 1981. Descriptive nomenclature and classification of pyroclastic deposits and fragments: Recommendations of the IUGS Subcommission on the Systematics of Igneous Rocks. *Geology* **9**:41-43.

Senft, Ferdinand. 1857. *Classification und Beschreibung der Felsarten.* Breslau: W. G. Korn.

Shipley, Robert M. 1974. *Dictionary of Gems and Gemology.* 6th ed. Santa Monica, Calif.: Gemological Institute of America.

Streckeisen, Albert L. 1967. Classification and Nomenclature of Igneous Rocks (Final Report of an Inquiry). *Neues Jahrb. Mineralogie Abh.* **107**:144-240.

Tarr, William A. 1936. *Terminology of the Chemical Siliceous Sediments.* Washington D. C.: National Research Council, Division of Geology and Geography.

Teall, Jethro J. H. 1888. *British Petrography.* London: Dulau.

Tomkeieff, Sergei I. 1954. *Coals and Bitumens and Related Fossil Carbonaceous Substances.* London: Pergamon Press.

Tröger, W. Ehrenreich. 1969. *Spezielle Petrographie der Eruptivgesteine; mit 1. Nachtrag.* Bonn: Deutschen Mineralogischen Gesellschaft.

Tucker, Maurice E. 1981. *Sedimentary Petrology: An Introduction.* New York: John Wiley & Sons (Halsted Press).

Wadsworth, Marshman E. 1893. A sketch of the geology of the iron, gold, and copper districts of Michigan. *Michigan Geological Survey, Report of the State Geologist for 1891-1892* 75-174.

Williams, Howel, Francis J. Turner, and Charles M. Gilbert. 1982. *Petrography.* 2nd ed. San Francisco: W. H. Freeman.

Zirkel, Ferdinand. 1893-1894. *Lehrbuch der Petrographie.* 2nd ed. 3 vols. Leipzig: W. Engelmann.

ABBREVIATIONS AND SYMBOLS

Abbreviations of Languages Used in Name Derivations

Afrik.	Afrikaans
Am. Ind.	American Indian
Arab.	Arabic
Celt.	Celtic
Corn.	Cornish
Dan.	Danish
Du.	Dutch
E. Pruss.	East Prussian
Ecuad.	Ecuadoran
Egypt.	Egyptian
Eng.	English
Finn.	Finnish
F.	French
G.	German
Gk.	Greek
Haw.	Hawaiian
Heb.	Hebrew
Icel.	Icelandic
It.	Italian
Jap.	Japanese
L.	Latin
L.L.	Late Latin
M.L.	Medieval Latin
M.Eng.	Middle English
M.H.G.	Middle High German
Mong.	Mongolian
Nor.	Norwegian
O. Eng.	Old English
O.F.	Old French
O.G.	Old German
O.H.G.	Old High German
O.N.	Old Norse
O. Sp.	Old Spanish
O. Sw.	Old Swedish
Peru.	Peruvian
Pol.	Polish
Port.	Portuguese
Russ.	Russian
Scot.	Scottish
Sp.	Spanish
Sp. Am.	Spanish American
Sw.	Swedish
Turk.	Turkish
dial.	dialect
dim.	diminutive

Scientific Labels

aschist.	aschistic
autometa.	autometamorphism
diaschist.	diaschistic
extru.	extrusive
hydrotherm.	hydrothermal
hypab.	hypabyssal
ig.	igneous rock
impact.	impactite
meta.	metamorphic rock
meteor.	meteorite
migmat.	migmatite
pluton.	plutonic
sed.	sedimentary rock
tekt.	tektite

The terms listed above are defined either in the main rock section or in the glossary.

Mineralogical Symbols

The symbols are those used in the International Union of Geological Sciences (*IUGS*) classification of igneous rocks. The numbers used in the definitions throughout this book refer to modal mineral percentages by volume.

Light-colored constituents:

Q	silica minerals (mainly quartz)
A	alkali feldspars, including albite An_{00-05}
P, pl	plagioclase An_{05-100}
F	feldspathoid
P/(A + P)	feldspar ratio (f.r.); plagioclase/ total feldspar

Dark-colored constituents:

M	mafic minerals
ol	olivine
opx	orthopyroxene
cpx	clinopyroxene
px	pyroxene
hbl	hornblende

SAMPLE ENTRY

Rock name —— **norite** *ig. pluton.* 1. Coarse-grained rock in which calcic-plagioclase (usually labradorite) and orthopyroxene (usually hypersthene) are the essential constituents; differs from gabbro by the presence of the orthopyroxene rather than clinopyroxene (augite). [Norge (Norway), original locality] (Esmark in 1823) 2. *IUGS*, plutonic rock satisfying the definition of gabbro, in which pl/(pl + px + ol) is 10-90 and the opx/(opx + cpx) is greater than 95.

First definition if more than one

Derivation of name

Definition proposed by International Union of Geological Sciences

Classification and occurrence

Originator of name; date of its introduction

Mineral abbreviations; given on page *10*

A

aa *ig. extru.* Rough, scoriaceous basaltic lava, consisting chiefly of sharp, angular fragments of compact lava and rough clinkers cemented together. Also named aphrolith. [Haw. *aā,* rough lava] (pub. by Dutton in 1883)

abraum salts *sed.* General term for the mixed salts overlying rock salt deposits at Stassfurt, Germany. [G. *Abraum,* refuse; the salts were originally thought to be of no value]

absarokite *ig. extru.* Porphyritic basaltic rock with phenocrysts of olivine and augite but no feldspar, in a groundmass of labradorite with orthoclase rims, olivine, augite, and some leucite. [Absaroka Indians, Absaroka Range, Yellowstone National Park, Wyoming-Montana] (Iddings in 1985)

acaustobiolith *sed.* General term for a noncombustible biolith; rocks like diatomite, radiolarite, phosphorite, and some limestones belong here. [Gk. *a,* not + *kaustos,* burning + biolith] (Grabau in 1924)

accretion *sed.* Concretion growing from the center outward in a regular manner. [L. *ad,* to + *crescere,* to grow] (Todd in 1903)

accumulative rock *ig.* = cumulate

achnahaite *ig. pluton.* Seldom used term for a biotite eucrite, esp. from Achnaha, Scotland. [Achnaha] (Niggli in 1936)

achnelith *ig. extru.* Pyroclastic particle whose external form is determined by surface tension on the ejected lava, rather than by fracture surfaces, e.g., Pele's tears and Pele's hair. [Gk. *achnē,* foam, froth] (Walker and Croasdale in 1971)

achondrite *meteor.* Aerolite (stony meteorite) lacking the characteristic chondrules of the chondritic stones; somewhat heterogeneous and more coarsely crystallized than the chondrites. [Gk. *a,* not, without + chondrule]

achondrite, augite *meteor.* Stony meteorite consisting of more than 90% augite with traces of olivine and troilite; no important feldspar. Also named angrite.

achondrite, diopside-olivine *meteor.* Stony meteorite containing about 75% diopside, 15% olivine, with a small amount of plagioclase (oligoclase-

andesine) and a little magnetite. Also named nakhlite.

achondrite, enstatite *meteor.* Stony meteorite consisting almost entirely of enstatite; often with accessory amounts of forsterite, diopside, oligoclase, kamacite; usually there is a brecciated structure with coarse enstatite occurring in crushed enstatite. Also named aubrite.

achondrite, hypersthene *meteor.* Stony meteorite consisting almost entirely of hypersthene, with only accessory plagioclase, olivine, troilite, and others; structurally brecciated, with coarse hypersthene occurring in crushed hypersthene. Also named diogenite.

achondrite, olivine *meteor.* Stony meteorite consisting of more than 95% olivine, with accessory chromite, plagioclase, and nickel-iron; very similar in composition and structure to terrestrial dunite. Also named chassignite.

achondrite, olivine-pigeonite *meteor.* Stony meteorite consisting of grains of olivine and pigeonite (clinopyroxene) in a black carbonaceous matrix; nickel-iron occurs as small granules around the grains of silicate minerals; specimens from Novo-Urei, Soviet Union, and Goalpara, India, contain very small diamonds. Also named ureilite.

achondrite, pyroxene-plagioclase *meteor.* Stony meteorite, the commonest of the achondrites, divided into two main types: eucrite, composed of pigeonite and anorthite, and howardite, composed of hypersthene and anorthite; consisting almost entirely of pyroxene and plagioclase, although pyroxene may somewhat exceed plagioclase; often brecciated but similar to terrestrial gabbros.

Ach'uaine-type hybrid *ig. pluton.* Series of unusual alkalic rocks from near Golspie, Sutherlandshire, Scotland; ranging from dark ultramafic to light-colored rocks and in composition including: olivine-hornblende biotite rock (scyelite), hornblende-biotite-pyroxene rock, orthoclase sodic-plagioclase rock (perthosite). [Ach-uaine, Scottish locality] (Read and Phemister in 1925)

acidic rock *ig.* = acidite

acidite *ig.* Collective name for all igneous rocks

high in silica (65-70% SiO_2 by weight). Also named acidic rock. [acid, from concept that it was high in silicic acid] (Cotta in 1864)

actinolite-magnetite rock *meta.* Somewhat schistose metamorphic rock rich in magnetite and actinolite and often accompanied by quartz or dolomite.

adamellite *ig. pluton.* Synonym for quartz monzonite; term originally used for an orthoclase-bearing tonalite. [Monte Adamello, Italy] (Cathrein in 1890)

adder stone *sed., meta.* = serpent stone

adinole *meta.* Dense albite-quartz-chlorite metamorphic rock with accessory epidote, actinolite, titanite, and others, formed by low-grade contact metasomatism by diabase of shale, argillite, or slate. [Possibly Gk. *adinos,* dense] (Haussmann in 1847)

adipocere *sed.* Waxy to greasy, brownish or light-colored natural substance consisting mainly of fatty acids, hydroxy acids, and their calcium and magnesium salts, formed by the decay of animal remains in damp areas or in fresh or salt water. Not to be confused with adipocerite and adipocire, synonyms for the mineral hatchettine. [L. *adeps,* fat + *cera,* wax]

adobe *sed.* Sandy, often calcareous, clay used in the American West and Southwest for making sun-dried bricks. [Sp. *adobar,* plaster] (pub. by Russel in 1889)

aegagropile *sed.* Lake ball consisting of radial, outgrowing, hairlike filaments formed by algae. [Gk. *aigagros,* goat + *pilus,* hair]

aegiapite, calcite *ig. hypab.* Carbonatite consisting of about equal amounts of aegirine-augite and apatite (30% each), calcite (22%), feldspars, and others; original rock from Cape Turij, Kola, Soviet Union. [aegirine + apatite]

aegirinolith *meta.?* Rock related to skarn and composed of aegirine-augite, titanite, and magnetite. [aegirine] (Kretschmer in 1917)

aegirite *sed.* Trade name for a bitumen allied to elaterite; not to be confused with the pyroxene mineral aegirine. [Possibly Aegir, god of the sea in Norse mythology]

aeolianite *sed.* = eolianite

aeonite *sed.* Trade name for a bitumen very similar to elaterite. [Possibly L. *aeon,* space, period of time] (pub. by Day in 1911)

aerinite *meta.?* Mixture of a calcium-bearing leptochlorite (blue in color) with pyroxene, quartz, and spinel. [Gk. *aerinos,* aerial, alluding to blue color] (Lasaulx in 1876)

aerolite *meteor.* 1. Meteorite composed of silicate minerals, generally with some nickel-iron alloy; the major types are achondrites and chondrites. Also named stones. (Maskelyne in 1863) 2. Synonym for any meteorite. (Blumenbach in 1804) [Gk. *aer,* air]

aerosiderite *meteor.* = siderite. [Gk. *aer,* air + siderite] (Maskelyne in 1863)

aerosiderolite *meteor.* = siderolite. [Gk. *aer,* air + siderolite] (Maskelyne in 1863)

aetite *sed.* = eaglestone. [Gk. *aetos,* eagle] (Pliny, *Natural History*)

agalmatolite *meta.* Term applied to massive varieties of muscovite (pinite), pyrophyllite, and steatite; in China small images are carved from these materials, often called Honan jade, Soochow jade. Pagodite is nearly identical. [Gk. *agalmatos,* image] (Klaproth in 1797)

agate *sed.* Rock composed essentially of chalcedony (a microfibrous variety of quartz) and characterized by a special color, or by colors arranged in bands, in irregular clouds, or in mosslike forms; commonly forms in cavities or vugs in sedimentary and volcanic rocks. Variety names are numerous and for the most part descriptive, e.g.: banded a., blood a., blue a., botryoidal a., brecciated a. (also named ruin a.), cat's-eye a. (opalescent), cer a. (chrome-yellow), chrysocolla a. (blue green), cloud a. (with large spots), coral a. (agatized coral), dot a., eye a. (with concentric bands), fire a. (iridescent), flower a., fortification a. (like drawings of ancient forts), frost a. (with snow white markings), glass a., grape a. (botryoidal), iris agate (with spectral colors), lace agate, landscape agate, milk a., mosaic a. (brecciated), moss a. (with dendritic inclusions), oolitic a., pigeon blood a., pipe a. (with tubelike inclusions), plume a. (featherlike), polka dot a., ribbon a., ring a., rose a., sagenite a. (with tiny needle inclusions), scenic a., seaweed a., shell a. (with silicified fossil shells), topographic a. (with maplike markings), tree a. (with treelike inclusions), tube a. (with hollow tubelike inclusions), wart a., wax a., white a., wood a. (silicified wood), zebra a., zigzag a. [Achates river, probably either modern Carabi or the Cannitello in southwestern Sicily] (Theophrastus, *On Stones*)

agate, Iceland *ig. extru.* 1. Obsidian from Iceland. 2. Grayish or brownish variety of obsidian.

agathocopalite *sed.* = kauri gum. [*Agathis australis,* a tall timber tree of New Zealand + copalite]

agglomerate *ig. extru.* 1. Consolidated pyroclastic deposit composed chiefly of volcanic bombs and subangular fragments lying in a tuff matrix. [L. *ad,* to + *glomus,* ball] (Lyell in 1831) 2. *IUGS,* pyroclastic rock or deposit whose average pyroclast size exceeds 64 mm and in which rounded pyroclasts predominate.

agglomerate, basalt *ig. extru.* Cemented coarse pyroclastics consisting of basaltic material, esp. bombs greater than 64 mm across.

agglomerate, volcanic *ig. extru.* = agglomerate

agglutinate *ig. extru.* 1. Rock similar to an agglomerate, but in which the volcanic bombs have been cemented only by the thin glassy skin of the elements at their points of contact; distinguished by the presence of glassy cement and the absence of a tuff matrix. Also spelled agglutinite. [L. *ad,* to + *glutinare,* to glue. 2. *IUGS,* pyroclastic rock like that defined above, but where the average pyroclast size exceeds 64 mm and rounded pyroclasts predominate.

agglutinate *lunar impact.* Term applied to certain particles on the lunar surface held together by and largely composed of glass, probably from spatter and melted ejecta formed by small high-velocity impacts on the lunar surface.

agglutinite *ig. extru.* = agglutinate (*ig. extru.*)

agmatite *migmat.* Migmatite with a breccialike structure. [Gk. *agma,* fragment] (Sederholm in 1923)

agpaite *ig. pluton.* Collective name for nepheline syenites with large amounts of sodium pyroxene and sodium hornblende, including naujaite, lujavrite, kakortokite, and others; distinguished by having (Na+K) greater than Al on a molecular or atomic basis. [Agpa, place of the auks, locality in southern Greenland] (Ussing in 1911)

agrite *sed.?* Brown, mottled calcareous rock sometimes cut as a gem.

agstein *sed.* = jet. [G. Gagates, ancient name of jet + *Stein,* stone]

aillikite *ig. dike.* Carbonatite; essentially an alnoite containing up to 50% fine-grained calcite that replaced earlier olivine, biotite, and possibly melilite. [Cape Aillik, Labrador] (Kranck in 1953)

ailsyte *ig. pluton.* Aplite containing sodium orthoclase and quartz with minor riebeckite (6%). [Ailsa Craig, Scotland] (Heddle in 1897)

aiounite *ig. hypab.* Rock related to kersantite, composed of augite (56%) and biotite (16%), in a cryptocrystalline groundmass (19%) that may represent plagioclase. [El Aioun, west of Oudjda, Morocco] (Duparc in 1926)

ajkaite *sed.* Pale yellow to dark reddish brown, sulfur-bearing fossil resin found in brown coal. Also spelled ajkite. [Ajka, Hungary] (Hlasiwetz in 1871)

akafoamite *ig., meta.?* Rarely used varietal name for charnockite.

akenobeite *ig. dike.* Light-colored aplitic granodiorite, consisting of oligoclase and orthoclase, with minor quartz and biotite altered to chlorite. [Akenobe district, Tajima, Japan] (Kato in 1920)

akerite *ig. pluton.* Syenite containing orthoclase, oligoclase, and often abundant biotite and green clinopyroxene (diopside or diopside-aegirine series), with a little quartz. [Åker parish, Oslo, Norway] (Brögger in 1890)

alabaster *sed.* 1. Massive, usually translucent, fine-grained gypsum rock, often used for statuary and decorative purposes. 2. = onyx marble. [Possibly Alabastron, a town in Egypt; the name was given to materials from which ointment vases called alabastra were made]

alabaster, Egyptian *sed.* Banded calcite onyx marble from near Thebes, Egypt.

alabaster, oriental *sed.* = onyx marble; cave onyx

aladzha *sed.* Ozokerite admixed with associated rock materials, found in the region of the Caspian Sea. [Tartar origin, possibly related to modern Turkish *alaca,* motley, variegated]

alaskite *ig. pluton.* Crystalline granular igneous rock characterized by essential alkali-feldspar and quartz, with little or no dark component; like a normal granite that is light in color because of the lack of mafics. [Alaska] (Spurr in 1900).

alaskite-quartz *ig. dischist.* = tarantulite. (Spurr in 1906)

albanite *ig. extru.* = leucitite. [Albano Hills, Italy] (Washington in 1920)

albanite *sed.* Black bituminous mixture from Albania. [Albania] (Istrati and Mihailescu in ca. 1912)

albertite *sed.* Jet black, brilliant, pitchlike asphalt rock; differs from other asphaltum in being only partially soluble in oil of turpentine and in its very imperfect fusion when heated. Also named albert coal. [Albert shale of Albert County, New Brunswick] (How in 1860)

albitite *ig. dike.* Aplitic rock composed entirely of a granular aggregate of albite; some specimens contain minor muscovite and quartz. [albite] (Turner in 1896)

albitite, quartz *ig. dike.* Igneous dike rock composed of a coarse aggregate of quartz (sometimes blue, opalescent) and albite that are not graphically intergrown. (Tilley in 1919)

albitophyre   *ig. dike.* Porphyritic rock containing albite phenocrysts in a groundmass of the same mineral. Also named albite porphyrite. |albite| (Coquand in 1857)

alboranite   *ig. extru.* Olivine-free hypersthene basalt, containing calcic plagioclase (bytownite-anorthite) and hypersthene phenocrysts in a groundmass of calcic-labradorite, clinopyroxene, and magnetite. |Island of Alboran, off Cabo de Gata, Spain| (Becke in 1899)

alboranite, quartz   *ig. extru.* Essentially identical to tridymite alboranite but containing quartz instead of tridymite.

alboranite, tridymite   *ig. extru.* Alboranite containing more than 12.5% tridymite by volume; additional minerals include calcic-plagioclase, clinopyroxene, hypersthene, and others. (Burri in 1937)

aleurite   *sed.* Unconsolidated sediment intermediate in grain size between sand and clay, consisting of particles in the range of 0.01-0.1 mm; commonly used in Russian literature and usually translated as silt. |Gk. *aleuron,* flour|

aleurolite   *sed.* Consolidated aleurite, intermediate in texture between sandstone and shale; nearly identical to siltstone. |Gk. *aleuron,* flour|

aleutite   *ig. extru.* Seldom used term for porphyritic extrusive igneous rocks, with aphanitic or fine-grained groundmasses, whose feldspars are intermediate in composition between those of andesites and basalts; belugite is the plutonic phaneritic equivalent. |Aleutian Islands, Alaska| (Spurr in 1900)

alexeyevite   *sed.* Waxlike, white to brown hydrocarbon, from near Kaluga, Soviet Union. |Vladimir Feodorovich Alekseev (=Alexeyev), 1852-1919, Russian chemist|

alexoite   *ig. pluton.* Rare pyrrhotite-rich dunite occurring as a border zone around serpentinized peridotite in Ontario. Miners' term. |Alexo mine, Dundonald Township, Ontario|

algarite   *sed.* Bitumen of unknown composition derived from algae. |alga| (Hackford in 1932)

algarvite   *ig. pluton.* Biotite melteigite; dark rock containing much more biotite and less nepheline than the original melteigite. |Algarve province, Portugal| (Lacroix in 1922)

algon   *sed.* Viscous, organic binding material of vase, consisting of remains of algae, or of land vegetation, mixed with iron principally in the form of FeS. |alga|

algovite   *ig.* Obsolete term applied to a group of augite-calcic-plagioclase rocks, including diabase, diabase porphyry, and gabbro, in the Algäuer Alps. Also spelled allgovite. |Algäuer Alps| (Winkler in 1895)

alios   *sed.* Hardpan, esp. an impervious ferruginous crust or iron pan. |F. *alios,* iron pan|

alkorthosite   *ig. pluton.* Syenite composed almost entirely of microperthitic orthoclase. |alkali + orthosite|

alkremite   *ig.* Rock composed of pyropic garnet and spinel with minor secondary minerals, often including kornerupine, chlorite, corundum, and sulfides; occurring as xenoliths. |aluminum + Russ. *kremei,* silicon; alluding to composition| (Ponomarenko in 1975)

allalinite   *meta.* Altered olivine gabbro, consisting of uralite, talc, saussurite, garnet, chlorite, and actinolite. |Allalin, Switzerland| (Rosenbusch in 1896)

allgovite   *ig.* = algovite

allingite   *sed.* Retinite containing no succinic acid but with considerable sulfur. |Allinges, Haute-Savoie, France| (Tschirch and Aweng in 1894)

allite   *sed.* Sediment from which silica has been largely removed and containing a high proportion of aluminum and iron compounds in the clay fraction, e.g., bauxite and laterite. |Gk. *allos,* other, altered|

allivalite   *ig. hypab.* Gabbro composed of olivine and anorthite, in which the latter is most abundant; related to harrisite and troctolite. |Allival Hill, Isle of Rum, Scotland| (Harker in 1908)

allochem   *sed.* In mechanically deposited limestones, carbonate aggregates that serve as the coarser framework grains in contrast to micrite (carbonate-mud matrix) and sparite (crystalline calcite cement); may include fossils, lumps, pellets, ooliths, sand, silt, and gravel. |Gk. *allos,* other + chemical| (Folk in 1959)

allochetite   *ig. dike.* Porphyritic dike rock with phenocrysts of labradorite, orthoclase, titanaugite, nepheline, magnetite, and apatite in a dense groundmass of augite, biotite, magnetite, hornblende, nepheline, and orthoclase; the groundmass is in the form of a microscopic felt. |Allochet Valley, Monzoni region, Italy| (Ippen in 1903)

alluvium   *sed.* General term for unconsolidated clay, silt, sand, or gravel deposited by running water. |L. *alluvium,* inundation| (Lyell in 1859)

almashite   *sed.* Green to black variety of amber

with a low oxygen content (2.5-3%). [Almas (Almash) Valley, near Piatra, Moldavia] (Murgoci in 1924)

alnoite   *ig. dike.* Lamprophyre composed of biotite (or phlogopite) and melilite as essential minerals, accompanied by olivine, augite, nepheline, and garnet. [Island of Alnö, Sweden] (Rosenbusch in 1887)

alnoite, biotite   *ig. dike.* Alnoite containing strongly colored pseudohexagonal plates of biotite as well as euhedral melilite (5 mm) crystals; from Isle Cadieux, between Montreal and Ottawa, Canada. (Stansfield in 1923)

alnoite, monticellite   *ig. dike.* Alnoite differing from the normal type in that it contains two kinds of olivine, normal and monticellite.

alomite   *ig. pluton.* Trade name for a fine blue sodalite rock quarried at Bancroft, Ontario, and used as an ornamental stone. [Charles Carrick Allom (not Alom), 1865-1947, of the firm White, Allom and Co., marble merchants, London] (pub. by Renwick in 1909)

alphitite   *sed.* Silt or clay composed largely of rock flour, such as the fine material produced by a glacier. [Gk. *alphiton,* barley meal] (Salomon in 1915)

alsbachite   *ig. aschist.* Granodiorite porphyry composed of quartz, sodic-plagioclase, some alkali feldspar, and often with rose-red garnets and shredded mica. [Alsbach, Germany] (Chelius in 1892)

alumyte   *sed.* Alum clay (bauxite) from County Antrim, Ireland. [alum] (Kinahan in 1889)

alunite rock   *ig., altered.* Alunite-rich rock often accompanied by kaolinite and formed by solfataric action on various volcanic rocks, e.g., trachyte, andesite, tuffs; formed by the action of sulfuric acid, resulting from the oxidation of pyrite in aluminous rocks.

alvikite   *ig. hypab.* Medium- to fine-grained hypabyssal calcite-carbonatite; nearly equivalent to sövite, which is coarser. [Alvik, Alnö Island, Sweden] (Eckermann in 1928)

amargosite   *sed.* = bentonite. [Amargosa River, Inyo County, California] (Melhase in 1926)

amatrix   *sed.* Trade name for green, blue-green, and bluish concretions of variscite, wardite, and other phosphates, in a matrix of chalcedony quartz. Also spelled amatrice [American + matrix] (pub. by Sterrett in 1909)

amausite   *ig. extru.* Felsite or petrosilex, very finely crystalline rock; some are apparently formed by the devitrification of glass. See obsidian, devitrified. [Possibly Gk. *amauros,* dark]

amber   *sed.* Yellow to orange to dark brown fossil resin exuded from prehistoric coniferous trees; occurs in irregular amorphous masses that, although brittle, can be easily worked or carved and take a good polish; insects, pine needles, buds, and flowers often occur as inclusions. The two major classes include succinite (with succinic acid) and retinite (without succinic acid). [Arab. *anbar,* amber, ambergris]

amber, black   *sed.* 1. Variety of rumanite that is not truly black but a deep ruby red, blue, or brown when viewed in strong light.   2. = jet

amber, blue   *sed.* Variety of osseous amber with a bluish tinge probably due to the presence of calcium carbonate.

amber, bone   *sed.* = amber, osseous

amber, earth   *sed.* 1. Amber that is mined; the term is used to distinguish mined amber from that obtained from the sea (sea amber).   2. Amber whose outer portion has deteriorated in luster, transparency, and color.

amber, flohmig   *sed.* Amber resembling goose fat; it is full of tiny bubbles but not as opaque as cloudy amber. [E. Pruss., *Flohmfett,* yellowish fat of goose or duck]

amber, osseous   *sed.* Cloudy to opaque amber containing numerous minute bubbles. Also named bone amber. [L. *osseus,* bony]

amber, sea   *sed.* Amber scooped from the ocean or found on beaches.

amberine   *sed.* Yellowish green chalcedony rock from Death Valley, California. [amber] (Ward, pub. by Sterrett in 1914)

amberite   *sed.* = ambrite

ambonite   *ig. extru.* Collective name for rocks of various compositions from Ambon Island that have phenocrysts in a fine-grained groundmass; included are bronzite-bearing andesite and dacite, mica-bearing andesite and dacite, hornblende-bearing andesite and dacite. [Ambon Island, Indonesia] (Verbeek in 1899)

ambrite   *sed.* Variety of retinite that is yellowish gray, subtransparent, and resembles amber; found in New Zealand. [amber] (Hochstetter in 1861)

ambrosine   *sed.* Yellowish to clove brown variety of amber with a conchoidal fracture and occurring in rounded masses in a phosphatic deposit near Charleston, South Carolina. [amber] (Shepard in 1870)

americanite   *ig. extru.* Natural glassy rock from

Peru or Colombia; originally thought to be akin to tektite, but now considered to be obsidian. Also spelled amerikanite. [South America] (Easton in 1921)

amherstite *ig. pluton.* Variety of anorthosite containing andesine-antiperthite (85%) with blue quartz, hypersthene, rutile, and others. [Amherst County, Virginia] (Watson and Taber in 1913)

amiatite *ig. extru.* Obsolete collective term used for rocks in which potassium is greater than sodium or calcium. [Monte Amiata, Italy] (Lang in 1891)

amiatite *sed.* = siliceous sinter. Also incorrectly spelled amatite. [Possibly Monte Amiata, Italy]

ampasimenite *ig.* Ijolite porphyry characterized by the presence of nepheline, titanaugite, hornblende, and magnetite in a fine-grained or glassy groundmass. [Ampasimena, Madagascar] (Lacroix in 1922)

ampelite *sed.* Black carbonaceous earth or shale that, because of enclosed pyrite, may yield alum on weathering; used by the ancients to kill insects on vines. Also named ampelite alumineux and alum shale. [Gk. *ampelos,* vine] (Brongniart in 1807)

amphibolide *ig. pluton.* General field term for any coarse-grained holocrystalline igneous rock composed nearly entirely of amphibole minerals, e.g., hornblendite. [amphibole]

amphibolite *meta.* Granoblastic to foliated metamorphic rock composed of essential amphibole and plagioclase; the major amphibole is usually hornblende and the plagioclase is commonly andesine, although compositions varying from oligoclase to bytownite have been observed; on the basis of origin there are two major types, orthoamphibolite, derived from an earlier igneous rock, and para-amphibolite, derived from an earlier sedimentary rock. [amphibole] (Brongniart in 1813)

amphibolite, anthophyllite *meta.* Amphibolite composed of the orthorhombic amphibole anthophyllite, plagioclase, and, at times, garnet.

amphibolite, crossite *meta.* Rare amphibolite in which the amphibole is crossite, occurring as large unoriented blades, and the plagioclase is usually albite.

amphibolite, cummingtonite *ig.?* Old name applied to a dike rock in Finland, composed of cummingtonite (an amphibole) and plagioclase; now considered to be an igneous dioritic rock. (Eskola in 1914)

amphibolite, feather *meta.* Amphibolite in which the crystals of amphibole (usually hornblende) form stellate or sheaflike groups on the foliation planes.

amphibolite, garnet *meta.* Normal amphibolite containing garnet (usually almandine).

amphibolite, glaucophane *meta.* Coarse-grained, poorly foliated metamorphic rock containing glaucophane as the major constituent, with variable subordinate amounts of albite, garnet, epidote, lawsonite, crossite, and others.

amphibolite, hornblende *meta.* Normal amphibolite in which hornblende is the major amphibole.

amphibolite, zoisite *meta.* Amphibolite containing appreciable zoisite.

amphibololite *ig. pluton.* = amphibolide. (Lacroix in 1894)

amphigenite *ig.* Obsolete term originally used for rocks composed of leucite, labradorite, and pyroxene and later used for rocks composed only of leucite. [F. *amphigène,* leucite] (Cordier in 1868)

amphoterite *meteor.* Obsolete name for an olivine-hypersthene chondrite; the presence of chondrules was not recognized at first. [Gk. *amphoteros,* both, in reference to the two major minerals] (Tschermak in 1883)

anabohitsite *ig. pluton.* Ilmenite-rich pyroxenite in which hypersthene and hornblende make up 65% of the rock, and ilmenite-magnetite make up 30%. [Anabohitsy, Madagascar] (Lacroix in 1914)

anagenite *sed.* Conglomerate in which the cementing material resembles mica schist; from the northern Apennines. [Gk. *ana,* on + *genes,* born] (Haüy in 1822)

analcime rock *sed.* Argillaceous sedimentary rock in which analcime is abundant, occurring as cement, as crystals, as ooliths with concentric structures, and as veinlets; analcime apparently precipitated by reactions of clay minerals with sodium-rich marsh and lacustrine waters.

analcimite *ig. extru.* Olivine-bearing basalt containing so much analcime in the groundmass that it makes up as much as two-thirds of the rock. [analcime] (Gemmellaro in 1829)

analcimite-tinguaite *ig. hypab.* Olive green aphanitic igneous rock whose composition is intermediate between analcimite and tinguaite, composed of anorthoclase, aegirine, nepheline, and analcime. (Johannsen in 1938)

analcimolith *ig., sed.* Monomineralic rock composed of analcime, whether primary or secondary in origin. [analcime] (Johannsen in 1938)

analcitite *ig. extru.* Augite-analcime basalt free

of important olivine and feldspar. |analcime| (Pirsson in 1896)

anamesite *ig. extru.* Rock in the basalt family lying in texture between dense basalt and coarse diabase; term seldom used. |Gk. *anamesos,* in the middle| (Leonhard in 1832)

anamesite, nepheline *ig. extru.* Nepheline basalt whose grain size is intermediate between a dense nepheline basalt and a coarse nepheline diabase. (Zirkel in 1894)

anatectite *migmat.* = anatexite

anatexite *migmat.* 1. Migmatite approaching a granite in composition and texture; metamorphic foliation may remain as relics of the original structure. 2. Rock formed by anatexis, that is, from the melting of a pre-existing rock. |Gk. *ana,* up, upwards + *tektos,* fused| (Sederholm in 1907)

anchorite *ig. pluton.* Nodular and veined variety of diorite, the normal facies of the rock being variegated with dark mafic segregation patches and light felsic contemporaneous veins. |Anchor Inn, near Caldecote, England| (Lapworth in 1889)

andalusite rock *meta., hydrotherm.* Hydrothermal rock whose major component is andalusite; esp. at Oreana, Nevada, and Mono County, California.

andelatite *ig. extru.* Obsolete term for an intermediate rock between andesite and latite; the extrusive equivalent of monzodiorite. |andesite + latite| (Johannsen in 1920)

andendiorite *ig.* Obsolete term for a Tertiary, quartz-augite diorite that occurs as islandlike areas in the midst of the volcanic rocks of the Chilean Andes. |Andes + diorite| (Stelzner in 1885)

andengranite *ig.* Obsolete term for a Tertiary, biotite-bearing hornblende granite that occurs as islandlike areas in the midst of the volcanic rocks of the Chilean Andes. |Andes + granite| (Stelzner in 1885)

andesinite *ig. pluton.* Coarse-grained igneous rock composed chiefly of andesine. |andesine| (Turner in 1900)

andesite *ig. extru., hypab.* Aphanitic, microcrystalline to glassy, igneous rock whose composition is close to diorite in that it consists of sodic-plagioclase (oligoclase to calcic-andesine) and one or more mafic minerals (hornblende, biotite, diopside, augite, hypersthene); most are porphyritic with feldspar and mafic phenocrysts; the rocks may be brown, reddish, gray, or greenish but are usually not black. |Andes| (von Buch in 1835)

andesite, analcime *ig. extru.* Extrusive rock composed of sodic-plagioclase and analcime, usually with some mafic.

andesite, augite *ig. extru.* Andesite in which the characterizing accessory mafic is augite, often accompanied by olivine, hornblende, biotite, and others.

andesite, basaltic *ig. extru.* Poorly defined term for rocks intermediate, in one or more respects, between a typical basalt and a typical andesite.

andesite, biotite *ig. extru.* Andesite in which the characterizing accessory mafic is biotite, often accompanied by hornblende.

andesite, hauyne *ig. extru.* Extrusive rock composed of sodic-plagioclase and hauyne, usually with some mafic.

andesite, hornblende *ig. extru., hypab.* Andesite in which the characterizing accessory mafic is hornblende, often accompanied by biotite and more rarely augite; this is the normal andesite.

andesite, hypersthene *ig. extru.* Andesite in which the characterizing accessory mafic is hypersthene, often accompanied by clinopyroxene (augite).

andesite, leucite *ig. extru.* Extrusive rock composed of sodic-plagioclase and leucite, usually with some mafic.

andesite, leucosodaclase *ig. extru., meta.* Light-colored andesite-like rock rich in sodic-plagioclase, esp. albite; e.g., sodaclase keratophyre.

andesite, nepheline *ig. extru.* Extrusive rock composed of sodic-plagioclase and nepheline, usually with some mafic.

andesite, pyroxene *ig. extru.* Andesite in which the characterizing accessory mafic is a pyroxene, usually either augite or hypersthene, with other minor minerals.

andesite, quartz *ig. extru.* = dacite

andesite, sodaclase *ig. extru., meta.* Andesite rock in which the plagioclase is very sodic (albite) and that also contains clinopyroxene; some have suggested that the albite formed from the metasomatic replacement of a more calcic-plagioclase. (Johannsen in 1937)

andesite, sodalite *ig. extru.* Extrusive rock composed of sodic-plagioclase and sodalite, usually with some mafic.

andesite, trachyte- *ig. extru.* = trachyandesite

andesite, vitrophyric *ig. extru.* = vitrophyre, andesite

anemoclastic rock *sed.* Rock consisting primarily of anemoclasts, fragments broken and more or less rounded by wind action. |Gk. *anemos,* wind + *klastos,* broken| (Grabau in 1904)

anemolite   *sed.* Helictite in which the eccentricity is ascribed to the action of air currents; originally noticed at Derbyshire, England. |Gk. *anemos,* wind| (Barnes and Holroyd in 1896)

anemosilicarenite   *sed.* Eolian (wind-borne) sand of siliceous composition. |Gk. *anemos,* wind + silica + arenite| (Grabau in 1904)

angrite   *meteor.* = achondrite, augite. |Angra dos Reis, Brazil|

anhydrite rock   *sed.* White, grayish or bluish evaporite rock in which the major constituent is anhydrite; the mineral may be in the form of interlocking grains or be as rectangular blades; often resembles a marble. Varieties include tripestone and vulpinite.

anhydrite rock, oolitic   *sed.* Anhydrite-cemented sandstone in which subrounded quartz grains are surrounded by encrustations of anhydrite.

anhydrite-carbonate rock   *sed.* Evaporite rock in which the major minerals are anhydrite and a carbonate mineral; varieties include anhydrite-calcite, anhydrite-dolomite, anhydrite-magnesite, and anhydrite-dolonite-magnesite.

anhydrite-dolomite rock   *sed.* Evaporite anhydrite-carbonate rock in which dolomite is significant.

anhydrite-gypsum rock   *sed.* Evaporite rock composed of anhydrite and gypsum; some of these form by the hydration of anhydrite to form gypsum and grade into gypsum rock.

anhydrite-halite rock   *sed.* Typical evaporite rock that, in addition to anhydrite and halite, may contain some accessory dolomite.

anhydrite-polyhalite rock   *sed.* Evaporite rock composed of anhydrite and polyhalite, with minor halite, dolomite, magnesite, and langbeinite; usually transitional between anhydrite-halite rock and anhydrite-dolomite rock.

anhydrock   *sed.* = anhydrite rock

animé   *sed.* Recent fossil resin often containing insects and sometimes mistaken for amber; copal. Also named gum animé. |F. *animé,* resin|

ankaramite   *ig. dike.* Olivine-bearing basalt, containing phenocrysts of titanaugite (very numerous) and olivine, in a fine-grained groundmass composed of titanaugite, labradorite, and others. |Ankaramy, Madagascar| (Lacroix in 1916)

ankaranandite   *ig., meta.?* Member of the charnockitic rock suite containing less than 10% quartz and whose feldspar is 0-40% plagioclase. (Giraud in 1964)

ankaratrite   *ig. extru.* Dark-colored olivine nephelinite lava composed of titanaugite (56%), nepheline (14%), olivine (10%), and biotite (8%). |Ankaratra, Madagascar| (Lacroix in 1916)

ankaratrite, melilite   *ig. extru.* Ankaratrite containing melilite in addition to the other minerals. (Lacroix in 1916)

anorthite rock   *ig. pluton., dike.* = anorthitite. (Irving in 1883)

anorthitfels   *ig. pluton., dike.* = anorthitite

anorthitissite   *ig. dike.* Hornblendite containing some anorthite (22%); from the middle Urals, Soviet Union. |anorthite + issite| (Tröger in 1935)

anorthitite   *ig. pluton., dike.* Igneous rock composed chiefly of anorthite; anorthosite composed of anorthite. |anorthite| (Turner in 1900)

anorthitite, quartz   *ig. pluton., dike.* Quartz-bearing anorthitite; the rock is composed of anorthite and quartz.

anorthoclasite   *ig. pluton.* Igneous rock composed entirely of anorthoclase. |anorthoclase| (Loewinson-Lessing in 1901)

anorthosite   *ig. pluton.* 1. Gabbroic igneous rock in which the mafics are absent or are only accessory constituents and the predominant mineral is calcic-plagioclase; most could be named labradoritite, bytownitite, anorthitite, or more generally plagioclasite. |anorthose, old name for the triclinic feldspars| (Hunt in 1863) 2. *IUGS,* a plutonic rock with 0-5 Q, P/(A + P) greater than 90, and M less than 10.

anorthosite, anorthite   *ig. pluton., dike.* = anorthitite

anorthosite, garnetiferous   *ig. pluton.* Garnet-bearing anorthosite; e.g., routivarite composed of labradorite (96%) with some almandine.

anorthosite, olivine   *ig. pluton.* Anorthosite in which olivine is a rather common accessory mineral; kelyphitic rims often form where the olivine and plagioclase come in contact.

anorthosite, quartz   *ig. hypab.* Anorthosite containing quartz; some consist of pure labradorite (85-90%) and quartz (10-15%). (Loughlin in 1912)

anthodite   *sed.* Speleothem composed of clusters of needle- or quill-like crystals; most are composed of aragonite, some are calcite. |Gk. *anthos,* flower|

anthracite   *sed., meta.* Hard, black coal of the highest rank, in which fixed carbon content is 92-98% (dry, mineral-matter-free basis); it has a semimetallic luster and a conchoidal fracture; it ignites with difficulty but burns with a short smokeless flame. [Gk. *anthrax,* coal] (Haüy in 1795)

anthraconite   *sed.* Black bituminous limestone or marble that usually emits a fetid odor when struck or rubbed. [Gk. *anthrax,* coal + *konis,* dust, powder] (Moll in 1806)

anthraxolite   *sed.* Black, coaly, lustrous variety of asphaltite with a high fixed-carbon content, occurring in veins and masses in sedimentary rocks, esp. oil shales. [Gk. *anthrax,* coal] (Chapman in 1871)

antiperthite   *ig.* Feldspar parallel or subparallel intergrowth in which sodic-plagioclase (albite, oligoclase, or andesine) appears to be the single-crystal host from which the potassic feldspar (usually orthoclase) has exsolved. [Gk. *anti,* opposite + perthite] (Suess in 1905)

antiperthite, myrmekite   *ig.* Myrmekitelike intergrowth of predominant plagioclase and vermicular orthoclase (or other potassic feldspar).

antsohite   *ig. hypab.* Lamprophyre related to hamrongite, but without feldspar; it is composed of biotite (42%), hornblende (39%), and quartz (19%). [Antsohy Creek, Tsaratanana, Madagascar] (Lacroix in 1922)

Apache tear   *ig. extru.* Rounded pebblelike obsidian that has weathered out of perlite flows in the American West, esp. Arizona; these are usually gray to gray-brown and are translucent; variety of marekanite. [Apache Indians and tearlike shape]

apachite   *ig. extru.* Gray, medium- to fine-grained phonolite with small phenocrysts of sanidine and nepheline in a matrix with aenigmatite and sodic amphibole about equal to pyroxene in abundance. [Apache Mountains, western Texas] (Osann in 1895)

apalhraun   *ig. extru.* = lava, block; aa [Icel. *apal,* rough + *hraun,* lava, lava field]

apaneite   *ig. pluton.* Nepheline-bearing apatitite; rock containing apatite (86%), nepheline (8%), and minor mafics (aegirine, biotite). [apatite + nepheline] (Vlodavetz in 1930)

apatitite   *ig. pluton.* Igenous rock in which apatite is the only essential mineral; apaneite is a variety. [apatite]

apenninite   *meta.* = besimaudite. [Apennines, Italy] (Gastaldi in 1878)

aphanide   *ig.* Informal field term designating a wholly or predominantly fine-grained (aphanitic) rock. [Gk. *aphanes,* not showing]

aphanite   *ig.* General term for any fine-grained igneous rock whose components are too small to be distinguished with the unaided eye. [Gk. *aphanes,* not showing] (Haüy in 1822)

aphanophyre   *ig.* Aphanite with phenocrysts; one which is porphyritic. [Gk. *aphanes,* not showing]

aphrolith   *ig. extru.* = aa. [Gk. *aphros,* foam] (Jaggar in 1917)

aplite   *ig. diaschist.* White, cream, yellowish, reddish, or gray dike rock having a fine-grained saccharoidal texture (called aplitic) and a granitic composition in which the major minerals are quartz, potassic feldspar, and sodic-plagioclase without important mafics. Aplite is a variety of granite; however, there are aplitic facies of many other plutonic rocks, and when appropriate the term may be modified to include other rocks, e.g., alkali-syenite a., basalt a., diorite a., gabbro a., granodiorite a., kaligranite a., monzonite a., nepheline syenite a., pulaskite a., syenite a., syenodiorite a., tonalite a., and others. [Gk. *haploos,* simple; long applied to rocks of simple composition] (Retz in ca. 1800)

aplite, alaskite   *ig. diaschist.* Aplitic alaskite differing from normal aplite in containing sodic-plagioclase, 5-50% of total feldspar, in addition to orthoclase, and from granite aplite, whose plagioclase is usually a bit more calcic. (Spurr in 1900)

aplite, foyaite   *ig. diaschist.* Nepheline syenite aplite; light-colored, saccharoidal rocks composed of orthoclase, nepheline, and other feldspathoids.

aplite, granite   *ig. diaschist.* = aplite. Some petrologists make a distinction between aplite and granite aplite, suggesting that the latter rock has more plagioclase (though less in amount than orthoclase) and more dark minerals.

aplite, hauyne   *ig. diaschist.* Dark-gray, saccharoidal dike rock composed of sanidine, aegirine-augite, and hauyne; an example occurs at Port Cygnet, Tasmania. (Twelvetrees and Petterd in 1900)

aplite, leucodiorite   *ig. diaschist.* Light-colored aplitic dike rock composed of oligoclase or andesine, sometimes with muscovite or hornblende but usually no quartz.

aplite, leucogabbro   *ig. diaschist.* Light-colored saccharoidal dike rock, related to gabbro in that the essential constituent is a calcic-palgioclase; accessories may be pyroxene and iron oxides.

aplite, nepheline   *ig. diaschist.* = nepheline syenite aplite. See syenite, nepheline.

aplite, ornöite *ig. diaschist.* Light red to white aplitic rock consisting almost entirely of feldspar (sodic-plagioclase 63%; microcline 36%); very minor hornblende and biotite may be present. (Högbom in 1910)

aplite, soda *ig. diaschist.* Quartz-albite aplite with a saccharoidal texture.

aplodiorite *ig. pluton.* Light-colored variety of biotite granodiorite containing little or no hornblende. [Gk. *haploos,* simple + diorite; alluding to its simple composition] (Bailey in 1916)

aplogranite *ig. pluton.* Light-colored granite composed of potassic feldspar and quartz, with minor biotite and some muscovite. [Gk. *haploos,* simple + granite; alluding to its simple composition] (Bailey in 1916)

apo- *ig.* Prefix often given to the names of various ancient volcanic rocks to indicate that metasomatic or metamorphic changes have occurred, e.g., devitrification or silicification, without the destruction of the original texture. The prefix eo- has an identical meaning but is used less frequently. [Gk. *apo,* from] (Bascom in 1893)

apobsidian *ig. extru., altered.* = obsidian, devitrified

apogranite *ig., autometa.* Albitized and greisenized granite, esp. located at the peripheral and apical parts of certain intrusives and commonly enriched in rare elements (Sn, W, Mo, Li, Nb, Ta, etc.). (Beus in 1962)

apogrit *sed.* = graywacke. [Gk. *apo,* from + grit]

aporhyolite *ig. extru., altered.* Metarhyolite showing clear evidence of an original fluidal texture (glassy) but having become completely crystalline through devitrification. Also named eorhyolite. (Bascom in 1893)

appinite *ig. pluton.* Group of dark-colored varieties of syenite, monzonite, or diorite, rich in prismatic hornblende phenocrysts and also containing hornblende in the groundmass. [Appin, Loch Linnhe, Scotland] (Bailey in 1916)

apricotine *sed.* Trade name for yellowish red, apricot-colored quartz pebbles from near Cape May, New Jersey; used as a gem. [apricot] (pub. by Sterrett in 1911)

apron *sed.* Type of cave flowstone covering sloping projections. [apron, alluding to its shape]

aquatillite *sed.* Glaciomarine or glaciolacustrine till-like deposit, such as one formed from a melting iceberg. [L. *aqua,* water + tillite] (Schermerhorn in 1966)

aquifer *ig., sed., meta.* Body of rock of any kind that is sufficiently permeable to conduct ground water and to yield significant quantities of water to wells and springs. [L. *aqua,* water + *ferre,* bear]

aquifuge *ig., sed., meta.* Impermeable rock body of any kind with no interconnecting openings and thus lacking the ability to conduct ground water and yield water to wells and springs. [L. *aqua,* water + *fugere,* to flee]

Arabicus lapis *ig.?* Term used by Theophrastus, Dioscorides, and Pliny; appears to have meant a cellular rock related to pumice; some descriptions say it resembles ivory. Also named Arabic stone.

aragonite rock *sed.* Aragonite-rich masses occurring in low-temperature, near-surface deposits, e.g., some varieties of calcareous sinter, stalactites, helictites, modern ooliths and pisoliths, coralloidal aggregates (flos ferri), and the skeletal parts of living and recent lower organisms; metastable relative to calcite and easily changes to calcitite with changes in the environment.

aragotite *sed.* Yellow bituminous substance found in the Sulphur Springs district of California. [Dominique François Jean Arago, 1786-1853, French astronomer and physicist] (Durand in 1872)

arapahite *ig. extru.* Black aphanitic magnetite-rich basalt composed of bytownite, pale-green augite, and magnetite (about 50%). [Arapaho Indians, Colorado] (Washington and Larsen in 1913)

ardmorite *sed.* Trade name for a bentonite clay. [Ardmore, South Dakota] (Spence in 1924)

arendalite *ig., meta.?* Variety of charnockite containing garnet. [Arendal, Norway]

arenite *sed.* 1. General term applied to clastic sedimentary rocks composed of sand-sized fragments irrespective of composition, e.g., sandstone, arkose, graywacke, calcarenite. 2. Nearly pure quartz sandstone containing less than 10% argillaceous matrix; nearly equivalent to sandstone. [L. *arena,* sand] (Grabau in 1904)

arenite, arkosic *sed.* Rock essentially identical to arkose but characterized by an abundance of unstable materials in which the feldspar grains exceed the rock fragments. (Williams, Turner, and Gilbert in 1954)

arenite, carbonate *sed.* = calclithite

arenite, chert *sed.* 1. Quartz arenite containing more than 25% quartz. (McBride in 1963) 2. Litharenite in which the main rock fragment is chert.

arenite, feldspathic *sed.* Sandstone containing abundant quartz, less than 10% argillaceous matrix, and 10-25% feldspar, and characterized by an abundance of unstable materials in which the feldspar grains exceed the rock fragments; less feldspathic and more mature than arkosic arenite; essentially identical to subarkose. (Williams, Turner, and Gilbert in 1954)

arenite, lithic *sed.* Sandstone containing abundant quartz, less than 10% argillaeous matrix, and more than 10% feldspar, and characterized by an abundance of unstable materials in which rock fragments exceed feldspar grains. (Williams, Turner, and Gilbert in 1954)

arenite, plagioclase *sed.* Arkose in which plagioclase is the main feldspar; some petrologists would require more than 25% plagioclase.

arenite, sandstone *sed.* Sedarenite composed chiefly of sandstone. (Folk in 1968)

arenite, schist *sed.* Lithic arenite containing abundant clasts of low-grade pelitic metamorphic rocks such as slate, phyllite, and sericite schist.

arenite, shale *sed.* Sedarenite composed chiefly of shale fragments. (Folk in 1968)

arenite, subfeldspathic lithic *sed.* Arenite with 10% or less of feldspar and a larger quantity of rock fragments. (Williams, Turner, and Gilbert in 1954)

arenite, volcanic *sed.* Lithic arenite composed chiefly of volcanic fragments and having a low quartz content. (Williams, Turner, and Gilbert in 1954)

argeinite *ig. pluton.* Hornblendite containing olivine (as much as 20%). [Argein, Haute Garonne, France] (Lacroix in 1933)

argil *sed.* White clay. esp. potters' clay. [Gk. *argillos,* white]

argillite *sed., meta.* 1. Massive, fine-grained rocks, usually thinly, even seasonally banded, containing quartz, feldspar, chlorite, and some clay minerals (chiefly illite); some are considered to be sedimentary, cemented by silica, while others appear to be weakly metamorphosed argillaceous rocks, intermediate between a claystone and a meta-argillite. 2. Claystone composed entirely of clay minerals. [argil]

argillite, varved *sed.* Argillite derived from a varved clay; a distinctly laminated lacustrine sediment showing yearly sequences of deposition.

argillith *sed.* = claystone. Also named argillyte. [argil] (Grabau in 1924)

argillutite *sed.* Pure lutite. Also spelled argillutyte. [argil + lutite] (Grabau in 1904)

argulite *sed.* Asphaltic sandstone from Utah. [Argyle Creek, Uintah County, Utah] (Abraham in 1942)

ariégite *ig.* Collective name for a group of spinel-containing pyroxenites; these are without any essential feldspar or olivine but may contain some garnet or amphibole. [Ariège, Pyrenees, France] (Lacroix in 1901)

arizonite *ig. dike.* Alaskite aplite in which quartz (80%) dominates over orthoclase (18%). [Helvetia, Arizona] (Spurr, pub. by Washington in 1917)

Arkansas stone *sed., meta.* Local term for a variety of novaculite found in the Ouachita Mountains of western Arkansas.

arkesine *ig. pluton., altered.* Talc- and chlorite-bearing hornblende granite from Mont Blanc, French Alps. [Gk. *archaios,* ancient] (Jurine in 1806)

arkite *ig. pluton.* Porphyritic ijolite containing melanite (14%), pseudoleucite (37%), nepheline (26%), diopside (11%), and minor orthoclase and apatite; large pseudoleucite phenocrysts occur in a fine-grained dark groundmass. [Arkansas] (Washington in 1901)

arkose *sed.* Light-colored (usually pink or reddish) feldspar-rich sandstone containing 25% or more feldspar (potassic or sodic) and not more than 20% phyllosilicate (clay, sericite, chlorite), cemented by calcite, limonite, or more rarely silica; the feldspars may be fresh or altered. [G. *archaios,* ancient] (Brongniart in 1823)

arkose, basal *sed.* Genetic type of arkose, represented by thin and local blankets of reworked material by seas encroaching on terrain underlain dominantly by felsic plutonic rocks; often underlie feldspathic and quartzose sandstones and resemble granite.

arkose, bentonitic *sed.* Sandy volcanic ash containing less than 25% bentonitic clay minerals. (Ross and Shannon in 1926)

arkose, impure *sed.* Variously defined term for an impure highly feldspathic sandstone or graywacke, e.g., micaceous arkose; specifically, sandstone containing 25-90% feldspar and igneous rock fragments, 10-50% mica and metamorphic rock fragments, and 0-65% quartz and chert. (Folk in 1954)

arkose, lithic *sed.* Arkose containing appreciable rock fragments; specifically, sandstone containing 10-50% fine-grained rock fragments, 25-90% feldspar, and 0-65% quartz and quartz rock. (McBride in 1963) arkose, micaceous   *sed.* Arkose containing appreciable mica; specifically, sandstone containing 25-90% feldspars and feldspar-rich rock fragments, 10-50% micas and mica-rich rock fragments, and 0-65% quartz and quartz rock. (Hubert in 1960)

arkose, plagioclase   *sed.* Arkose in which the chief feldspar is plagioclase, generally sodic (albite or oligoclase).

arkose, quartzitic   *sed.* = quartzite, arkose

arkose, quartzose   *sed.* Quartz-rich arkose containing 50-85% quartz and quartz-rock fragments, 15-25% feldspars and feldspar-rock fragments, and 0-25% micas and mica-rock fragments. (Hubert in 1960)

arkose, residual   *sed.* = arkose, basal

arkose, tectonic   *sed.* Genetic type of arkose, representing materials of fluvial and torrential origin associated with coarser fanglomerates; usually thicker than basal or residual arkose deposits.

arkosite   *sed.* = quartzite, arkose. [arkose]

arkositite   *sed.* Arkose so well-cemented that the particles are interlocking. [arkose] (Tieje in 1921)

Armenian stone   *meta., etc.* Usually, lapis lazuli; has been applied to other blue materials, e.g., the mineral azurite.

Armenian whetstone   *meta.* = emery. (Theophrastus, *On Stones*)

aromatite   *sed.* Bituminous rock resembling myrrh in color and odor. [aroma]

arsoite   *ig. extru.* Trachyte consisting of phenocrysts of sanidine, andesine, diopside, and some olivine, in a groundmass of sanidine, oligoclase, diopside, magnetite, and sodalite. Also named Arso-type trachyte. [Arso flow of 1302 at Epomeo, Ischia, Italy] (Reinisch in 1912)

arterite   *migmat.* Migmatite formed by the injection of magma from outside. [L. *arteria,* artery] (Sederholm in 1897)

arthrolite   *sed.* Cylindrical concretion with transverse joints, sometimes found in clays or shales. [Gk. *arthron,* joint] (Tschersky in 1887)

articulite   *sed., meta.* = itacolumite. [L. *articulus,* joint; alluding to its flexibility] (Wetherell in 1867)

aschaffite   *ig. dike.* Old term for certain kersantites containing large quartz and plagioclase areas, originally thought to be primary but later believed to be foreign inclusions (xenoliths) brought up by the intrusion of the dike. [Aschaffenburg, Bavaria] (Gümbel in 1865)

ash, volcanic   *ig. extru.* 1. Fine-grained, unconsolidated volcanic pyroclastic material, of any composition; some petrologists use 4 mm diameter for the upper limit, others use 2 mm; likewise, fine ash is under 0.25 mm or 0.063 mm. 2. *IUGS,* pyroclasts with mean diameters smaller than 2mm. Also named ash grains. Coarse ash grains range from 1/16 mm to 2 mm, while fine

Arkose. St. Die, Vosges, France. 12 cm.

ash grains (dust grains) are smaller than 1/16 mm.

ashstone   *ig. extru.* Indurated deposit of fine volcanic ash; fine-grained tuff.

asiderite   *meteor.* Old term for stony meteorites lacking metallic components; stony meteorite. [Gk. *an,* without + *sideros,* iron] (Daubrée in 1867)

asperite   *ig. extru.* Collective field name for rough cellular lavas whose chief feldspar is plagioclase, but whose other components cannot be determined without a microscope. [L. *asper,* rough] (Becker in 1888)

asphalt   *sed.* Dark brown to black solid or semisolid bitumen; strictly one derived from an oil rich in cycloparaffin hydrocarbon. [Gk. *an,* without + *sphallein,* to slip; meaning nonslip or sticky] (Theophrastus, *On Stones*)

asphalt, lake   *sed.* Soft natural asphalt, rich in bitumen, from the pitch lake of Trinidad.

asphalt, land   *sed.* Hard natural asphalt, containing more impurities than lake asphalt, from areas outside the pitch lake of Trinidad.

asphalt, native   *sed.* Natural liquid or semiliquid asphalt in exudations or seepages including surface flows and lakes; there are ocurrences in Trinidad, in the Uinta Basin of Utah, near the Dead Sea, and elsewhere.

asphalt, natural   *sed.* = asphalt, native

asphalt rock   *sed.* Any porous sedimentary rock that is impregnated naturally with asphalt. Also named asphalt stone.

asphalte   *sed.* Originally, limestone saturated with bitumen, e.g., that from Seyssel, France; later, sandstone and sand saturated with bitumen. Varieties can be designated as calcareous asphalte (like that from Seyssel and from Oklahoma), siliceous asphalte (examples in Oklahoma), and earthy asphalte. [asphalt]

asphaltite   *sed.* Any one of several naturally occurring black solid bitumens soluble in carbon disulfide and fusing above 230°F, e.g., uintahite and grahamite. Also named asphaltum; essentially the same as asphalt.

assyntite   *ig. pluton* Obsolete name for a nepheline syenite containing titanite. [Assynt, Scotland] (Shand in 1910)

astite   *meta.* Variety of hornfels in which mica and andalusite are the characterizing minerals. [Cima d'Asta, Italian Alps] (Salomon in 1898)

astridite   *meta.?* Ornamental stone consisting mainly of chromian jadeite; from New Guinea. [Astrid, 1905-1935, wife of Leopold III, King of the Belgians] (Willems in 1934)

atatschite   *ig., altered?* Vitrophyric rock characterized by sillimanite and cordierite; there are microscopic crystals of orthoclase, augite, and biotite in a glassy matrix. [Atatsch Mountain, southern Urals, Soviet Union] (Morozewicz in 1902)

Asphalt. Trinidad, West Indies. 14.5 cm.

ataxite *ig. extru.* Taxite in which the aggregation of components resembles a breccia. [Gk. *ataxia,* disorder] (Loewinson-Lessing in 1888)

ataxite *meteor.* Iron meteorite generally containing more than 10% nickel and lacking the structure of either a hexahedrite or an octahedrite. [Gk. *ataxia,* disorder] (Brezina in 1896)

ataxite, nickel-poor *meteor.* Iron meteorite essentially like a hexahedrite but having a fine-grained structure.

ataxite, nickel-rich *meteor.* Iron meteorite containing over 12% nickel and consisting essentially of plessite (a fine-grained intergrowth of kamacite and taenite); some with excessive nickel may contain taenite with small inclusions of kamacite.

atlantite *ig. extru.* Nepheline basanite or tephrite in which the dark minerals predominate over the light. [Atlantic petrographic province] (Lehmann in 1924)

atmoclastic rock *sed.* Sedimentary rock consisting of atmoclasts (fragments broken off in place by chemical or mechanical atmospheric weathering) that have been recemented without rearrangement by wind or water. [Gk. *atmos,* vapor + *klastos,* broken] (Grabau in 1924)

atmosilicarenite *sed.* Siliceous sand formed from the weathering and disintegration of a parent rock, resulting from the passive action of the atmosphere rather than from the atmosphere in motion. [Gk. *atmos,* vapor + silica + arenite] (Grabau in 1913)

attritus *sed.* In banded coal, the finely divided, dull plant residue composed of the more resistant plant products. [L. *attritus,* worn away]

aubrite *meteor.* = achondrite, enstatite. [Aubres, France]

auganite *ig. extru.* Basalt with calcic-plagioclase and pyroxene but no olivine. [augite andesite; later classed as a basalt] (Winchell in 1912)

augitite *ig. extru.* Tephrite composed of 40% clinopyroxene (augite) with small amounts of amphibole, magnetite or ilmenite, apatite, and occasionally nepheline and other feldspathoids and feldspar, in a glassy groundmass that may be analcime; the name incorrectly suggests the rock is composed entirely of augite. [augite] (Doelter in 1882)

augitophyre *ig. dike.* Obsolete term for a diabase porphyry having phenocrysts of augite. Also named augite porphyry. [augite] (Zirkel in 1894)

australite *tekt.* Term applied to tektites from Australia and Tasmania. Earlier named obsidian bombs. [Australia] (Suess in 1900)

austro-indomalaysianite *tekt.* General term for tektites from the Far East, including Australia, Idonesia, Malaysia. (Beyer in 1933)

autoarenite *sed.* Sand produced by crushing due to tectonic movements; like an autoclastic breccia but with sand-size clasts. [Gk. *autos,* self + arenite]

autobreccia *ig. extru.* Breccia formed by some process that is contemporaneous with the formation of the rock unit from which the fragments are derived; e.g., brecciated structures shown by some andesites when there is a spalling of the viscous lava as gases in vesicles expand into dilated joints. [Gk. *autos,* self + breccia]

autosite *ig. dike.* Lamprophyric igneous rock similar in composition to kersantite but without feldspar. [Possibly Monte Auto, Italy]

avasite *sed.?* Brittle, black massive rock that is apparently a siliceous limonite, from Hungary. [Avasthal, Hungary] (Krenner in 1881)

aventurine *meta.* Quartzite exhibiting strongly colored reflections from included platy minerals; some mica produces silvery, brassy, or golden reflections, fuchsite (chromian muscovite) and chlorite produce green reflections, and hematite or goethite produce metallic reddish or brownish reflections [It. *avventura,* chance; from the accidental discovery of a synthetic material with similar characteristics]

avezacite *ig. dike.* Perknite intermediate between hornblendite and pyroxenite, in which hornblende dominates; ilmenite makes up about one-fifth of the rock. [Avezac-Prat, near Lannemezan, Pyrenees] (Lacroix in 1900)

aviolite *meta.* Variety of hornfels composed essentially of mica and cordierite. [Monte Aviolo, Italian Alps] (Salomon in 1898)

Aztec stone *ig., sed., meta.* = chalchiuhatl

azurchalcedony *sed.* Chalcedony rock colored blue by chrysocolla; sometimes used as a gem; found in Arizona. [azure + chalcedony] (Kunz in 1907)

azure stone *meta.* = lapis lazuli

azurlite *sed.* = azurchalcedony. [azure] (Kunz in 1907)

azurmalachite *altered ore.* Mixture of azurite and malachite in concentric bands; used as a gem; found esp. in Arizona copper deposits. (Kunz in 1907)

B

bacalite   *sed.* Yellow to white variety of amber from Baja California, Mexico. [Baja California] (Buddhue in 1935)

bagotite   *ig., altered.* = lintonite. [Bagot, Ontario, Canada] (pub. by Egleston in 1889)

bahamite   *sed.* Lithocalcarenites, varying from calcisiltites to calcirudites, in which the grains are accretionary and commonly composite, consisting of smaller granules bound together by precipitated material (often aragonite) into aggregate grains. [Bahama Islands] (Beales in 1958)

bahamite, dolomitic   *sed.* Bahamite in which the primary texture is obliterated by the formation of an anhedral mosaic of dolomite crystals. (Beales in 1958)

bahiaite   *ig. pluton.* Brownish black coarsely granular hornblende hypersthenite containing minor amounts of olivine and ferroan spinel (pleonaste). [Bahia, Brazil] (Washington in 1914)

baikerite   *sed.* Variety of ozokerite. Also spelled baikalite. [Lake Baikal, Soviet Union] (Hermann in 1858)

baldite   *ig. dike.* Hypabyssal equivalent of an analcime basalt, composed of pyroxene phenocrysts in a groundmass of analcime, augite, and iron oxides. [Big Baldy Mountain, Montana] (Johannsen in 1938)

balkhashite   *sed.* Rubbery bitumen (elaterite) similar to coorongite, formed by algae in Lake Balkhash, Siberia. [Lake Balkhash, Soviet Union] (Kumpan in 1931)

ball, accretionary lava   *ig. extru.* Rounded mass formed on the surface of a lava flow by the molding of viscous lava around a core of already solidified lava.

ball, algal   *sed.* = biscuit, algal

ball, armored clay   *sed.* = ball, armored mud

ball, armored mud   *sed.* Large subspherical mud ball coated or armored with gravel; clay chunks, released by rapid river-bank erosion, upon rolling down stream may acquire a gravel armor.

ball, clay   *sed.* Piece of clay released by erosion of a clayey bank and rounded by wave action; essentially the same as a mud ball.

ball, coal   *sed.* Spherical concretions, composed of calcite or dolomite, that occur in coal deposits and often contain well-preserved plant fossils of the species from which the coal was divided.

ball, lava   *ig. extru.* Globular mass of lava, scoriaceous on the inside and compact on the outside; it is formed by the coating of scoria fragments by fluid lava. An accretionary lava ball is a variety.

ball, load   *sed.* Bulbous to pouchlike structure formed by the downward protrusion of sands into underlying mud (a load structure); when these load pouches are detached after induration they are named load balls and resemble concretionary sandstones.

ball, mud   *sed.* Spherical mass of mudstone or clay, formed by the weathering and breakup of clay deposits; they may develop from clay chunks that develop a globular shape upon rolling down grade.

ball, shale   *meteor.* Term applied to a meteorite partly or wholly converted to iron oxides by weathering and consisting of laminated iron shale. Also named oxidite.

ball, volcanic   *ig. extru.* = ball, lava

ballstone   *sed.* Large rounded nodule in a stratified unit, e.g., an ironstone nodule in coal.

baltimorite   *meta.* Serpentinite (picrolite type) of a grayish-green color and silky luster, from Bare Hills, Maryland. [Baltimore, Maryland] (Thomson in 1843)

banakite   *ig. extru.* Basaltic rock with olivine and clinopyroxene phenocrysts in a groundmass of labradorite with alkali feldspar rims, olivine, clinopyroxene, some leucite, and possibly quartz. [Bannock or "Robber" Indians, Idaho] (Iddings in 1895)

banatite   *ig. pluton.* Obsolete term for a group of dioritic rocks, varying from orthoclase-bearing quartz diorite to quartz-free augite diorite. [Banat district, Hungary] (von Cotta in 1865)

bandaite   *ig. extru.* Obsolete term for a labradorite- or bytownite-dacite whose classification was based on a chemical analysis. [Bandai San, Japan] (Iddings in 1913) banket   *sed.* Siliceous conglomerate consisting of vein-quartz pebbles embedded in a quartzitic matrix; originally applied to the gold-bearing mildly metamorphosed conglomerates of Witwatersrand area of South Africa. [Afrik. *banket,* confectionary]

baramite   *meta.* Magnesite serpentinite formed from the alteration of a lherzolite or marchite, consisting of serpentine (43%) and magnesite (36%), opal, chalcedony, and iron oxides. [Baramia quarry, southeastern Arabian desert, upper Egypt] (Hume in 1935)

barite rock   *sed.* Massive, earthy, fibrous, platy, or well-crystallized barite occurring widely in clay, siltstone, sandstone, marl, limestone, and dolomite; it may be in the form of nodules, spherulites, concretions, veins, lenses, cavity fillings, and cement. Varieties include barite concretions, barite nodules, barite roses, barite sand crystals. See also quartz-barite rock.

barolite   *sed.* Heavy rock composed either of barite or celestite. [Gk. *barus,* heavy] (Wadsworth in 1891)

barshawite   *ig. pluton.* Pinkish, even-grained ijolite, composed of barkevikite, nepheline, orthoclase, titanaugite, analcime, and andesine. [Barshaw, near Paisley, Scotland] (Johannsen in 1938)

basalatite   *ig. extru.* Obsolete general term for an alkali-rich basalt with orthoclase or anorthoclase in addition to labradorite. [basalt + latite] (Johannsen in 1920)

basalt   *ig. etru.* Volcanic, and locally intrusive (dikes, sills), aphanitic equivalent of gabbro, whose essential mineral is calcic-plagioclase usually with mafics (clinopyroxene and olivine); a very common lava. There are many varieties depending upon variations in texture (porphyritic, glomeroporphyritic, glass), structures (scoria, aa, pahoehoe, vesicular, cellular, columnar), mineralogy (presence of orthopyroxene, olivine, biotite, hornblende, native iron, graphite, quartz, etc.). [L. *basaltes,* dark marble; possibly from Gk. *basanos,* touchstone, or Egypt. *bechen,* hard dark rock, or Heb. *barzel,* iron] (Agricola in 1546)

basalt, alkali   *ig. extru.* Critically silica-undersaturated basalt, containing normative nepheline, diopside, and olivine, with no normative hypersthene. (Hibsch in 1910; redefined by Yoder and Tilley in 1962)

basalt, alkali-olivine   *ig. extru.* basalt, alkali. (Tilley in 1950)

basalt, alkaline   *ig. extru.* basalt, alkali. (Chayes in 1964)

basalt, amygdaloidal   *ig. extru.* Vesicular basalt in which the vesicles are filled with secondary minerals, such as quartz, jasper, calcite, epidote, feldspars, zeolites, and others. [Gk. *amygdale,* almond; alluding to the shape]

basalt, analcime   *ig. extru.* Extrusive or dike rock composed of analcime, augite, and olivine; not a true basalt. (Lindgren in 1880)

basalt, anemousite   *ig. extru.* = pacificite

basalt, anorthite   *ig. extru.* Basalt in which the plagioclase is anorthite; however, rocks described as such apparently have bytownite, rather than anorthite, and seem to be too rich in silica. (Wada in 1882)

basalt, calciclase   *ig. extru.* = basalt, anorthite

basalt, columnar   *ig. hypab., extru.* Structural form of basalt in which large polygonal prismatic columns of the rock form during crystallization; usually the columns are vertical, but there are cases where they are inclined, bent, or horizontal, e.g., the Giant's Causeway, on the northern coast of Ireland.

basalt, Deccan   *ig. extru.* Nonporphyritic tholeiitic basalt covering an area of about 200,000 sq. mi. in the Deccan region of southeastern India.

basalt, enstatite   *ig. extru.* Orthopyroxene basalt whose mafic is enstatite.

basalt, feldspathoidal   *ig. extru.* = basalt, foid

basalt, foid   *ig. extru.* Extrusive, usually holocrystalline and aphanitic, equivalent of foid gabbro. Also named feldspathoidal basalt and foidal basalt.

basalt, foidal   *ig. extru.* basalt, foid

basalt, Fra Mauro   *ig. lunar.* Basaltic rocks in the lunar highlands, differing from mare basalts primarily by their higher plagioclase content; several varieties have been distinguished on the basis of chemical composition, e.g., KREEP. [Fra Mauro, the Apollo 14 landing site on the moon where rock was collected]

basalt, glimmer   *ig. extru.* Mica basalt in which biotite is an essential constituent although it does not entirely proxy for pyroxenes. [G. *Glimmer,* mica] (Möhl in 1874)

basalt, graphite   *ig. extru.* Rare rock in which graphite, as fine particles or in little clusters, occurs distributed throughout a basalt. (Steenstrup in 1882)

basalt, hauyne   *ig. extru.* Olivine-bearing feldspar-free rock in which hauyne, a feldspathoid, is accompanied by pyroxene. (Trimmer in 1841)

basalt, hauyne-nepheline   *ig. extru.* Extrusive rock intermediate between hauyne basalt and nepheline basalt, which, in addition to pyroxene and olivine, contains the feldspathoids hauyne and nepheline.

basalt, high-alumina   *ig. extru.* Nonporphyritic basalt distinguished by a higher content of Al_2O_3 (generally higher than 17%) than that of tholeiite with similar SiO_2 and total alkalies and by lower alkali content than that of alkali basalt; generally transitional between tholeiite basalt and alkali basalt. (Kuno in 1960)

basalt, hornblende   *ig. extru.* Basalt with hornblende, usually as large phenocrysts in a holocrystalline groundmass of calcic-plagioclase, pyroxene, and sometimes olivine. (Sommerlad in 1881)

basalt, hypersthene   *ig. extru.* Olivine basalt in which the usual clinopyroxene (augite) is proxied by hypersthene, an orthopyroxene. (Diller in 1884)

basalt, iron   *ig. extru.* Extrusive basalt that, in addition to plagioclase and pyroxene, contains grains, flakes, lumps, and large masses of telluric iron; occurs on Disco Island, Greenland. (Stienstrup in 1876).

basalt, leucite   *ig. extru.* Extusive ash- to dark-gray basaltic rock composed of leucite, pyroxene, and olivine; rock in which the feldspathoid leucite takes the place of feldspar.

basalt, leucite-nepheline   *ig. extru.* Extrusive rock intermediate between leucite basalt and nepheline basalt, which, in addition to pyroxene and olivine, contains the feldspathoids leucite and nepheline.

basalt, leucitoid   *ig. extru.* Obsolete term applied to rocks that appeared to be leucite basalt but in which the leucite was irregular and undeterminable by microscopy. (Bořický in 1874)

basalt, lunar   *ig. lunar.* Basalt from the moon that is composed of nearly equal amounts of augite, highly calcic-plagioclase (bytownite), and ilmenite; chemically it contains more titanium dioxide, rare-earth elements, zirconium, and less nickel than terrestrial basalt.

basalt, melilite   *ig. extru.* Fine-grained or porphyritic extrusive rock composed of melilite, pyroxene, and olivine; rock in which melilite takes the place of feldspar. (Stelzner in 1882)

basalt, melilite-nepheline   *ig. extru.* Extrusive rock intermediate between melilite basalt and nepheline basalt, which, in addition to pyroxene and olivine, contains melilite and nepheline.

basalt, mica   *ig. extru.* Basalt in which biotite is essential; the labradorite is usually sodic, and there may be hornblende and augite; nearly the same as glimmer basalt.

basalt, nepheline   *ig. extru.* Fine-grained or porphyritic extrusive rock composed of nepheline, pyroxene, and olivine; rock in which nepheline takes the place of feldspar. (Naumann in 1850)

basalt, nosean   *ig. extru.* Olivine-bearing, feldspar-free rock in which nosean, a feldspathoid, is accompanied by pyroxene. (Lindgren in 1880)

basalt, oligoclase   *ig. extru.* Misnomer for a rock that is probably augite andesite; the term basalt implies calcic-plagioclase while oligoclase is sodic. (Bořický in 1874)

basalt, olivine   *ig. extru.* Basalt containing olivine as an essential constituent; it differs from the normal basalt only in the texture as modified by the presence of olivine, which is usually fresh and is present among the phenocrysts. (Johannsen in 1937)

basalt, orthoclase   *ig. extru.* Rare variety of basalt in which orthoclase is abundant; it is normally present only as a minor accessory.

basalt, picrite   *ig. extru.* Poor name for an olivine-rich, mafic basalt; essentially equivalent to masafuerite, a picrite.

basalt, pillow   *ig. extru.* = lava, pillow

basalt, quartz   *ig. extru.* Basalt carrying primary quartz; some examples are unusual in that they can also contain primary olivine. Foreign quartz (e.g., from sand) may be abundant in some basalts; these grains are often rimmed by shells of glass and augite prisms. (Diller in 1887)

basalt, quartz-bearing   *ig. extru.* Quartz basalt in which quartz is less than 5% of the total light-colored components.

basalt, sanidine   *ig. extru.* Rare variety of basalt containing sanidine; related to orthoclase basalt.

basalt, sodalite   *ig. extru.* Fine-grained or porphyritic extrusive rock composed of sodalite, pyroxene, and olivine; rock in which sodalite takes the place of feldspar.

basalt, subalkaline   *ig. extru.* Basalt with less sodium (without nepheline and sodic pyroxene) than alkali basalt. The term was proposed to replace tholeiite. (Chayes in 1964)

basalt, tholeiitic   *ig. extru.* = tholeiite

basalt, vesicular   *ig. extru.* Basalt lava containing abundant vesicles (rounded cavities or cells) formed as a result of the expansion of gases during the fluid stage. [L. *vesicula*, bladder]

basalt, VHA   *ig. lunar.* Very high alumina basalt found on the moon; since the term is based on the chemical composition, it may include a mixture of other rock types. [acronym for very high alumina] basalt, vitrophyric *ig. extru.* = hyalobasalt. A misnomer, since these glassy rocks seldom contain phenocrysts, a feature implied by the term vitrophyre.

basaltite *ig. extru.* Term sometimes used to refer to basalt without olivine. [basalt]

basanite *ig. extru.* Dark basaltlike rock characterized by the combination of calcic-plagioclase, feldspathoid, and olivine; without olivine the rock is a tephrite. Variations in the feldspathoids determine the following varieties: hauyne b., leucite b., nepheline b., nosean b. [Gk. *basanos,* touchstone] (Brongniart in 1813)

basanite *sed.* Velvety-black variety of cryptocrystalline quartz, siliceous shale, or similar substance, slightly tougher and finer grained than jasper; used by jewelers for testing the purity of precious metals, esp. gold. Also named Lydian stone or touchstone. [Gk. *basanos,* touchstone]

basanitoid *ig. extru.* Seldom used name for rock intermediate between olivine basalt and igneous basanite. [basanite] (Bücking in 1880)

basic rock *ig.* = basite

basite *ig.* Collective name for all mafic-rich (basic) igneous rocks with SiO_2 45-50% by weight. Also named basic rock. [basic, alluding to relatively low silicic acid] (von Cotta in 1864)

bastite *meta.* Altered orthopyroxene (enstatite or bronzite) having the composition of serpentine; green to brownish, occuring in foliated form in certain granular extrusive igneous rocks, and characterized by a bronzelike metalloidal luster or schiller. [Baste near Harzburg, Germany] (Haidinger in 1845)

Bath stone *sed.* Soft, cream-colored, oolitic limestone, easily quarried and used for building purposes. [Bath, England]

bathvillite *sed.* Amorphous, opaque, very brittle brown woody resin occurring as porous lumps in torbanite at Torbane Hill, in Bathville, Scotland. [Bathville] (Williams in 1863)

batukite *ig. extru.* Dark leucitite lava consisting of over 80% mafic minerals; phenocrysts of augite and some olivine occur in a groundmass of augite, magnetite, and leucite. [Batuku, Celebes] (Iddings and Morley in 1917)

bauxite *sed.* Residual sedimentary rock composed of one or more of the aluminum hydroxide minerals, gibbsite, boehmite, diaspore, usually with variable amounts of clay (usually kaolinite) and some iron and titanium oxides; although often pisolitic or oolitic, specimens may be massive, fragmental, or cellular. [Les Baux or Beaux, France] (Dufrenoy in 1847)

bauxite, lateritic *sed.* Term sometimes applied to bauxite formed from and upon crystalline igneous (or more rarely metamorphic) rocks under tropical conditions. [laterite]

bauxite, terra rossa *sed.* Term sometimes applied to bauxite associated with carbonate rocks and reportedly formed from clay minerals. [It. *terra rossa,* red earth]

Bayard egg *sed.* Smooth red, yellow, and orange jasper pebbles, stained jet-black at the surface. [Bayard, Nebraska] (Mitchell in 1946)

bayate *sed.* Local name for a brown ferruginous type of jasper occurring with manganese ores of Cuba. [Possibly Bayamo, Cuba] (Burchard in 1920)

beachrock *sed.* Friable to well-cemented sand or gravel rock (detrital or skeletal), formed in the intertidal zone in a tropical or subtropical region, cemented with calcium carbonate.

beaconite *meta.* Fibrous variety of talc resembling asbestos; found in Michigan. [Beacon, Michigan] (Wadsworth in 1893)

bebedourite *ig. pluton.* Medium- to coarse-grained pyroxenite containing biotite; the original specimens consisted essentially of diopside and biotite, with accessory perovskite, and others. [Bebedouro, Mina Geraes, Brazil] (Tröger in 1928)

beckerite *sed.* Dark-brown, soft, dense, nonfusible variety of amberlike fossil resin, occurring in lumpy, opaque to cloudy masses in the Baltic area. Also named brown resin (in German, Braunharz). [Mority Becker, a merchant, who with W. Stantien dredged for amber in the Baltic in the last half of the 19th century] (Pieszczek in 1881)

beckite *sed.* Common incorrect spelling of beekite.

bediasite *tekt.* Name often applied to tektites from Texas. [Bedias, Grimes County, Texas, after the Bedias Indians] (Barnes in 1940)

bedrock *ig., sed., meta.* General term for the solid rock that underlies surface gravel, soil, and similar materials; may be igneous, sedimentary, or metamorphic.

beekite *sed.* Chalcedonic chert formed by the replacement of limestone; often it takes the form of shells, coral, or other fossils, and frequently is subspherical, doughnut-shaped, or botryoidal with bands. [Henry Beeke, 1751-1837, dean of Bristol, England, who called attention to the rock] (Dufrénoy in 1847)

beerbachite *meta., ig. dike.* Saccharoidal hornfels derived from xenolithic inclusions in gabbro and composed of labradorite, clinopyroxene (diallage), hornblende, hypersthene, and others. The term also has been applied to igneous rocks having a similar composition, and the following varieties have been named: pyroxene beerbachite, amphibole beerbachite, and olivine beerbachite. [Beerbach, Odenwald, Germany] (Chelius in 1892)

beetle stone   *sed.* 1. Nodule of coprolite-bearing ironstone, so named from the resemblance of the enclosed coprolite to the body and limbs of a beetle.   2. = septarium, esp. turtle stone.

beforsite   *ig. pluton* Dolomite carbonatite; the original examples contained primary dolomite (60%), sodium-rich biotite (30%), barite (6%), iron oxides, and pyrite. [Bergeforseu, Alnö, Sweden] (Eckermann in 1948)

bekinkinite   *ig. pluton.* Feldspar-poor plutonic igneous rock composed of barkevikite, nepheline, and olivine; related to a theralite. [Mt. Bekinkina, Madagascar] (Rosenbusch in 1907)

beldongrite   *sed.* Black, pitchlike material closely allied to psilomelane; apparently from the alteration of spessartine. [Beldongri, Nagpur district, India] (Fermor in 1909)

beloeilite   *ig. pluton.* Light gray, granular igneous rock composed of abundant sodalite, less potassic feldspar (orthoclase), and a moderate amount of dark minerals (aegirine), sodic-plagioclase, magnetite, and zeolite. [Beloeil, old name for Mt. St. Hilaire, Quebec] (Johannsen in 1938)

belugite   *ig. pluton.* Unnecessary name for rocks whose feldspars are intermediate (andesine and/or labradorite) between diorite and gabbro. [Beluga River, Alaska] (Spurr in 1900)

benmoreite   *ig. extru.* Silica-saturated to undersaturated rock intermediate between trachyte and mugearite, with a differentiation index of 65-75 and $K_2O:Na_2O$ less than 1:2. [Ben More, Mull, Scotland] (Tilley and Muir in 1964)

bentonite   *sed.* Soft, highly colloidal, plastic, light-colored clay rock, essentially consisting of minerals of the smectite (montmorillonite) group; it can swell to several times its original volume when placed in water; it is usually formed by the alteration of volcanic ash in situ. [Fort Benton formation, Rock Creek district, Wyoming] (Knight in 1898)

bentonite, arkosic   *sed.* Bentonite containing 25-75% sandy impurities; the detrital grains usually represent unaltered crystalline grains from an original volcanic ash. (Ross and Shannon in 1926)

bentonite, potassium   *sed.* Potassium-bearing clay of the illite group, formed from the alteration of volcanic ash; a metabentonite consisting of randomly interstratified layers of montmorillon-

ite and illite with a ratio of 1:4 (potassium occupying about 80% of the exchangeable-cation positions of the mica portion). Also named potash bentonite and K-bentonite.

berengelite   *sed.* Asphaltlike, dark-brown resin from Peru. [St. Juan de Berengela, Peru] (Johnston in 1838)

beresite   *ig., autometa.* Dikelike rock altered to a material that resembles greisen, consisting of quartz, white mica (from alteration of feldspar), and minor pyrite. [Beresovsk gold mine, Urals] (Rose in 1837)

bergalite   *ig. hypab.* = bergalith

bergalith   *ig. hypab.* Lamprophyre with phenocrysts of melilite, hauyne, biotite, and rare augite in a dense groundmass composed of nepheline, magnetite, perovskite, apatite, secondary calcite, and glass. [Oberbergen, Kaiserstuhl, Baden, Germany] (Söllner in 1913)

beringite   *ig. extru.* Dark, barkevikite andesite in which the feldspars are albite and subordinate orthoclase. [Bering Island, Bering Sea] (Starzynski in 1912)

bermudite   *ig. extru.?* Poorly defined light-colored lamprophyric rock consisting of numerous small flakes of biotite in a felsic groundmass, possibly a mixture of nepheline, sanidine, and analcime. [Bermuda Island] (Pirsson in 1914)

bernstein   *sed.* = amber. [O. G. *börnen,* to burn + G. *Stein,* stone; alluding to its ability to ignite when heated]

berondrite   *ig. pluton.?* Theralite similar to luscladite, but the mafics are long hornblende prisms and titanaugite. [Berondra Valley, Madagascar] (Lacroix in 1920)

beschtauite   *ig. extru.* Sodium-rich rhyolitic porphyry containing sanidine, oligoclase, and quartz, with hornblende, and very minor biotite, augite, and apatite. Also spelled beschtaunite. [Mt. Beschtau, Caucasus] (Bayan in 1866)

besimaudite   *meta.* Gneissoid talc rock, originally described in Italy. [Besimauda, Piemonte, Italy] (Zaccagna in 1887)

bielenite   *ig. pluton.* Fine-grained, blackish-gray peridotite containing olivine and various pyroxenes (diallage, enstatite), with accessory chromite and magnetite; it differs from lherzolite in containing more pyroxene than olivine. [Biele River, Moravia, Czechoslovakia] (Kretschmer in 1917)

bielzite   *sed.* Massive, resinous, brownish-black hydrocarbon from Transylvania. [Eduard Albert Bielz, 1827-1898] (Benkö and Jahn in 1887) bigwoodite *ig. pluton.* Medium-grained alkali syenite composed essentially of microcline, microcline-microperthite, albite, and hornblende (bright green), with calcite as a common accessory. [Bigwood Township, Sudbury district, Ontario, Canada] (Quirke in 1936)

billitonite *tekt.* Name applied to tektites from Billiton (Belitung) Island, Indonesia. (Suess in 1900)

bimstein *ig. extru.* = pumice. Also spelled bimsstein. [G. *Bimsstein,* pumice stone]

binghamite *sed.* Chalcedony-rich, red, brown, or cream colored rock containing embedded fibers of goethite; when polished the chatoyancy produces fine cat's-eye gems; found in Minnesota. Also named cuyunite. [William J. Bingham, St. Paul, Minnesota]

biocalcarenite *sed.* Calcarenite in which fossil organic fragments of any kind (or products of organic activity), dominate. [Gk. *bios,* life + calcarenite]

biocalcilutite *sed.* Calcilutite in which biochemical materials play an important role; because these rocks are often entirely recrystallized, evidence of biochemical materials may be obliterated. [Gk. *bios,* life + calcilutite]

biocalcilyte *sed.* Calcareous biogenic clastic rock, such as shell rock, coral rock, or calcareous ooze. Also spelled biocalcilite. [Gk. *bios,* life + calcite or calcareous] (Grabau in 1924)

biocalcirudite *sed.* Calcirudite in which fragments of reef-building organisms, such as stromatoporoids, branched corals, shells, and calcareous algae, broken away and worn to a variable degree, are aggregated. [Gk. *bios,* life + calcirudite]

biocalcisiltite *sed.* Calcisiltite containing abundant fossils or fossil fragments. [Gk. *bios,* life + calcisiltite]

bioclastic rock *sed.* 1. Sedimentary rock consisting of the broken remains of organisms, such as a limestone composed of shell fragments. 2. Rock consisting of fragments broken from pre-existing rocks or pulverized or arranged, by the action of living organisms, such as plant roots or earthworms; the rock need not consist of organic materials. [Gk. *bios,* life + *klastos,* broken] (Grabau in 1904)

biogenetic rock *sed.* = biogenic rock

biogenic rock *sed.* Organic rock produced directly by the physiological activities of either plant or animal organisms, e.g., coal, peat, coral reefs, shelly limestone, and many others. [Gk. *bios,* life + *genes,* born] (Grabau in 1924)

bioherm *sed.* Organic reef or mound rock built by corals, gastropods, stromatoporoids, foraminifera, and other organisms, and composed almost exclusively of their calcareous remains. [Gk. *bios,* life + *eremites,* hermit] (Cumings and Shrock in 1928)

bioherm, crinoidal *sed.* Bioherm formed by a thickly populated crinoid colony; these restricted accumulations of crinoidal limestone have been noted to form reservoir rock in the Todd Oil Field in Texas.

biohermite *sed.* Limestone composed of debris broken from a bioherm and forming pocketfillings or talus slopes associated with reefs. [bioherm] (Folk in 1959)

biolite *sed.* 1. Old term for a concretion formed through the action of living organisms. 2. = biolith. [Gk. *bios,* life]

biolith *sed.* Rocks of organic origin that can be divided into two separate groups: acaustobiolith (noncombustible, inorganic in composition) and caustobiolith (combustible, organic in composition); biogenic rock. [Gk. *bios,* life]

biolithite *sed.* Limestone formed in situ, such as a reef rock or stromatolite. [Gk. *bios,* life]

biomicrite *sed.* Limestone composed of skeletal grains (bioclasts) in a matrix of micrite (carbonate mud). The major organism is often specified, e.g., brachiopod biomicrite [Gk. *bios,* life + micrite] (Folk in 1959)

biomicrite, packed *sed.* Biomicrite in which the skeletal grains make up over 50% of the rock.

biomicrosparite *sed.* Biomicrite in which the carbonate-mud matrix has recrystallized to microspar; a microsparite containing fossils. [Gk. *bios,* life + microsparite]

biomicrudite *sed.* Biomicrite containing fossils more than 1 mm in diameter. [biomicrite + rudite]

biopelite *sed.* Pelite containing organic materials, e.g., black (carbonaceous) shale. [Gk. *bios,* life + pelite]

biopelmicrite *sed.* Limestone intermediate between a biomicrite and a pelmicrite; specifically, limestone containing less than 25% intraclasts and less than 25% ooliths, with a volume ratio of fossils to pellets ranging between 3:1 and 1:3, and with the carbonate-mud matrix more abundant than the sparry-calcite cement. [Gk. *bios,* life+ pelmicrite] (Folk in 1959)

biopelsparite *sed.* Limestone intermediate in content between biosparite and pelsparite; it has the same compositional requirements as a biopelmicrite, except that the sparry-calcite cement is more abundant than the carbonate-mud matrix. |Gk. *bios,* life + pelsparite| (Folk in 1959)

biosparite   *sed.* Limestone composed of fossil skeletal grains (bioclasts) cemented by sparite (clear equant calcite); the major organism is often specified, e.g., pelecypod biosparite. Also named shelly biosparite. |Gk. *bios,* life + sparite| (Folk in 1959)

biosparrudite   *sed.* Biosparite containing fossil fragments (bioclasts) more than 1 mm in diameter. |biosparite + rudite|

biostrome   *sed.* Stratiform rock deposit, such as a coral bed or shell bed, consisting of, and built mainly by, organisms or fragments of organisms, and without a moundlike or lenslike form |Gk. *bios,* life + *stroma,* bed| (Cumings in 1932)

biotitite   *ig. pluton.* Perknite composed almost exclusively of biotite. |biotite| (Washington in 1927)

birkremite   *ig. pluton.* Plagioclase-poor hypersthene-bearing kalialaskite.|Birkrem (Bjerkreim), Egersund district, Norway| (Kolderup in 1896)

birmite   *sed.* = burmite

biscuit, algal   *sed.* Hemispherical or disklike calcareous mass (up to about 20 cm in diameter) produced in fresh water as a result of precipitation by various blue-green algae.

biscuit, marl   *sed.* Algal biscuit consisting of a hard, rounded flat concretion of marl formed around shell fragments; found on the shore or shallow bottom of a lake, esp. in the northern United States.

bistagite   *ig., meta.?* Rock composed of diopside, occurring in a serpentine deposit in Siberia; it is uncertain whether it formed from a magma, like a diopsidite, or is metamorphic. |Bis-Tag, Jenissei, district in Siberia| (Jaczewski in 1909)

bitumen   *sed.* Term loosely used for any natural hydrocarbon material in sediments; specifically, the term is used for liquid or solid hydrocarbons soluble in organic solvents. Bitumens range from petroleum (liquid) to anthraxolite (solid). |L. *bitumen,* traced to L. *pix,* pitch, and *tumere,* to swell, thus *pixtumens,* swelling pitch|

bitumen, eu-   *sed.* Collective name for various fluid, viscous, or solid bitumens that are easily soluble in organic solvents, including elaterite, asphalt, ozokerite, petroleum. (Tomkeieff in 1954)

bitumen, vein   *sed.* Any of several bitumens that give off a pitchy odor, burn readily with a smoky flame, and occupy fissures in rocks or less frequently form basin-shaped deposits on the surface.

bitumenite   *sed.* = torbanite. (Traill in 1857)

bizardite   *ig. hypab.* Nepheline-containing alnoite, consisting of olivine and mafic phenocrysts (augite and biotite) with melilite, nepheline, perovskite, apatite, and secondary calcite, serpentine, chlorite, and cancrinite. |Isle Bizard, Montreal, Quebec| (Stansfield in 1923)

bjerezite   *ig. hypab.* Porphyry showing phenocrysts of nepheline, pyroxene with aegirine borders, long andesine plates, and orthoclase in a fine-grained groundmass of pyroxene, brown mica, andesine, potassic feldspar, nepheline, analcime, and zeolites. |Bjerez River, Atschinsk, Siberia| (Erdmannsdörffer in 1928)

björnsjöite   *ig. hypab.* Sodium-rich porphyritic nordmarkite containing albite (as phenocrysts), aegirine, quartz, apatite, titanite, and calcite. |Island of Storo, Björn Lake, Norway| (Brögger in 1932)

blackmorite   *sed.* Yellow variety of opal from Montana. |Mt. Blackmore, Montana| (Peale in 1873)

blackstone   *ig., sed., meta.* General term applied to black rocks, e.g., basalt, basanite, oil shale, and others.

blaes   *sed.* Scottish name for gray blue carbonaceous shales associated in the Lothians with oil shales; they have a low content of bituminous matter and are more brittle than the oil shale. |Scot. *blae,* blue|

blairmorite   *ig. extru.* Phonolite porphyry characterized by dominant phenocrysts of analcime in a matrix of analcime, alkali feldspar, alkali pyroxene, with minor nepheline, melanite, and titanite. |Blairmore, southwest Alberta, Canada| (Knight in 1904)

blastite   *meta.* General term for a metamorphic rock composed of crystalloblasts (blasts), crystals grown during metamorphism; e.g., plagioclase blastite, consisting of rounded plagioclase grains in a biotite-rich matrix (sometimes referred to as a pearl gneiss). |Gk. *blastos,* sprout| (Becke in 1903)

blastocataclasite   *meta.* Cataclasite showing some recrystallization effects. |Gk. *blastos,* sprout + cataclasite|

blastomylonite   *meta.* Mylonitic rock in which some recrystallization or new mineralization has occurred. |Gk. *blastos,* sprout + mylonite|

blaviérite   *meta.* Contact metamorphic rock formed by the intrusion of granitic magma upon sericite schist; although the schistose structure is preserved, in addition to the fine micaceous components of the schist, there are crystals of quartz, orthoclase, and oligoclase. [Blaviér, Mayenne, France] (Munier-Chalmas in 1862)

block, glacial   *sed.* Large angular rock fragment not greatly modified during glacial transport.

block, volcanic   *ig. extru. IUGS,* pyroclast with a mean diameter exceeding 64 mm, whose commonly angular or subangular shape indicates that during its formation it was in the solid state. Also named block.

bloodstone   *sed.* Dark green jasper with bloodlike red spots; heliotrope is related but is usually defined as chalcedony with red jasper spots.

blue ground   *ig. hypab.* Hard, unaltered kimberlite found below the decomposed yellow ground in the diamond pipes of Africa. Miners' term.

blueschist   *meta.* Schist with a blue color from the presence of the amphiboles glaucophane or crossite; usually a glaucophane schist.

bluestone   *sed., ig.* 1. Highly argillaceous sandstone of even texture and bedding, usually formed in a lagoon or lake near the mouth of a stream.   2. Common local and commercial term for various bluish colored rocks, including sandstone (flagstone), shale, limestone, hard clay, and basalt.

boakite   *sed.* Local name for a brecciated green and red jasper found in Nevada. [Possibly Mr. C. C. Boak, Tonopah, Nevada]

bobrovkite   *ig.?* Alloy of nickel and iron, found as fine scales in platiniferous sands. [Bobrovka River, Nijni-Tagil, Urals] (Wyssotzky in 1913)

bog iron ore   *sed.* Hard, oolitic, pisolitic, and concretionary poorly crystalline bodies, composed of goethite (most common), siderite, vivianite, and magnanese oxides; presently they are forming in swamps and lakes of middle to high latitudes.

bogusite   *ig. pluton.* Igneous rock closely related to teschenite but lighter in color. [Boguschowitz, Silesia, Czechoslovakia] (Johannsen in 1938)

bojite   *ig. pluton.* Gabbro in which primary hornblende proxies for the usual augite; hornblende gabbro, not to be confused with a uralite gabbro in which the hornblende is secondary in origin. [Boii, a Celtic tribe that settled in Germany] (Weinschenk in 1899)

bole   *sed.* Any of several fine, compact, soft clays, frequently red, yellow, or brown from the presence of iron oxide and consisting essentially of hydrous silicates of aluminum or less often of magnesium; often forms from the decomposition of basaltic rocks. [Gk. *bolos,* clod]

Bologna stone   *sed.* Reddish gray nodular (spherulitic) barite, found in a clay bed in Mt. Paterno, near Bologna, Italy; an early source of wonder because of its strong phosphorescence.

boloretin   *sed.* Resin from fossil firwood. [Gk. *bolos,* clod, earth + *rhetine,* resin; alluding to its earthy luster] (Forchhammer in 1840)

bomb   *ig. extru.* 1. Lava ejected while still viscous and receiving a rounded shape while in flight; the actual form or shape varies greatly; some petrologists require a diameter greater than 32 mm, and others require 64 mm; most bombs are basaltic in composition. Also named volcanic bomb.   2. *IUGS,* pyroclast with a mean diameter commony exceeding 64 mm, wholly or partly molten during its formation and subsequent transport.

bomb, bread-crust   *ig. extru.* Light and porous volcanic bomb covered with a skin resembling the crust of bread; as the bomb swells, from the expansion of included gases during its flight, the crust is often cracked open.

bomb, cored   *ig. extru.* Volcanic bomb with a core of nonvolcanic rock or already solidified lava, around which lava has molded itself. Also named perilith.

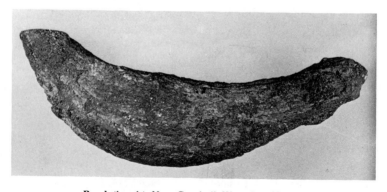

Bomb (basalt). Near Graybull, Wyoming. 22 cm.

bomb, cow-dung *ig. extru.* Volcanic bomb with a flattened shape (and a somewhat scoriaceous surface due to its impact while still viscous.

bomb, impact *ig. extru.* Volcanic bomb flattened by its impact while still viscous. A cow-dung bomb is a variety.

bomb, impact *impact.* Porous mass of impactite formed by splattering and exhibiting sculpturing while in flight.

bomb, obsidian *tekt.* Early name for tektites, esp. for those found in Australia and Tasmania. (Darwin in 1844?)

bomb, ribbon *ig. extru.* Volcanic bomb with a long, thin, sometimes tapering and twisted, flattened tail.

bomb, spindle *ig. extru.* Volcanic bomb with a tapering projection on the end of a football-shaped body; usually formed by the rotational or spiral motion of the lava clot as it was hurled through the air.

bomb, volcanic *ig. extru.* = bomb

bombite *ig extru.?* Dark, blackish gray material from near Bombay, India, resembling Lydian stone (basanite); probably a glassy igneous rock, perhaps hyalobasalt. |Bombay, India| (de Bournon in 1859)

bone bed *sed.* Phosphorite deposit, often bituminous, appearing mostly as a microbreccia containing bones, shark's teeth, fish scales, and various coprolites; collophane is major, but calcite, quartz, opal, dolomite, barite, and pyrite may occur. Also named bone phosphorite.

bone turquoise *sed.* = odontolite

boninite *ig. extru.* Glass-rich, nearly feldspar-free olivine-bronzite andesite with phenocrysts of olivine, bronzite, and augite. |Bonin Islands, Japan| (Petersen in 1891)

borate rock *sed.* A general term for rocks composed of borate minerals in lake-bed strata, as well as in brines and as incrustations around saline lakes. The major mineral associations include colemanite and ulexite (more abundant) and kernite and borax; esp. important in the Great Basin of California and Nevada.

borengite *ig. pluton., altered.* Ultra-potassic syenite (K_2O as high as 14%); essentially a nepheline syenite that has been sericitized metamorphically with almost complete substitution of potassium for sodium; occurs at Alnö Island, Sweden, and in Virginia. (Eckermann in 1960)

borolanite *ig. hypab.* Nepheline syenite consisting essentially of orthoclase, melanite, natrolite, and biotite; the rock has a pseudoporphyritic tex-

ture due to rounded masses thought to be altered aggregates of nepheline and orthoclase. |Loch Borolan, Sutherlandshire, Scotland| (Horne and Teall in 1892).

boryslavite *sed.* Hard and brittle variety of ozokerite. Also spelled boryslawite. |Boryslav, Galicia, Poland| (Hintze in 1933)

bostonite *ig. dike.* Aplitic dike rock of the pulaskite family, differing from normal aplites in having a texture characterized by clusters of rough irregular feldspar plates within a trachytoid groundmass. |Boston, Massachusetts| (Hunter and Rosenbusch in 1890)

bostonite, lime *ig. dike.* Variety of bostonite containing a notable amount of actual or normative anorthite and calcium pyroxene. (Brögger in 1894)

bostonite, quartz *ig. dike.* Syenitic aplite dike rock containing a small amount of quartz and having a rough trachytic texture characteristic of a bostonite.

bottle stone *tekt.* = moldavite

boulder *sed.* Rounded, subrounded, or subangular fragment of any rock more than 256 mm (about 10 in) in diameter. |M. Eng. *bulderston,* related to Sw. *bullersten,* rumbling stone in a stream|

boulder, faceted *sed.* Boulder ground flat on one or more sides by a natural agent, such as a glacier, or by wind or water.

boulder, glacial *sed.* Boulder moved by a glacier and modified by abrasion, not necessarily rounded.

boulderstone *sed.* Large rock lying on the surface of the ground or embedded in the soil, differing from the country rock of the region.

boundstone *sed.* Limestone bound together during deposition and remaining substantially in the position of growth; corals, e.g., often provide the framework which binds the rock together.

bouteillenstein *tekt.* = moldavite |F. *bouteille,* bottle + G. *Stein,* stone|

bowenite *meta.* Massive, fine grained, apple-green to greenish white serpentinite; resembles nephrite; originally from Smithfield, Rhode Island. (George T. Bowen, 1803–1828, American chemist and mineralogist, who made the first analyses) (Dana in 1850)

bowralite *ig. diaschist.* Coarse-textured syenite pegmatite composed of sanidine tablets, subordinate alkali amphibole, and aegirine; quartz, perovskite, zircon, and ilmenite are accessories. |Bowral, New South Wales, Australia| (Mawson in 1906) box stone   *sed.* Ferruginous concretion, often of rounded rectangular or boxlike form, having a hollow interior in which white sand is sometimes present; found in Jurassic and Tertiary sands in Great Britain. Also named box.

braccianite   *ig. extru.* Extrusive rock closely related to cecilite but without melilite. [Lake Bracciano, Italy] (Lacroix in 1917)

brandbergite   *ig. diaschist.* Aplitic rock composed of large white orthoclase (Carlsbad twins), grains of quartz, and aggregates of biotite, in a fine-grained groundmass that also contains arfvedsonite, zircon, and magnetite; the phenocrysts are characteristically frayed in outline. [Brandberg, South West Africa] (Cloos and Chudoba in 1931)

brazilite   *sed.* Oil-bearing rock from Bahia, Brazil; term also used as a synonym for the mineral baddeleyite. [Brazil] (pub. by Fletcher in 1893)

brea   *sed.* Uncommon term used for a viscous asphalt formed by the evaporation of volatile components from oil in seepages. (Sp. *brea,* pitch)

breccia   *ig., sed., meta.* Consolidated angular clastic fragments of any kind of rock or mineral whose size is larger than sand (greater than 2 mm diameter, or for some petrologists, 4 mm); they may be of sedimentary origin (e.g., talus accumulation), of igneous origin (e.g., pyroclastic volcanic breccia), or of metamorphic origin (e.g., cataclastic friction breccia). [It. *breccia,* breach, fragments of stone]

breccia, ablation   *sed.* = breccia, solution. [L. *ab,* away + *latus,* carried]

breccia, alloclastic   *ig.* Breccia formed by the disruption of nonvolcanic rocks by volcanic processes beneath the Earth's surface. [Gk. *allos,* other + *klastos,* broken]

breccia, anhydrite   *meta.* Cataclastic anhydrite rock showing extreme brecciation and flowage with dense fine-grained anhydrite pieces in a matrix of coarser anhydrite.

breccia, autoclastic   *meta.* Breccia formed in the place where it is found as a result of crushing, shattering, dynamic metamorphism, orogenic forces, or other mechanical processes; fault breccia is an example. [Gk. *autos,* self + *klastos,* broken]

breccia, bajada   *sed.* Wedge-shaped, imperfectly stratified accumulation of coarse, angular, poorly sorted rock clasts mixed with mud and formed in an arid region by an intermittent stream or a mudflow containing considerable water. [Sp. *bajada,* descent, slope]

breccia, beach   *sed.* Breccia formed on a beach where wave action is inefficient and angular blocks are supplied from cliffs; produced under conditions of rapid submergence. (Norton in 1917)

breccia, bone   *sed.* = bone bed

breccia, carnallite   *sed.* Carnallite rock containing angular to rounded fragments of halite, kieserite, anhydrite, and clay in a predominantly carnallitic matrix.

breccia, cataclastic   *meta.* Metamorphic rock formed principally by mechanical deformation and composed of angular rock fragments of varying size, set in a normally subordinate matrix made up of smaller rock pieces, mineral fragments, and powder; a wide variety of rocks may be affected, including sandstone, limestone, quartzite, marble, gneiss, plutonic igneous rocks, and others; essentially the same as autoclastic breccia. Also named crush breccia. [Gk. *kata,* down, very + *klastos,* broken]

breccia, cave   *sed.* Breccia composed of the angular fragments of limestone and other rocks that have fallen to the floor from the roof and sides of a cave and are cemented with calcium carbonate or occur in cave earth.

breccia, chert   *sed.* Breccia composed of chert fragments residual from the weathering of limestones.

breccia, clastic   *sed.* Breccia whose fragments have been formed by erosion. [Gk. *klastos,* broken]

breccia, clayey   *sed.* Breccia containing at least 10% clay and 80% rubble.

breccia, coal   *sed.* Coal in a seam that has been naturally fragmented and often shows slickensided surfaces.

breccia, collapse   *sed., ig.?* Breccia formed by the collapse of rock overlying an opening, e.g., the collapse of a cave roof or roof of country rock above an intrusion.

breccia, contact   *ig., meta.?* Breccia formed by wall-rock fragmentation around an igneous intrusion; consists of both wall-rock fragments and igneous materials.

breccia, crackel   *sed.* Incipient breccia in which fragments have parted by planes of rupture but have little or no displacement. (Norton in 1917)

breccia, crush   *meta.* = breccia, cataclastic

breccia, desiccation   *sed.* Breccia composed of polygons formed from mud-cracked sediments.

breccia, earthy   *sed.* Breccia in which rubble, sand, and silt (with clay) each constitute more than 10% of the rock. (Woodford in 1925)

breccia, eruptive   *ig. extru.* = breccia, pyroclastic

breccia, evaporite-solution　*sed.* Solution breccia formed where soluble evaporite minerals (halite, anhydrite, gypsum, and others) have been removed. (Sloss and Laird in 1947)

breccia, explosion　*ig. extru.* = breccia, pyroclastic

breccia, fallback　*ig. extru., impact.* Breccia composed of fragments that fall inside a crater (fallback materials), either from an igneous explosion crater or an impact crater.

breccia, fallout　*ig. extru., impact.* Breccia composed of fallout (deposited on the outside) materials, either derived from an igneous explosion crater or an impact crater.

breccia, fault　*meta.* Tectonic breccia composed of fragments resulting from the breaking of rocks during movements along a fault, from friction between the walls of the fault; essentially equal to a friction breccia.

breccia, flow　*ig. extru.* Breccia formed from the fragmentation of the surface of lava flows as a consequence of the crystallization of the surface upon cooling and its inability to deform plastically with the still fluid moving parts.

breccia, fold　*sed.* Breccia composed of fragments resulting from the sharp folding of brittle rock layers between more ductile beds, e.g., a chert breccia formed where interbedded chert and shale are sharply folded.

breccia, friction　*meta.* = breccia, fault

breccia, friction　*ig. hypab.* Volcanic breccia formed where a rising mass of nearly crystallized lava is shattered by friction against the walls of the volcanic vent and later cemented by newly rising magma.

breccia, ice　*sed.* Breccia composed of fragments of glacier ice that are of different ages or of comparable ages.

breccia, igneous　*ig.* 1. Breccia composed of fragments of igneous rocks.　2. Breccia produced by igneous processes, e.g., flow breccia.

breccia, impact　*impact.* Breccia whose origin is a result of a natural impact at the earth's surface, e.g., from a meteorite.

breccia, injection　*ig., hydrothermal.* Breccia formed by the introduction of largely foreign rock fragments into veins and fractures within the host rock. (Speers in 1957)

breccia, intraformational　*sed.* Rock formed by brecciation of partly consolidated material, followed by practically contemporaneous sedimentation; similar in nature and origin to an intraformational conglomerate.

breccia, intrusion　*ig., sed.* = breccia, intrusive

breccia, intrusive　*ig., sed.* Mixture of angular to rounded fragments in a matrix of other clastic material, mobilized and intruded into its present position along a pre-existing structure. (Bryant in 1968)

breccia, lunar　*impact.* Moon rock composed of fragments formed by shock lithification of fine-grained lunar materials during meteorite impact.

breccia, monomict　*ig., sed., meta., meteor.* Breccia in which all the fragments have essentially the same composition; esp. applied to brecciated meteorites. [Gk. *monos,* single + *miktos,* mixed]

breccia, osseous　*sed.* = bone bed. [L. *osseus,* bony]

breccia, peat　*sed.* Peat broken up and redeposited by water.

breccia, pillow　*ig. extru.* Rock composed of lava pillows and fragments in a tuff matrix.

breccia, polymict　*ig., sed., meta., meteor.* Breccia in which the fragments have a variety of compositions; esp. applied to brecciated meteorites. [Gk. *pur,* fire + *klastos,* broken]

breccia, pyroclastic　*ig. extru.* 1. Breccia consisting chiefly of angular to subangular blocks (not bombs) of more than 32 mm in diameter, lying in a matrix that is usually tuffaceous but that may be vesicular and pumiceous.　2. *IUGS,* pyroclastic rock whose average pyroclast size exceeds 64 mm and in which angular pyroclasts predominate. [Gk. *pur,* fire + *klastos,* broken]

breccia, raft　*sed.* Breccia containing fragments unworn during transportation, e.g., while being transported on floating vegetation or by an iceberg.

breccia, reef　*sed.* Breccia composed of limestone fragments broken off a reef by the action of waves and tides.

breccia, rubble　*meta.* 1. Breccia in which matching fragments are not separated by initial planes of rupture; they are in touch. (Norton in 1917)　2. Tectonic breccia characterized by prominent relative displacement of fragments and by some rounding.

breccia, salt-dome　*sed.* Brecciated shale associated with the dome of a salt plug.

breccia, sandy　*sed.* Breccia containing at least 80% rubble and 10% sand, and no more than 10% of other material. (Woodford in 1925)

breccia, sedimentary　*sed.* Breccia formed by sedimentary processes, e.g. talus breccia.

breccia, shale-pebble　*sed.* Desiccation breccia consisting of thin shale pieces in a sandy matrix.

breccia, shatter   *meta.* Tectonic breccia composed of angular fragments that show little rotation.

breccia, shoal   *sed.* Submarine breccia, usually of limestone, formed by the action of waves and tides on a shoal. (Norton in 1917)

breccia, silty   *sed.* Breccia containing at least 80% rubble and 10% silt, and no more than 10% of other material. (Woodford in 1925)

breccia, slump   *sed.* Breccia formed by slumping (usually subaqueous) brought about by simple gravitational stress.

breccia, solution   *sed.* Collapse breccia formed where soluble material has been dissolved out, thus allowing the overlying rock to settle and become fragmented; it is esp. related to the solution of salt beds.

breccia, talus   *sed.* Breccia composed of rock fragments derived from and lying at the base of a cliff or very steep, rocky slope. [L. *talus,* ankle; alluding to the base of a cliff]

breccia, tectonic   *meta.* Breccia formed as a result of crustal movements, e.g., fault breccia. [Gk. *tekton,* carpenter; often used in reference to Earth deformations]

breccia, tuff   *ig. extru.* Intermediate rock between tuff and pyroclastic breccia; it is a mixture of ash and blocks.

breccia, tuffaceous   *ig. extru. IUGS,* breccia with average clast size greater than 64 mm and containing 25-75% by volume of pyroclasts, the remaining components being sedimentary clasts.

breccia, volcanic   *ig. extru.* = breccia, pyroclastic

brecciola   *sed.* Well-graded, intraformational breccia consisting of small, angular limestone fragments in well-defined beds separated by dark shale. [It. dim. of *breccia*]

broggite   *sed.* Variety of brown asphalt from Peru. [Jorge Alberto Broggi, of Lima, Peru, founder of the Geological Society of Peru in 1925] (Fester and Cruellas in 1935)

brontolite   *meteor.* = stony meteorite. Also spelled brontolith. [Gk. *bronte,* thunder]

bronzitfels   *ig. pluton.* = bronzitite. (Lang in 1877)

bronzitite   *ig. pluton.* Pyroxenite composed almost exclusively of bronzite; small amounts of olivine, chromian spinel (picotite), chromite, hornblende, and secondary serpentine also may be present. [bronzite] (Williams in 1890)

brown rock   *sed.* Local Tennessee term for dark brown to black phosphorite that results from the weathering of phosphatic limestone; also, any rock of brown color.

brownstone   *sed.* Ferruginous sandstone (or sometimes arkose) whose grains are coated by iron oxide films; often applied to the dark reddish-brown Triassic sandstone quarried in the Connecticut River valley for use in building.

bucaramangite   *sed.* Pale yellow retinite from Colombia resembling amber; insoluble in alcohol and yielding no succinic acid. [Bucaramanga, Colombia] (Boussingault in 1842)

buchite   *meta.* Vitrified hornfelslike rock formed when aluminous or siliceous rocks are subjected to intense thermal metamorphism by contact with mafic or ultramafic magmas; often consisting of various combinations of cordierite, mullite, sillimanite, tridymite, corundum, spinel, calcic-plagioclase, pyroxenes, and magnetite set in an abundant matrix of brownish glass. A related rock type is basalt jasper. [Christian Leopold von Buch, 1774-1853, German geologist]

buchnerite   *ig. pluton.* = lherzolite. [Christian Ludwig Otto Buchner, 1828-1897, authority on meteorites]

buchonite   *ig. extru.* Dark orthoclase-bearing tephrite, consisting of zoned labradorite crystals and nepheline, with analcime, nosean, magnetite, pyroxene, and others. [Buchonia, Fulda, Germany] (Sandberger in 1872)

bugite   *ig., meta.?* Hypersthene-bearing quartz diorite, related to charnockite, consisting of andesine (60%), hypersthene (18%), and quartz (16%), and also with biotite, apatite, zircon, and iron oxide. [Bug, near Schkurinotz, Podolia, Ukraine] (Bezborodko in 1931)

buhrstone   *sed.* Siliceous rock suitable for use as a millstone; some are sandstones with angular grains, and others may be silicified fossiliferous limestone with abundant cavities formerly occupied by fossil shells. Also spelled burstone and burrstone. [M. Eng. *bur,* whetstone]

bullion   *sed.* 1. Concretion or nodule that generally encloses a fossil. 2. Carbonate or silica-rich concretion (sometimes over 1 m across) found in some types of coal; stained brown by humic derivatives and often with plant structures at the nucleus. [L. L *bullio,* mass of gold or silver; apparently alluding to its rounded massive form]

burl   *sed.* Small nodule or oolith in fireclay, usually with a high content of iron oxide or alumina. [O. F. *bourle,* from L. *burra,* tuft of wool; alluding to its resemblance to a knot or lump in cloth]

burmite   *sed.* Amberlike resin found in Upper Burma; generally pale yellow but also reddish and dark brown. [Burma] (Helm in 1892) burnite   *altered ore.* Mixture of silicates, oxides, and carbonates of copper, from near Battle Mountain, Nevada; sometimes used as a gem. [Frank Burnam, its discoverer in 1952]

burnt rock   *meta., altered.* In some altered deposits, rocks reddened by the introduction of hematite as disseminated dust and veinlets so that the rock looks burned. Prospectors' term.

burstone   *sed.* = buhrstone. Also spelled burrstone.

busorite   *ig. pluton.* Cancrinite syenite in which the cancrinite is primary in origin, from Lueshe area, Congo. [Busoro, Kivu, Congo] (de Béthune in 1956)

bustite   *meteor.* = achondrite, enstatite. [Bustee, India] (Tschermak in 1883)

byerite   *sed.* Jet-black bituminous coal resembling albertite, from Middle Park, Colorado. [William Newton Byers, 1831-1903, American surveyor and pioneer from Denver, Colorado] (Mallet in 1875)

bytownitfels   *ig. pluton.* = bytownitite. (Kolderup in 1902)

bytownitite   *ig. pluton.* Anorthosite composed almost exclusively of bytownite; minor pyroxene and iron oxides have been noted. [bytownite] (Johannsen in 1920).

C

cacholong   *sed.* Pale bluish-white, opaque to semi-translucent, porcelainlike opal, sometimes banded with chalcedony. Also spelled cachalong. [Mong. *qus* (*khas, khash*), jade, precious object + *chuluun* (chilagun), stone; possibly from Cach (Kash) River, Chinese Turkestan, a jade-producing area.] (Used in English as early as 1791)

cahemolith   *sed.* = coal, humic. Incorrect spelling of chameolith.

calbenite   *sed.* = myrickite. [Possibly California + San Benito County]

calc-   Common prefix used with rock names to indicate calcareous, calcite, or calcium.

calc-aphanite   *ig., altered.* Diabase or related rock largely replaced by calcium carbonate minerals. [calcareous + aphanite]

calc-dolomite   *sed.* = dolomite, calcitic

calc-flinta   *meta.* Very fine-grained, flinty thermal metamorphic rock, often banded, and derived from a calcareous mudstone; composed of calcium-rich silicates and may contain vesuvianite, grossular, andradite, clinozoisite, quartz, feldspars, and others. [calc- + Sw. *flinta,* flint] (Barrow and Thomas in 1908)

calc-sapropel   *sed.* Sapropel containing calcareous algae.

calc-schist   *meta.* Foliated metamorphic rock, formed by the metamorphism of an argillaceous limestone, in which calcite has recrystallized in elongated or platy forms, rather than in the commoner granular forms, thus giving to the rock, with the other products of metamorphism, a schistose structure. There are many varieties depending upon the composition of the original sediment and upon the grade of metamorphism, e.g., diopside c.-s., epidote c.-s., hornblende c.-s., quartz c.-s., sericite c.-s., talc c.-s., tremolite c.-s., or combinations of these and other silicate minerals. (Brongniart in 1827)

calc-silicate rock   *meta.* Metamorphic rock consisting mainly of calcium-bearing silicates, such as wollastonite, grossular, andradite, tremolite, diopside, titanite, and formed by the metamorphism of impure limestone; some are monomineralic, and others have structures and mineralogies allowing them to be described as c.-s. gneiss, c.-s. hornfels, c.-s. schist, and others.

calc-sinter   *sed.* = travertine

calc-tufa   *sed.* = tufa

calcarenite   *sed.* General designation for mechanically deposited carbonate rocks of sand-grain size (1/16 mm to 2 mm in diameter) composed of 50% or more of carbonate fragments. Varieties include: arenaceous c. (mixed with quartz grains), dolomite c. (dolomite rather than calcite), oolitic c. (composed of ooliths). (Grabau in 1903)

calcdolomite   *sed.* = dolomite, calcitic

calcibreccia   *sed.* Calcirudite whose clasts are angular in form; limestone breccia, or consolidated calcareous rubble.

calciclasite   *ig. pluton.* Anorthosite composed nearly entirely of anorthite; also called calciclase. (Johannsen in 1937)

calciclasite, quartz   *ig. pluton.* Quartz-bearing anorthite anorthosite (calciclasite). (Johannsen in 1937)

calcicrete   *sed.* = calcrete

calcigranite   *ig. pluton.* Granite whose accessory plagioclase is calcic (labradorite or bytownite). (Johannsen in 1919)

calcigravel   *sed.* Unconsolidated equivalent of calcirudite.

calcikersantite   *ig. dike.* Kersantite in which the plagioclase is calcic; most contain andesine, and some have labradorite or bytownite (calcic-plagioclase). (Johannsen in 1937)

calcilith   *sed.* 1. = limestone. (Grabau in 1924)   2. Rock composed almost entirely of the calcareous remains of organisms.

calcilutite   *sed.* Carbonate rocks composed (more than 50%) of the finest-grained products of mechanical processes; because of recrystallization, the coarseness varies from about 2 microns to 15 microns, and the original detrital origin of the rock is difficult to ascertain in most cases. (Grabau in 1903)

calcimicrite   *sed.* Limestone in which the particles are less than 20 microns in diameter and in which the micrite component exceeds the allochem component. (Schmidt in 1965)

calcimixtite *sed.* Mixtite that is dominantly calcareous. (Schermerhorn in 1966)

calcipelite *sed.* = calcilutite

calciphyre *meta.* Marble containing conspicuous calcium and/or magnesium silicate minerals such as garnet, forsterite, pyroxene, wollastonite, and others. (Brongniart in 1813)

calcipulverite *sed.* Old term for an aphanitic limestone formed by precipitation, to be used in contrast to those emplaced by currents. |L. *pulvis,* dust|

calcirhyolite *ig. extru.* Aphanitic extrusive equivalent of calcigranite, whose accessory plagioclase is calcic (labradorite or bytownite), rather than the more common sodic types. (Johannsen in 1932)

calcirudite *sed.* General designation for mechanically deposited carbonate rocks composed of 50% or more of angular to rounded carbonate fragments over 2 mm in diameter. (Grabau in 1903)

calcisiltite *sed.* General designation for mechanically deposited carbonate rocks composed of 50% or more of original carbonate fragments between 1/256 mm and 1/16 mm across; difficult to identify because many calcilutites recrystallize to grain sizes within this range.

calcisphere *sed.* In some limestones, spherical objects, up to 0.5 mm in diameter, composed of calcite (usually sparite), often with a micritic wall; often ascribed to algae, although an affinity for foraminifera has been suggested.

calcisyenite *ig. pluton.* Syenite in which the small amount of plagioclase present is calcic (labradorite or bytownite) instead of the usual oligoclase or andesine. (Johannsen in 1937)

calcitite *sed., meta.* Any rock in which calcite is the major component, e.g., limestone, marble, and others. [calcite] (Kay in 1951)

calclithite *sed.* Transported terrigenous sands containing a large quantity of detrital limestone and dolomite particles; used to distinguish this type of sand from those produced by chemical or biochemical precipitation, such as skeletal particles, ooliths, and other materials formed within the basin of deposition. (Folk in 1968)

calcrete *sed.* 1. Calcareous duricrust, e.g., some caliche. 2. Conglomerate formed by the cementation of superficial gravels by calcium carbonate. Also named calcicrete. [calcareous concrete] (Lamplugh in 1902)

calcsinter *sed.* = travertine. Also spelled calc-sinter.

calctufa *sed.* = tufa. Also spelled calc-tufa.

caldasite *ig.? altered.* Rock consisting mainly of baddeleyite or a mixture of zircon and orvillite (poorly defined hydrated zirconium silicate), from the Caldas district, Brazil. |Caldas district, Minas Geraes, Brazil| (Derby pub. by Lee in 1917)

caliche *sed.* 1. Calcitic crust (duricrust) that may also contain clay and sand, produced on and near the ground surface of some semiarid regions by evaporation of ground water drawn upward under capillary action. 2. Clastic materials cemented with soluble salts of sodium, esp. soda niter, in the nitrate deposits of the Atacama Desert of northern Chile and southern Peru. Also named soda niter rock. 3. Term applied in Spanish American countries to, e.g., clayey soil on a gold vein, clay in the selvage of veins, bank composed of clay or other detrital substance encountered in placer mining. |Sp. *caliche,* from L. *calx,* lime|

californite *meta.* Closely compact variety of idocrase (vesuvianite), with an olive green to grass green color, from Siskiyou, Fresno, and Tulare Counties, California; often substituted for jade. Also named California jade or vesuvianite jade. |California| (Kunz in 1903)

calm *sed.* Scottish name for light-colored shale or mudstone; light-colored blaes. Also spelled caulm. |Possibly Scot. *calm,* mold or frame used for casting metal|

caltonite *ig. extru.* Very compact, bluish black analcime basanite, containing microphenocrysts of olivine and augite in a groundmass of platy labradorite, augite, analcime, and iron oxides. |Calton Hill, Derbyshire, England| (Johannsen in 1938)

campanite *ig. extru.* Originally described as a leucite-bearing tephrite with large crystals of leucite; later called the extrusive equivalent of a pseudoleucite-bearing nepheline syenite. |Vesuvius, Campania, Italy| (Lacroix in 1917)

camptonite *ig. hypab.* Lamprophyre similar in composition to a nepheline diorite; composed of plagioclase (usually andesine or labradorite), titanaugite, and brown hornblende (usually barkevikite). |Campton Falls, New Hampshire| (Rosenbusch in 1887)

camptospessartite *ig. diaschist.* Dark spessartite, containing titanaugite. [camptonite + spessartite, from its similarity to camptonite] (Tröger in 1931) canadite   *ig. pluton.* Nepheline syenite with calcium-sodium feldspar present in the norm but not in the mode; recognized variety is cancrinite canadite, with 10-20% cancrinite. [Canada] (Quensel in 1913)

cancarixite   *ig. extru.* Quartz-bearing aegirine fortunite, a silica-rich end-member of the lamproite series; contains sanidine (51%), aegirine augite (36%), quartz (9%), biotite, titanite, and iron oxides, and also secondary calcite. [Cancarix, Sierra de las Cabras, Spain] (Parga-Pondal in 1935)

candelite   *sed.* = coal, cannel. [candle] (Gümbel in 1883)

canga   *sed.* Tough, well-consolidated, unstratified ferruginous breccia or conglomerate composed of fragments of hematite and the rock itabirite cemented together by limonite or hematite, and occasionally by other lateritic constituents. [Port. *canga,* yoke]

cangagua   *sed.* Widespread tufalike deposit composed of clay, with gypsum and alum, occurring in Ecuador. [Ecuad. *cangagua,* clay for making bricks]

cannelite   *sed.* = coal, cannel

cannonball   *sed.* Large, dark concretion resembling a cannonball.

cantalite   *ig. extru.* Sodic rhyolite pitchstone. [Cantal, France] (Leonhard in 1821)

carbonaceous rock   *sed.* Any sedimentary rock either consisting of, or containing an appreciable amount of, original or introduced organic material, including plant and animal residues and organic derivatives greatly altered from the original remains, e.g., carbonaceous shale and limestone, coal, asphalt, bituminous materials, and many other materials. Also named carbonolite and carbonolith. [carbon]

carbonate rock   *ig., sed., meta.* Any rock consisting chiefly of carbonate minerals (calcite, dolomite, siderite, aragonite, and others), e.g., limestone, dolostone, marble, and carbonatite; term sometimes restricted to a sedimentary rock composed of more than 50% carbonate minerals by weight.

carbonatite   *ig. hypab.* Carbonate-rich rock of apparent magmatic (igneous) derivation or descent; in addition to the carbonate minerals (calcite, dolomite, ankerite, or siderite), they also may contain biotite, pyroxene, amphibole, feldspar, feldspathoid, apatite, titanite, or other minerals. [carbonate] (Brögger in 1921)

carbonatite, aegirine   *ig. hypab.* Carbonatite in which aegirine is the major mafic mineral; a good example is the calcite carbonatite ringite with as much as 20% aegirine.

carbonatite, calcite   *ig. hypab.* Carbonatite in which calcite is the characterizing carbonate mineral, e.g., sovite (a coarse-grained rock) and alvikite (a medium- to fine-grained rock.)

carbonatite, dolomite   *ig. hypab.* Carbonatite in which dolomite is the characterizing carbonate mineral, e.g., beforsite and rauhaugite.

carbonatite, ferro-   *ig. hypab.* Carbonatite in which either siderite, ankerite, or iron-rich dolomite is the characterizing carbonate mineral.

carbonatite, magnesio-   *ig. hypab.* Carbonatites in which MgO is greater than (FeO + Fe_2O_3 + MnO); if the nature of the carbonate phase is known the rock may be referred to as dolomite carbonatite or ankerite carbonatite. (Woolley in 1982)

carbonite   *meta.* = coke, natural. (Heinrich in 1875)

carbonolite   *sed.* = carbonaceous rock. (Wadsworth in 1891)

carbonolith   *sed.* = carbonaceous rock. (Grabau in 1924)

carburan   *ig.?* Pitchy carbonaceous material containing uranium, occurring in a pegmatite in Karelia, Soviet Union; related to thucholite. [carbon + uranium] (Labuntzov in 1934)

Carlsbad spring stone   *sed.* = Karlsbad spring stone

carmazul   *altered ore.* Oxidized copper-rich rock showing red, brown, blue, and green colors, and composed of jasper, chalcedony, quartz, hematite, chrysocolla, and malachite; from Lower California, Mexico; sometimes cut as a gem. Also named chrysocarmen. [Possibly Carmen Island + Sp. *azul,* blue]

carmeloite   *ig. extru.* Olivine-bearing andesite, containing iddingsite (a reddish brown mixture of secondary ferric iron, magnesium, and calcium silicates) as an alteration product of olivine phenocrysts. [Carmelo Bay, California] (Lawson in 1893)

carnallite rock   *sed.* Saline evaporite rock in which carnallite is the major constituent; associated minerals may include halite, kieserite, sylvite, anhydrite, clays, hematite, quartz, and others.

carnallite-kieserite rock   *sed.* Saline evaporite rock containing abundant carnallite and kieserite; in the Stassfurt, Germany, deposits a zone of this rock up to 40 ft in thickness occurs.

carnelian   *sed.* Red, orange red, brownish red, or brownish orange translucent to semitranslucent form of chalcedony. [Possibly L. *carneus* fleshy, alluding to the color]

carpolite *sed.* Ellipsoidal concretion or nodule about 1-2 cm in diameter, originally believed to be a fossil seed. [Gk. *karpos,* fruit]

carstone *sed.* Local British (East Anglia) term for a hard, firmly cemented ferruginous sandstone used in grinding (quernstone) and building. [Possibly alluding to use in making steps for mounting carts and horses (horse blocks)]

carvoeira *ig. dike.* Obsolete name for a quartz-tourmaline rock, now named tourmalite. [Carvoeiro, Brazil] (Eschwege in 1832)

cascadite *ig. dike.* Dark colored, sodium-rich minette, containing sodium orthoclase, biotite, diopside, and others. [Cascade Creek, Highwood Mountains, Montana] (Pirsson in 1905)

Cassel brown *sed.* = earth, black. [Cassel, Germany]

Cassel earth *sed.* = earth, black. [Cassel, Germany]

cassianite *sed.* = kassianite

cataclasite *meta.* Any metamorphic rock formed principally by tectonic (mechanical) deformation or cataclastic metamorphism, e.g., metamorphic breccia (friction breccia), phacoidal rock, mylonite, and phyllonite. Also named cataclastic rock. [Gk. *kata,* down, very + *klastos,* broken] (Kjerulf in 1885)

catalinaite *sed.* Local name for jasper beach pebbles from Santa Catalina Island, California. Also named catalinite. [Catalina] (Sterrett in 1911)

catarinite *meteor.* Obsolete term for a Brazilian iron meteorite containing about 33% nickel. [Santa Catharina, Brazil] (Damour in 1877)

catawbarite *meta.* Intimate mixture of talc and magnetite; associated with itacolumite in South Carolina. Also incorrectly spelled catawbirite and catawberite. [Possibly town of Catawba, or Catawba River, South Carolina] (Lieber in 1857)

catlinite *sed.* Red, siliceous argillite, used by the Dakota Indians in Minnesota for pipe bowls. A variety of pipestone. [George Catlin, 1796-1872, American painter of Indians] (Jackson in 1839)

caucasite *ig. hypab.* = kaukasite. [Caucasus]

caustobiolith *sed.* Biolith that is combustible and usually derived from plant materials, e.g., the various coals and bitumens. [Gk. *kaustikos,* burning + biolith]

caustolith *sed.* Rock with the property of combustibility; although it is usually of organic origin (caustobiolith), it may be inorganic, e.g.,

sulfur rock. [Gk. *kaustikos,* burning] (Grabau in 1924)

cavalorite *ig. pluton.* Granular rock related to diorite in which there is an excess of orthoclase over oligoclase and an abundance of hornblende, with accessory quartz, apatite, magnetite, and others. [Monte Cavaloro, near Bologna, Italy] (Capellini in 1877)

cave bacon *sed.* Thin, translucent drapery speleothem with parallel color banding.

cave balloon *sed.* Thin-walled pearly speleothem resembling a small inflated sack.

cave beard *sed.* = cave rope

cave blade *sed.* Any broad, flat speleothem; usually composed of flat bladelike calcite crystals arranged in a rosette fashion.

cave blister *sed.* Hollow, white hemispherical cave wall deposit.

cave bubble *sed.* Hollow sphere of calcium carbonate deposited on the surface of a gas bubble in a cave.

cave cauliflower *sed.* Subaqueous coralloidal speleothem with radiating branches and nodules on the tips of the branches.

cave cone *sed.* Upward-pointing conical speleothem composed of many sunken cave rafts.

cave coral *sed.* Coralloidal speleothem.

cave cotton *sed.* White, mounded fibrous sulfate speleothem.

cave cup *sed.* Hollow, semicircular, rimmed speleothem that projects upward and outward from a cave wall.

cave fill *sed.* Detrital material washed into a cave or derived from solution of the cavern bedrock.

cave flower *sed.* Sulfate (gypsum, epsomite, mirabilite) speleothem growing from the base and whose "petals" curve radially outward from the center. Also named oulopholite.

cave grape *sed.* Speleothem shaped like a bunch of grapes; botryoidal.

cave grass *sed.* In caves, bundles of long gypsum needles resembling grass.

cave hair *sed.* Single fibrous sulfate crystals, occurring in caves.

cave ice *sed.* Speleothems composed of ice; sometimes incorrectly applied to calcitic dripstone and other speleothems.

cave marble *sed.* = onyx, cave

cave nest *sed.* = Carbonate bowl-shaped speleothem containing cave pearls.

cave onyx *sed.* = onyx, cave

cave orange *sed.* Round, sometimes colored, speleothem formed when a floating nucleus becomes layered with calcium carbonate.

cave pearl *sed.* Carbonate pisolith, often spheroidal in shape, which may form in a shallow cave pool.

cave pendant *sed.* Hanging projection of cave bedrock.

cave popcorn *sed.* Popular term for nodular, coralloidal speleothems.

cave puffball *sed.* = cave snowball

cave raft *sed.* Thin layer of crystalline carbonate that floats on the surface of a quiet cave pool.

cave rope *sed.* Flexible speleothem with parallel fibers of gypsum aligned in a ropelike manner. Also named cave beard.

cave snowball *sed.* Speleothem very similar to a mounded cave cotton formation. Also named cave puffball.

cave sword *sed.* Long, wide, bladed gypsum crystal occurring in caves.

cave terrace *sed.* Gently sloping series of rimstone dams, found in caves.

cave velvet *sed.* Speleothem with a velvety surface luster.

cayeuxite *sed.* Pyrite nodules, rich in arsenic, antimony, germanium, molybdenum, nickel, silicon, and others, from Lower Cretaceous shales in the Carpathians [Lucien Cayeux, 1864-1944, French sedimentary petrologist, Paris]

cecilite *ig. extru.* Very dark gray and dense rock with rare leucite and augite phenocrysts in a holocrystalline matrix composed of leucite, augite, melilite, nepheline, olivine, anorthite, magnetite, and apatite; essentially a melilite leucitite in which leucite comprises about 50% of the total rock. [Cecilia Metella, daughter of Quintus Caecilius Metellus Creticus, Roman consul ca. 69 B.C., whose tomb is at Capo di Bove, Italy] (Cordier in 1868)

cedarite *sed.* = chemawinite. [Cedar Lake, Manitoba, Canada] (Klebs in 1896)

cedricite *ig. extru.* Leucite-rich lava containing larger crystals of leucite, diopside, and phlogopite in a matrix of priderite, perovskite, magnophorite (titanian potassian richterite), phlogopite, zeolites, and often serpentine pseudomorphs after olivine. [Mt. Cedric, Western Australia] (Wade and Prider in 1940)

celestialite *meteor.* Substance related to ozoke-rite found in some iron meteorites [L. *caelestis,* heavenly] (Smith in 1875)

celestite rock *sed.* = strontium rock

cement rock *sed.* Argillaceous limestone in which the proportion of clay to calcite (about 1:3) approximates that required for the manufacture of Portland cement.

cementstone *sed.* = cement rock

cenuglomerate *sed.* 1. Rock formed by the consolidation of mudflow material. (Harrington in 1946) 2. Coarse breccia formed by the accumulation of material resulting from rockfalls, landslides, or mudflows. (Dunbar and Rodgers in 1957) [L. *coenum,* mud + *glomerare,* to wind into a ball]

ceramicite *ig. extru.* Collective name for rocks resembling cordierite-bearing glassy rhyolite and occurring as a silica-rich differentiate in necks of andesite volcanoes; some are holocrystalline, others are glassy. [ceramic] (Koto in 1916)

ceraunite *meteor.* = meteorite. [Gk. *keraunos,* thunderbolt]

ceraunite *sed.* Poorly defined variety of jasper. Also named thunder stone. [Gk. *keraunos,* thunderbolt] (Allan in 1814)

cerolite *meta.* Compact, massive, yellow or greenish mixture of serpentine minerals and stevensite. [Gk. *keros,* wax] (Breithaupt in 1823)

cerulene *sed., altered ore?* Form of calcium carbonate stained green and blue by malachite and azurite and used as a gemstone; from Bimbowrie, South Australia. [L. *caeruleus,* sky blue]

ceyssatite *sed.* = randanite. [Ceyssat, Auvergne, France] (Gonnard in 1875)

chalcedonite *sed.* = chalcedony rock

chalcedony rock *sed.* Rock in which chalcedony, the massive, commonly translucent and microscopically fibrous variety of quartz, occurs as a major component (e.g., agate). The numerous varieties of chalcedony are described elsewhere in this dictionary under specific names, esp. agate. [Chalcedon or Calchedon, ancient maritime city of Bithynia, on the Sea of Marmara, Asia Minor]

chalcedony, psilomelane *sed.* Black, banded chalcedony rock, heavily impregnated with psilomelane (black manganese oxides), and occasionally cut into gems; esp. from Mexico.

chalchihuitl *ig., sed., meta.* General Mexican name for jade, as well as other green rocks, e.g., turquoise, smithsonite, serpentine, green porphyries, that can be carved. There are several variations in the spelling of the term. [Aztec *chalchihuitl,* blood or water of precious stones; apparently the wearing of precious stones was believed to enrich the blood and promote health].

chalchihuitl, iztac *sed.* White or less commonly greenish Mexican onyx. [Aztec *iztac,* white]

chalite *sed.* Obsolete term for a conglomerate containing pebbles intermediate between chalcedony and flint. [Possibly Gk. *chalix,* pebble] (Pinkerton in 1811)

chalk *sed.* Porous, fine-textured, and somewhat friable variety of limestone, normally light-colored; usually formed by shallow-water accumulation of calcareous tests of floating micro-organisms (chiefly foraminifers). The following varieties have been noted: argillaceous c. (with clay minerals), dolomitic c. (somewhat replaced by dolomite), glauconitic c. (abundant glauconite grains), magnesian c. (hard layers of magnesian calcite in chalk), phosphatic c. (somewhat replaced by phosphate minerals). [O. Eng. *cealc,* lime]

chalk, black *sed.* Bluish black, carbonaceous clay, shale, or slate, used as a pigment or crayon.

chalk, French *meta.* Soft, compact, whitish masses of steatite (talc rock); used in pencil form for marking cloth or in powder form for cleaning cloth.

chalk, Spanish *meta.* Soft, compact variety of steatite (talc rock) from the Aragon region of Spain.

chalk rock *ig., sed., meta.* 1. Any soft white rock resembling chalk, e.g., steatite (soapstone), diatomite, tripolite, tufa, limestone, volcanic tuff, and many others. 2. = chalk

chameolith *sed.* = coal, humic. Also incorrectly spelled chaemolith and cahemolith. [Gk. *chamai,* low, on the ground]

chapinite *sed.* Brown and yellow recemented brecciated jasper occurring west of Tiefort Village, Camp Irwin Military Reservation, San Bernardino County, California. [Dr. Roy E. Chapin, Barstow, California]

chapopote *sed.* Viscous bitumen related to maltha and nearly synonymous with brea; usually considered to be a little more solid than brea. Also spelled chapapote. [Aztec *chapopote,* tar, asphalt]

charnockite *ig., meta.?* Orthopyroxene (hypersthene)-bearing rock close to granite or granulite; contains 10-60% quartz, and the proportion of alkali feldspar to total feldspar is 35-90%; although its origin is debated (igneous versus metamorphic), high temperature and pressure are generally thought to be necessary to its formation. [Job Charnock, died 1693, British founder of Calcutta, India, from whose tombstone the rock was first described] (Holland in 1893)

charnockite, alkali *ig., meta.?* Rock in the charnockite series containing 20-60% quartz of the felsic constituents and in which the ratio of alkali feldspar to total feldspar is greater than 90%. (Tobi in 1971)

charnockite, m- *ig., meta.?* Charnockite containing mesoperthite (about equal amounts of potassic feldspar and plagioclase in the perthite) as the only feldspar. (Tobi in 1971)

charoite rock *meta.* Lilac to violet, massive charoite occurring in potassic-feldspar metasomatites at the contact of nepheline- and aegirine-syenites with limestone; the charoite-rich rock is important in the gem-cutting industry, where it is carved and cut. [Charo River area, Murun massif, northwest Aldan, Yakutsk, Soviet Union] (Rogova and others in 1978)

chassignite *meteor.* = achondrite, olivine. [Chassigny, France] (Rose in 1863)

chemawinite *sed.* Pale yellow to dark brown variety of retinite, in fragments the size of a pea to that of a robin's egg, in decayed wood at Cedar Lake in Manitoba. Also named cedarite. [Chemahawin or Chemayin, Indian name of a Hudson Bay post near its occurrence] (Harrington in 1891)

chemical rock *sed.* Any sedimentary rock composed primarily (more than 50%) of material formed directly by precipitation from solution or colloidal suspension (as from evaporation) or by the formation of insoluble precipitates (as by the mixing of solutions of two or more soluble salts); used in contrast to the term detrital rock. (Krynine in 1948)

cheremchite *sed.* Variety of sapropelic coal composed of a mixture of structureless humic sapropel and algal remains. Also spelled tcheremkhite and tscheremkhite. [Cheremchovo district, Irkutsk, Siberia] (Zalessky in 1917)

cherokite *sed.* Dense, brown, residual sand making up the cement of the chert breccias in the zinc-mining district around Joplin, Missouri. [Cherokee Mississippian limestone formation] (Jenney in 1893)

chert *sed.* Tough, compact, dull to semivitreous, opaque quartz rock composed predominantly of impure fine-grained chalcedony or micro-granular quartz or of combinations of the two; often white or light gray but may be green, blue, pink, red, yellow, brown, or black and occurs as nodular segregations in limestones and dolomites and less commonly as extensive beds. Variety names are numerous: arenaceous c. (sand grains in the rock), bedded c., black c., calcareous c. (with calcite inclusions), carbonaceous c., nodular c., oolitic c. (replaced calcite ooliths), red c., red shaly c., and others listed below. [English term, origin is unknown] (in geological reports as early as 1680)

chert, bone   Weathered, residual chert that appears somewhat porous and chalky; usually white but may be stained red, brown, and other colors.

chert, chalcedonic   *sed.* Vitreous to waxy smooth chert, transparent to milky, generally buff or blue-gray, sometimes mottled; generally the chalcedony component dominates over the cryptocrystalline (microgranular) component.

chert, chalky   *sed.* Dull or earthy, soft to hard variety of chert resembling chalk; the fracture is generally rough or uneven.

chert, chrome   *meta., metasomatic.* Chertlike rock in which quartz has replaced the silicate minerals of a chromite peridotite, the more resistant chromite grains remaining unaltered in the secondary siliceous groundmass. (Fermor in 1919)

chert, cotton   *sed.* = chert, chalky

chert, diatomaceous   *sed.* Opaline-rich variety of chert derived from diatoms.

chert, dolocastic   *sed.* Spongy chert whose porosity has resulted from the removal of dolomite rhombs, originally in the rock, by solution. [dolomite cast]

chert, granular   *sed.* Compact, weathered chert composed of distinguishable and relatively uniform-sized grains, with a saccharoidal appearance characterized by an uneven fracture surface and dull glimmering luster.

chert, granulated   *sed.* Type of granular chert characterized by rough, irregular grains or granules of chert tightly or loosely held together in small masses or fragments.

chert, novaculitic   *sed.* Any gray chert that breaks into slightly rough, splintery fragments; usually less vitreous and somewhat coarser-grained than chalcedonic chert. [L. *novacula,* razor]

chert, pinhole   *sed.* Weathered chert pierced by minute holes or pores.

chert, porcelaneous   *sed.* Chert having a smooth fracture surface and a typically china-white appearance, resembling glazed porcelain or chinaware.

chert, quartzose   *sed.* Variety of chert containing very tiny quartz crystals that give the rock a vitreous, sparkly appearance; term also applied to chert containing anhedral secondary quartz.

chert, radiolarian   *sed.* Chert essentially composed of recrystallized opaline radiolarian capsules. Varieties include black radiolarian chert (dark carbonaceous pigments) and red radiolarian chert (red iron-oxide pigments).

chert, rhomb-bearing   *sed.* Chert containing perfectly formed, small sharp-edged rhombs of a carbonate mineral (dolomite, ankerite, siderite, or calcite); rhombs are sometimes replaced by iron oxides, iron silicates, pyrite, or crystalline quartz or may be weathered out.

chert, sandy   *sed.* Chert with an oolitelike structure,

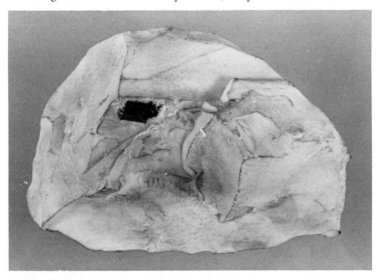

Chert. Arkansas. 13 cm.

formed when silica replaces cement or fills pore spaces in sandy beds and incorporates large, rounded sand grains in a cherty matrix.

chert, sapropelic *sed.* Dark, red brown, massive or banded chert containing dark opaque carbonaceous and bituminous material and often pyrite. [Gk. *sapros,* rotten + *pelos,* mud]

chert, smooth *sed.* Ordinary hard, dense, homogeneous chert, characterized by a conchoidal to even fracture surface that is devoid of roughness and exhibits no crystallinity, granularity, or other distinctive structure.

chert, spiculitic *sed.* Chert essentially derived from the crystallization of opaline sponge spicules and composed of sponge spicules, opal, and chalcedony with clay, pyrite, organic matter, and other materials.

chian *ig., sed.?* Ancient term for a rock whose identity is now uncertain; perhaps for obsidian or for black marble. [Chios, Greek island] (Theophrastus, *On Stones*)

chibinite *ig. pluton.* Granular eudialyte-bearing nepheline syenite, differing from lujavrite in containing less dark-colored constituents, which are in compact aggregates of thick crystals instead of in fine needles, and in having eudialyte as interstitial patches instead of well-formed individuals. [Umptek (Chibinä), Kola Peninsula, Soviet Union] (Ramsay in 1894)

china-clay rock *sed.* Weathered granite composed chiefly of quartz and kaolinite, with muscovite and tourmaline as possible minor accessories; crumbles easily in the fingers.

china stone *sed.* 1. Partially weathered granite containing quartz, kaolinite, and minor mica and fluorite and used as a glaze in the manufacture of china; harder than china-clay rock. 2. Fine-grained, compact Carboniferous limestone or mudstone found in England and Wales.

chinarump *sed.* Seldom used name for the petrified wood occurring in Arizona. [Possibly Triassic Shinarump conglomerate formation, in which much of the wood occurs]

chirvinskite *sed.* High-carbon bitumen near shungite or anthraxolite in composition. [Petr Nikolaevich Chirvinsky, 1880-1955, Russian mineralogist and petrologist] (Platonov in 1941)

chladnite *meteor.* Group of achondrite meteorites including both enstatite achondrites as well as hypersthene achondrites. [Ernst Florence Friedrich Chladni, 1756-1827, German physicist, one of the first to demonstrate that meteorites could not have originated on earth] (Rose in 1863)

chlorastrolite *ig., altered.* Variety of pumpellyite occurring as small rounded pebbles showing a finely radiated or stellated structure and light bluish green color; derived from an amygdaloidal trap rock and occurring on the shores of Isle Royale, Lake Superior, Michigan. [Gk. *chloros,* green + *astron,* star] (Jackson and Whitney in 1847)

chloromelanite *meta.* Dark green to nearly black jadeite jade [Gk. *chloros,* green + *melanos,* black] (Damour in 1865)

chloropal *sed.* Massive, opal-like, greenish-yellow to pistachio-green clay composed of nontronite; it has a chonchoidal fracture and feebly adheres to the tongue. [Gk. *chloros,* green + opal] (Bernhardi and Brandes in 1822)

chlorophyre *ig. pluton.* Term applied to certain greenish porphyritic quartz diorites near Quenast, Belgium. [Gk. *chloros,* green] (Dumont in 1847)

chondrite *meteor.* Aerolite (stony meteorite) characterized by the presence of chondrules, which are spheroidal aggregates of olivine or pyroxene usually about 1 mm in diameter; the most common of all meteorites and, depending upon mineralogical and chemical compositions, with several varieties (listed below). [Gk. *chondrion,* granule]

chondrite, carbonaceous *meteor.* Relatively rare chondrite characterized by a dull black color, friability, generally low density, and lack or almost total lack of free nickel-iron. Some contain amorphous hydrated silicates and serpentine; although the amount of organic material in these in relatively small, the nature of these compounds is a subject of great interest.

chondrite, crystalline *meteor.* Hard, crystalline stony meteorite containing firm round radial chondrules that break with the matrix.

chondrite, enstatite *meteor.* Relatively rare chondrite meteorites characterized by the orthorhombic pyroxene enstatite; some diopside may be present, but olivine is normally absent.

chondrite, olivine-bronzite *meteor.* Relatively abundant chondrite composed of olivine, orthorhombic pyroxene (bronzite), with nickel-iron, plagioclase (often oligoclase), troilite, and others.

chondrite, olivine-hypersthene *meteor.* The most abundant chondrite; contains olivine, orthorhombic pyroxene (hypersthene), with lesser amounts of plagioclase (usually oligoclase), nickel-iron, troilite, and others.

chondrite, olivine-pigeonite *meteor.* Small group of chondrites characterized by the predominance of olivine (about 70%), pigeonite in accessory amounts (about 5%), and plagioclase, troilite, and small amounts of nickel-iron. Some contain carbonaceous compounds and are nearly black.

chorismite   *migmat.* Megascopically composite rock (migmatite) consisting of two or more petrographically different parts of uncertain or doubtful origin. See also migmatite. [Gk. *choris,* apart] (Niggli and Huber in 1943)

christianite   *ig. pluton.* Obsolete term for a type of granitic magma in which K_2O equals Na_2O and is greater than CaO. [Possibly Christiania, Oslo, Norway] (Lang in 1891)

chromitite   *ig. pluton.* Igneous rock composed of 95% or more of chromite; similar rocks with too large a percentage of ferromagnesian constituents to be pure chromitite have been named bronzite chromitite, hypersthene chromitite, and olivine chromitite. [chromite] (Johannsen in 1938)

chrysanthemum stone   *ig. extru.* Glomeroporphyry consisting of feldspar phenocryst clusters (1 in across) in a fine-grained, gray-green to black basaltic matrix; occurs around Victoria, Vancouver Island, Canada. Also named chrysanthemum rock and flowering gabbro.

chrysocarmen   *altered ore.* = carmazul. [Possibly chrysocolla + Carmen Island in Gulf of California, Mexico]

chrysojasper   *sed., altered ore?* Jasper colored green with chrysocolla. [chrysocolla + jasper]

chrysoprase   *sed.* Yellow green variety of chalcedony that can resemble jade. [Gk. *chrysos,* gold + *prason,* leek]

chrysoquartz   *meta.?* Uncommon name for golden aventurine. [Gk. *chrysos,* gold + quartz]

ciminite   *ig. extru.* Trachydolerite containing orthoclase with labradorite, augite or diopside, and olivine. [Cimino volcano, Italy] (Washington in 1906)

cinder   *ig. extru.* Highly vesicular vitric pyroclastic fragment with size 4-32 mm; often like scoria. Also named volcanic cinder. [O. Eng. *sinder*]

cinerite   *ig. extru.* Deposit of volcanic cinders; cindery lapilli tuff. [L. *cineris,* ash] (Cordier in 1816)

cipolin   *meta.* Variety of marble containing mica (usually phlogopite) and forming a transition between marble and mica schist. Also spelled cipolino and cipollin. [It. *cipolla,* onion; because it is stratified like an onion]

clarain   *sed.* In banded coal, thin bands characterized by bright color and silky luster; composed largely of translucent attritus. Also named clarite. [L. *clarus,* clear] (Stopes in 1919)

clasmoschist   *sed.* Obsolete term originally introduced to replace "graywacke" (arenaceous rock in the lower part of the Secondary, roughly Mesozoic, strata). [Possibly Gk. *klastos,* broken + schist] (Conybeare, date unknown)

clasolite   *sed.* General term for a rock composed of fragments of other rocks; clastic rock. [Gk. *klastos,* broken]

clast   *sed.* Individual grain, particle, or fragment of a rock or mineral, produced by the disintegration (mechanical weathering) of a larger rock mass. [Gk. *klastos,* broken]

clay   *sed.* Unconsolidated or poorly indurated natural earth, containing an excess of particles of clay size (1/256 mm or less in diameter), and consisting predominantly of the clay minerals; in practice the name clay is applied to material containing as little as 10% clay minerals. Important mineralogical varieties include kaolinite c. (kaolin), illite c., montmorillonite c. (bentonite), diaspore c., and halloysite c. [O. Eng. *claeg,* clay]

clay, acid   *sed.* Any clay yielding hydrogen ions in a water suspension.

clay, ball   *sed.* Light buff to gray, highly plastic, sometimes refractory, clay, commonly characterized by the presence of organic matter; used as a bonding constituent of ceramic wares. [from early English practice of rolling the clay into balls of about 10 in diameter]

clay, book   *sed.* Clay occurring as thin, leaflike laminae.

clay, boulder.   *sed.* Term used in Great Britain, nearly equivalent to till; used esp. where boulders of various sizes are embedded in stiff, hard, pulverized clay or rock flour.

clay, brick   *sed.* 1. = earth, brick.   2. Impure clay containing iron, calcium, magnesium and other ingredients.

clay, brown   *sed.* = clay, red.

clay, burley   *sed.* Clay containing burls; more specifically, a diaspore-bearing clay in Missouri, averaging 45-65% alumina.

clay, calcareous   *sed.* Clay containing calcium carbonate, e.g., marl.

clay, china   *sed.* Clay consisting essentially of the mineral kaolinite and suitable for use in the manufacture of chinaware.

clay, diaspore   *sed.* Name applied to clay composed of diaspore (with or without boehmite) as well as those containing diaspore with considerable or even predominating kaolinite; some are fireclays.

clay, flint   *sed.* Smooth, flintlike clay, composed

of kaolinite, that breaks with a good conchoidal fracture and resists slaking in water.

clay, glacial   *sed.* = flour, rock

clay, halloysite   *sed.* Clays whose major component is halloysite but that may also contain alunite, allophane, and phosphatic materials; esp. important in Indiana at the base of the Mansfield (Pennsylvania) sandstone.

clay, illite   *sed.* Clay rich in minerals of the illite group esp. deep-sea clay; often occurs with montmorillonite, kaolinite, and chloritic clays; underclays often consist of kaolinite-illite mixtures or, if they are calcareous, of illite alone.

clay, kaolinite   *sed.* = kaolin

clay, montmorillonite   *sed.* 1. = bentonite. 2. Transported clays, esp. those derived from bentonite, usually mixtures of montmorillonite and kaolinite with some halloysite and allophane; up to 25% clastic grains of quartz, potassic feldspars, sodic-plagioclase, muscovite, and other minerals may be present and gypsum and barite crystals, opal, pyrite, zeolites, and other minerals may occur.

clay, pelagic   *sed.* Clay occurring in deep-ocean marine sediments. [Gk. *pelagos,* sea]

clay, red   *sed.* Marine deposit, fine-grained and red to reddish-brown, formed by the slow accumulation of material a long distance from the continents and at depths generally greater than 3,500 m; consists of dust (meteoric and volcanic), pumice, skeletal materials, and others.

clay, residual   *sed.* Clays that form in place and make up soil or are products of soil-forming processes; the character of these deposits depends upon climate, drainage, and the composition of the parent rock.

clay, salt   *sed.* Clay containing appreciable amounts of saline minerals, esp. halite; these can grade into silty halite rock (rock salt).

clay, sandy   *sed.* Unconsolidated sediment consisting of a mixture of sand and clay particles; various percentages have been suggested, e.g., 40-75% clay, 12.5-50% sand, 0-20% silt (Shepard in 1954); 10-50% sand and having a ratio of silt to clay less than 1:2 (Folk in 1954).

clay, silty   *sed.* Unconsolidated sediment consisting of a mixture of silt and clay particles; various percentages have been suggested, e.g., 40-75% clay, 12.5-50% silt, and 0-20% sand (Shepard in 1954).

clay, varved   *sed.* = varve

claycrete   *sed.* Weathered argillaceous material forming a layer immediately overlying bedrock. [clay + L. *crescere,* to grow]

claypan   *sed.* Dense subsurface layer of soil whose hardness and relatively slow permeability to water are chiefly due to high clay content.

claystone   *sed.* 1. Indurated clay lacking the thin-bedded character of shale; essentially the same as mudstone. 2. Concretionary clay found in alluvial deposits in the form of rounded disks, often coalesced to give curious shapes.

cleftstone   *sed.* = flagstone

cliachite   *sed.* Ferruginous bauxite from Dalmatia (Yugoslavia); also the colloidal aluminum hydroxides occurring in bauxite. Also spelled kliachite and kljakite. [Cliache (Kljake), Dalmatia] (Breithaupt in 1847)

cliffstone   *sed.* British term for a hard chalk used in paint, as a filler for wood, and in the manufacture of rubber. [Possibly alluding to chalk cliffs]

clinker   *ig. extru.* Irregular porous fragment of lava, often formed by the fragmentation of the surface of lava flows like aa. [Du. *Klinker,* a brick that resounds sonorously, from *klinken,* to clink]

clinker   *meta.* Masses of coal altered by an igneous intrusion; vitreous to porous fragments of ash or natural coke. [Du. *Klinker,* a brick that resounds sonorously]

clinkertill   *meta.* Glacial till baked by the burning of lignite beds.

clinkstone   *ig. extru.* = phonolite

clinopyroxenite   *ig. pluton.* 1. Pyroxenite in which the pyroxene is monoclinic. 2. *IUGS,* plutonic rock with M equal to or greater than 90 and cpx/(ol+opx+cpx) greater than 90. [clinopyroxene]

clinopyroxenite, olivine   *ig. pluton.* 1. Olivine-bearing pyroxenite in which the pyroxene is monoclinic. 2. *IUGS,* plutonic rock with M equal to or greater than 90, 5-40 ol/(ol+opx+cpx), opx/(ol+opx+cpx) less than 5, and cpx/(ol+opx+cpx) less than 90.

closterite   *sed.* Dense, laminated, brownish-red oil shale from the Irkutsk River basin of Siberia; the organic material includes *Closterium,* a variety of algae. [Closterium] (Zalessky in 1917)

cloustonite   *sed.* Rock related to asphalt, jet-black with a brilliant luster like obsidian; soluble in benzene and occurs in blue limestone at Inganess, Orkney. [Charles Clouston, 1800-1884, Scottish minister who wrote in 1839 an account of the geology and mineralogy of Orkney]

coal   *sed.* Brown to black, brittle, compact, amorphous combustible rock containing more than

50% by weight and more than 70% by volume of carbonaceous material, including inherent moisture; formed by the partial decomposition of vegetable matter without free access to air, under the influence of moisture, and in many cases, of increased pressure and temperature. There are various ranks of coal depending upon the degree of change from the original vegetable matter, e.g., lignite, bituminous, and anthracite coals. [O. Eng. *col,* coal]

coal, Albert   *sed.* = albertite

coal, algal   *sed.* = coal, boghead

coal, alum   *sed.* Argillaceous brown coal containing alum as a weathering product of pyrite.

coal, anthracite   *sed., meta.* = anthracite

coal, anthraxylous   *sed.* Bright coal with a ratio of anthraxylon (vitreous lustrous bands) to attritus (dull gray to black bands) greater than 3:1.

coal, anthraxylous-attrital   *sed.* Rather bright coal with a ratio of anthraxylon (vitreous lustrous bands) to attritus (dull gray to black bands) between 3:1 and 1:1.

coal, attrital   *sed.* Coal with a ratio of attritus (dull gray to black bands) to anthraxylon (vitreous lustrous bands) greater than 3:1.

coal, attrital-anthraxylous   *sed.* Coal with a ratio of anthraxylon (vitreous lustrous bands) to attritus (dull gray to black bands) between 1:1 and 1:3.

coal, ball   *sed.* Coal occurring in spheroidal masses, probably formed by jointing fractures; not to be confused with a coal ball.

coal, banded   *sed.* Heterogeneous coal, usually bituminous, containing bands of varying luster.

coal, bastard   *sed.* 1. Impure shaly coal occurring in the lower part of the shale strata immediately overlying a coal seam.   2. Any coal with a high ash content.

coal, bird's-eye   *sed., meta.* Anthracite with numerous small areas displaying a semiconchoidal fracture pattern resembling eyes.

coal, bituminous   *sed.* Dark brown to black, usually banded, coal, containing more than 14% volatile matter (dry, ash-free basis) and, on average, 84% carbon; the most abundant coal, between lignite and anthracite in rank.

coal, blind   *sed., meta.* Any coal (e.g., anthracite) that burns without a flame.

coal, bog   *sed.* Earthy variety of brown coal.

coal, boghead   *sed.* Clean, compact, blocky sapropelic coal of massive structure related to cannel coal but containing large masses of oil algae.

coal, bone   *sed.* Very impure coal high in ash (33% or more).

coal, bright   *sed.* Banded coal in which translucent matter predominates.

coal, brown   *sed.* Brown to brownish black coal in which original plant structures may usually be seen; intermediate in rank between peat and lignite; term commonly used outside the United States.

coal, candle   *sed.* Cannel coal burning with a steady flame like a candle.

coal, cannel   *sed.* Nonbanded dull, black coal of bituminous rank with a conchoidal fracture; burns readily with a long smoky flame. [Eng. dial. *cannel,* candle]

coal, carbonaceous   *sed.* Coal intermediate in composition between metabituminous coal and anthracite.

coal, cinder   *meta.* = coke, natural

coal, coke   *meta.* = coke, natural

coal, cone-in-cone   *sed.* Coal exhibiting a cone-in-cone structure, consisting of a set of inter-

Coal (cannel). New Zealand. 14 cm.

penetrating right circular cones packed closely together.

coal, crystallized *sed.* = coal, cone-in-cone. |the cones resemble crystals|

coal, eye *sed.* Coal showing circular or elliptical structural disks, either parallel or normal to the bedding, with concentric bending rims and radiating striations; they reflect light in a mirrorlike way and resemble eyes, e.g., bird's-eye coal.

coal, gas *sed.* Bituminous coal suitable for the manufacture of flammable gas because of its high content of volatile matter.

coal, glance *sed.* Bright coal composed primarily of anthraxylon (bright material derived chiefly from the woody parts of plants). |M. Eng. *glacen,* to slide, influenced by M. Eng. *glinten,* to shine|

coal, hard *sed., meta.* = anthracite

coal, humic *sed.* Any coal derived from peat by the process of humification (the development of humus or humic acids), essentially by slow oxidation; most coal is of this type. |L. *humus,* earth| (Potonié in 1904)

coal, humic-cannel *sed.* = coal, pseudocannel

coal, intermediate *sed.* Type of banded coal defined microscopically as consisting of 40-60% of bright ingredients such as vitrain, clarain, and fusain.

coal, lean-cannel *sed.* Cannel coal low in hydrogen and transitional to bituminous in rank.

coal, lignite *sed.* = lignite

coal, matte *sed.* = coal, splint. |G. *matt,* dull|

coal, metabituminous *sed.* Coal of slightly higher rank than bituminous, containing about 90% carbon (analyzed on a dry, ash-free basis).

coal, metacannel *sed.* Cannel coal of higher than usual rank.

coal, parrot *sed.* Cannel coal from near Edinburgh, Scotland; burns with a crackling noise, hence the name parrot.

coal, peacock *sed.* Iridescent coal exhibiting variegated colors on fracture surfaces due to thin films. (Plot in 1686)

coal, peat *sed.* Transitional coal between peat and lignite; brown coal.

coal, pitch *sed.* Bituminous or lignite coal, brittle and with a conchoidal fracture and pitchy luster.

coal, pseudocannel *sed.* Cannel coal containing much humic matter.

coal, sapropelic *sed.* Dull massive coal derived from fine-grained organic residues in stagnant or standing bodies of water; formed through putrefaction under anaerobic conditions rather than peatification. Examples include cannel coal and boghead coal. |Gk. *sapros,* rotten + *pelagos,* sea| (Potonié in 1904)

coal, semibituminous *sed.* Coal ranking between bituminous coal and semianthracite; harder and more brittle than bituminous coal.

coal, semibright *sed.* Type of banded coal defined microscopically as consisting of 61-80% of bright ingredients such as vitrain, clarain, and fusain.

coal, semicannel. *sed.* = coal, lean-cannel

coal, semidull. *sed.* Type of banded coal defined microscopically as consisting of dull ingredients such as durain with 21-40% bright ingredients such as vitrain, clarain, and fusain.

coal, semisplint *sed.* Banded coal intermediate in composition and character between banded bright coal and splint coal.

coal, soft *sed.* 1. = bituminous coal. 2. Term sometimes used for brown coal or lignite, esp. outside the United States.

coal, splint *sed.* Type of grayish-black coal, dull with a rough, uneven granular texture and composed of more than 30% opaque attritus. Also spelled splent. |splintery|

coal, stellar *sed.* = stellarite

coal, stone *sed., meta.* = anthracite

coal, subbituminous *sed.* Black coal intermediate in rank between lignite and bituminous coal and distinguished from lignite by a higher carbon and lower moisture content.

coal, subcannel *sed.* Cannel coal of subbituminous rank.

coal, tar *sed.* Resinous brown coal rich in bitumen.

coal, xyloid *sed.* = lignite, woody. |Gk. *xylon,* wood|

coal, yellow *sed.* = tasmanite. (Selwyn in 1855)

cobble *sed.* Rounded, subrounded, or subangular fragment of any rock 64-256 mm in diameter. |dim. of *cob,* of obscure origin, possibly O. Eng. *cop, copp,* head, related to G. *Kopf,* head|

cobble, glacial *sed.* Till stone of cobble size, often striated, with a faceted flat-iron shape.

cobblestone *sed.* 1. Rounded water-worn stone; may be used for paving; esp. a stone of cobble size. 2. Rock composed of cobble-sized stones, e.g., a cobble conglomerate.

cocite *ig. dike.* Yellowish green lamprophyre showing phenocrysts of olivine and diopside and smaller amounts of olivine, pyroxene, magnetite, biotite, and leucite, in a groundmass of augite, orthoclase, and biotite. [Coc Pia, Upper Tonkin, Vietnam] (Lacroix in 1933)

coconucite   sed. Variety of travertine from Colombia. [Coconuco, Colombia]

coerulene   sed. Trade name for calcium carbonate (usually calcite) colored green and blue by malachite and azurite. [L. coeruleus, blue]

coke   sed., altered. Carbon-rich substance used as a fuel and obtained by distilling the volatile constituents from coal, petroleum, and the like, by heating in ovens or retorts. [Eng. dial. cokes, coaks, cinders; perhaps M. Eng. colk, core]

coke, native   meta. = coke, natural

coke, natural   meta. Coke formed by the natural carbonization of coal under the influence of an igneous intrusion (contact metamorphism) or by natural combustion of coal in mines. Also named native coke, cinder coal, and cokeite.

cokeite   meta. = coke, natural. [coke] (Lacroix in 1910)

collobriérite   meta. Grunerite eulysite composed of fayalite and grunerite, with magnetite, spessartine-almandine, and accessory zircon, apatite, hornblende, and others. [Collobrièu, France] (Lacroix in 1917)

colloclast   sed. Accretionary aggregate of mud or silt, lobate-shaped with surficial irregularities; aggregate is usually the size of sand or gravel and resembles a mineral or rock fragment (clast). [Gk. kolla, glue + clast] (Sander in 1967)

colloidstone   sed. Indurated sedimentary rock composed of colloid-size particles. [Gk. kollodes, gluelike] (Alling in 1943)

colluvium   sed. General term for any loose mass of rock and soil materials deposited by rainwash, sheetwash, or slow continuous downslope creep, usually collecting at the base of a gentle slope. [L. colluvies, washings] (Used in English as early as the 17th century)

colombianite   ig. extru. Variety of obsidian (americanite) from near Cali, Colombia; once thought to be a tektite. [Colombia]

coloradoite   ig. extru. Quartz trachyandesite containing orthoclase (28%), andesine (28%), quartz (15%), with biotite, diopside, hematite, and others. Not to be confused with the mineral of the same name. [San Cristobal quadrangle, Colorado] (Niggli in 1923)

columbretite   ig. extru. Type of phonolite consisting of sanidine and altered hornblende prisms in a dense groundmass of corroded oligoclase micro-

lites whose embayments are filled with analcime, small augite prisms, and small magnetite grains. [Columbrete Islands, Spain] (Johannsen in 1938)

column   sed. Speleothem formed by the union of a stalactite and stalagmite in caves.

column, basalt   ig. extru. = basalt, columnar

comendite   ig. extru. Very light-colored pantellerite consisting of phenocrysts of quartz, alkali feldspar, and mafics (aegirine, arfvedsonite, riebeckite, biotite) in a gray, bluish, or yellowish groundmass of quartz and alkali feldspar. [Comende, San Pietro Island, south of Sicily] (Bertolio in 1895)

composite rock   ig., sed., meta. General term for a rock consisting of two or more different parts, each having an apparently different origin; e.g., a migmatite.

comptonite   ig., altered. Opaque variety of thomsonite from Somma, Italy, and the Lake Superior region; often cut as a gem. [Spencer Joshua Alwyne Compton, 1790-1851, Marquis and Earl of Northampton, once president of the Geological Society of London, who brought the material from Italy] (Brewster in 1821)

conachatae   sed. Variety of agate containing inclusions of cacholong opal arranged as conic patches. [Gk. konos, cone + and G. Achat, agate]

conchilite   sed. Small, rusty, bowl-shaped objects of goethite (limonite) that grow in an inverted position on mineralized bedrock in lakes, esp. noticed in Canada. [Gk. konche, shell] (Tanton in 1944)

concrete, ice   sed. Solid frozen mixture of rock fragments, sand, other clasts, and ice. Also named icecrete. [L. concretus, grown together]

concretion   sed. 1. Regular to irregular spherical, ellipsoidal to rounded flattened, compact bodies, commonly composed of calcite, siderite, barite, gypsum, or similar minerals, together with some of the original sediment in which they occur; they are diagenetic and form by localized concentrated cementation within sediments and sometimes grow around a nucleus, e.g., a fossil.   2. Collective term including concretions, septaria, spherulites, sand crystal rosettes, geodes, and the like. [L. concretus, grown together]

concretion, calcareous   sed. Concretion in which the major cementing material is calcite; term includes concretions of sand (kugelsandstein), silt, and clay (shale).

concretion, iron carbonate   sed. = concretion, siderite

concretion, phosphatic   sed. Concretion in which the cementing material is a calcium phosphate (usually an apatite); usually phosphatic objects of concretionary shape are more accurately described as nodules or spherulites, since they incorporate very little of the clastic material in which they occur.

concretion, sandstone   *sed.* Concretions composed of sand-sized grains usually cemented by calcite; they vary in size from less than 1 cm to as much as 9 m in diameter; variations in particle sizes can result in siltstone concretions, conglomerate concretions, and others. Also named kugelsandstein.

concretion, siderite   *sed.* Concretion in which the cementing material is siderite; often occurs in dark shale related to coal deposits and often has a fossil nucleus. Also named iron carbonate concretion.

concretion, voidal   *sed.* Iron-oxide concretion, usually tube-shaped, with a central hollow surrounded by a hard, dense limonitic rim; apparently formed by the weathering (oxidation) of a siderite concretion. Box stone is related.

cone-in-cone   *sed.* Unusual structure consisting of a succession of concentric right cones fitting one into another in inverted position (base upward, apex downward) rarely exhibited by calcareous shales, clay rocks, and coals; the cone axes are parallel, and their bases are on an approximately parallel surface; the cones usually contain fibrous calcite or, more rarely, siderite or gypsum.

conglomerate   *sed.* Cemented clastic sedimentary rock consisting of rounded fragments of older rocks and/or minerals (larger than 2 mm across, including granules, pebbles, cobbles, or boulders),

embedded in a finer matrix; the older rocks can be of any kind, hard rocks naturally predominating; calcite, quartz, and limonite are the common cementing materials. [L. *con,* with, together + *glomerare,* to wind into a ball]

conglomerate, arkosic   *sed.* Orthoconglomerate in which most of the clasts consist of granite, aplite, or pegmatite, so that the composition is like an arkose.

conglomerate, basal   *sed.* Well-sorted, lithologically homogeneous conglomerate forming the bottom stratigraphic unit of a sedimentary sequence and resting on a surface of erosion, thereby marking an unconformity.

conglomerate, boulder   *sed.* Conglomerate consisting of boulder-size fragments (greater than 256 mm in diameter).

conglomerate, breccia   *sed.* Sedimentary rock intermediate between a conglomerate and breccia in that it consists of both angular and rounded rock fragments. (Norton in 1917)

conglomerate, breccio-   *sed.* = conglomerate, breccia

conglomerate, cataclastic   *meta.* = conglomerate, crush

conglomerate, cobble   *sed.* Conglomerate consisting of cobble-size fragments (64-256 mm in diameter).

conglomerate, crush   *meta.* Cataclastic metamorphic rock closely related to a crust breccia, but in this case there has been considerable rotation of the particles, the corners becoming rounded off through mutual abrasion, so that the frag-

Concretion (barite) with some septarian cracks. Pryor Mountains, Montana. 15 cm.

ments are subangular to rounded and in an abundant finer matrix. (Lamplugh in 1895)

conglomerate, desiccation *sed.* Conglomerate consisting of fragments of mud-cracked sediment rounded by transportation; mudstone conglomerate.

conglomerate, dolomitic *sed.* Conglomerate consisting of limestone pebbles and dolomite cement. (Nelson and Nelson in 1967)

conglomerate, edgewise *sed.* Conglomerate exhibiting an edgewise structure, one in which flattened fragments are set at steep angles to the bedding, e.g., an intraformational conglomerate containing elongated calcareous pebbles transverse to the bedding.

conglomerate, granite-pebble *sed.* = conglomerate, arkosic. (Krumbein and Sloss in 1963)

conglomerate, graywacke *sed.* Conglomerate close to graywacke in composition.

conglomerate, intraformational *sed.* Conglomerate in which the fragments are essentially contemporaneous with the matrix in origin; developed by the breaking up and rounding of fragments of a newly formed or partly consolidated sediment and their nearly immediate incorporation in new sedimentary deposits; often composed of shale or limestone.

conglomerate, mudstone *sed.* 1. = conglomerate, desiccation.   2. Any conglomerate composed of mudstone fragments.

conglomerate, oligomictic *sed.* Conglomerate in which the clasts are lithologically homogeneous, e.g., all pebbles being limestone or quartzite or granite. [Gk. *oligos,* few + *miktos,* mixed]

conglomerate, orthoquartzitic *sed.* Well-sorted oligomictic conglomerate consisting of quartz-rich fragments (vein quartz, chert, quartzite), derived from eroded granitic or metamorphic terrain, with the removal of less stable and finer materials by weathering or long transport.

conglomerate, pebble *sed.* Conglomerate consisting of pebble-size fragments (4-64 mm in diameter).

conglomerate, petromict *sed.* Conglomerate characterized by a mixture of metastable rock fragments, pebbles, or cobbles of plutonic, eruptive, sedimentary, or metamorphic rocks. [Gk. *petros,* rock + *miktos,* mixed]

conglomerate, polymictic *sed.* Conglomerate in which the clasts are not lithologically homogeneous; there is a broad mixture of rock types in the pebbles. [Gk. *polys,* many + *miktos,* mixed]

conglomerate, quartz-pebble *sed.* = conglomerate, orthoquartzitic

conglomerate, roundstone *sed.* Any conglomerate of sedimentary origin consisting of rounded clasts; in contrast, sharpstone conglomerate is a sedimentary breccia.

conglomerate, sandy *sed.* 1. Conglomerate containing 30-80% sand and having a ratio of

Conglomerate (Potomac marble). Point of Rocks, Maryland. 11.5 cm.

sand to mud (silt and clay) greater than 9:1. (Folk in 1954) 2. Conglomerate containing more than 20% sand. (Krynine in 1948)

conglomerate, second-cycle *sed.* Conglomerate containing fragments showing evidence of having been derived from an older conglomerate.

conglomerate, sharpstone *sed.* Sedimentary breccia; term used in contrast to roundstone conglomerate, a true sedimentary conglomerate.

conglomerate, tuffaceous *ig. extru., sed.* IUGS, conglomerate whose average clast size is greater than 64 mm and containing 25-75% by volume of pyroclasts, the remaining components being sedimentary clasts.

conglomerate, turbidite *sed.* Petromict conglomerate deposited in deep water.

conglomerate, volcanic *ig. extru., sed.* Conglomerate composed of over 50% water-laid volcanic pyroclastic boulders and pebbles showing some effects of erosion.

conglomerite *sed., meta.* Conglomerate with the same state of induration as a quartzite; if the rock is a result of metamorphic changes it is more commonly named a metaconglomerate. (Willard in 1930)

congressite *ig. pluton.* Coarse-grained, usually foliated, urtite composed of much nepheline (73%) and small amounts of orthoclase, albite, sodalite, muscovite, biotite, and magnetite. [Congress Bluff, Craigmont Hill, Ontario] (Adams and Barlow in 1913)

coniatolite *sed.* Hard, sheetlike crust of aragonite occurring in supratidal saline environments in the Persian Gulf area. [L. *konia,* powdered lime] (Purser and Loreau in 1973)

contactite *meta.* Term applied to any rock formed by contact metamorphism. [contact]

conulite *sed.* In caves, a hollow conical carbonate speleothem formed as a drill-hole lining in mud; subsequent erosion isolates the lining. [L. *conus,* cone, cone-shaped]

coombe rock *sed.* Mass of unstratified rock debris of any type that has accumulated as a result of the slow somewhat viscous downslope flow of waterlogged materials; esp. applied to unrolled and unweathered flint mixed with chalk fragments, partly filling a dry valley (coombe), as in southwestern England. [O. Eng. *cumb,* narrow valley]

coorongite *sed.* Brown, soft, elastic variety of elaterite, forming from deposits of *Elaeophyton* algae in salt water bodies. [Coorong district, South Australia] (Morriss in 1877)

copal *sed.* General term for a wide variety of hard, brittle, somewhat transparent, yellowish to reddish fossil resins; term also used for modern resinous exudations. [Sp. Am. *copalli,* resin]

copal, Congo *sed.* Hard, yellowish to colorless fossil resin used in making varnish; derived from trees of the genus *Copaifera,* found in the Congo.

copalite *sed.* Fossil, pale yellow to dirty brown, resin resembling copal; emits a resinous aromatic odor when broken and burns easily with a yellow flame; the original specimens came from the blue clay of Highgate Hill, near London. Also spelled copaline. [copal] (Hausmann in 1847)

coppaelite *ig. extru.* Olivine-free melilitite lava showing small phenocrysts of diopside in a holocrystalline groundmass of melilite, diopside, and phlogopite. [Coppaeli di Sotto, near Rieti, Umbria, Italy] (Sabatini in 1903)

coprolite *sed.* 1. Brown to black fossilized excrement of fish, reptiles, and mammals, measuring up to 20 cm in length; since some of these are composed of calcium phosphates, the term is often extended to include phosphate nodules of any origin. 2. Term also applied to colorful agate nodules apparently derived from true coprolites; occurs in the American West, esp. in Utah, and may be used by lapidaries. [Gk. *kopros,* dung]

coquina *sed.* Unconsolidated, or poorly cemented, biocalcirudite containing at least 50% of mechanically deposited mollusc shells or shell fragments, of size greater than 2 mm. [Sp. *coquina,* shellfish, cockle]

coquinite *sed.* Fully cemented coquina limestone. [coquina]

coral rag *sed.* Well-cemented, rubbly limestone breccia or conglomerate, composed of broken and rolled fragments from coral reef deposits.

coral reef *sed.* Ridge or mound of in-place coral colonies (accompanied by the accumulation of skeletal fragments, carbonate sands, algae) resulting from the organic secretion of calcium carbonate; potentially a wave- and surf-resistant framework. Various types have been named, e.g., patch reef (small and circular), barrier reef (separated from a coast by a lagoon), pinnacle reef (conical), fringing reef (attached to the coast), and atoll (enclosing a lagoon).

coral rock *sed.* = limestone, coral-reef

coralgal *sed.* Solid limestone formed by an intergrowth of frame-building corals and algae. [coral + algae]

corallite *sed.* Fossil coral rock. [L. *corallum,* coral] (Phillips in 1815) corcovadite   *ig. hypab.* Fine-grained porphyritic rock having the composition of granodiorite, in which phenocrysts of plagioclase (oligoclase-andesine), green hornblende, and some quartz and biotite, are in a groundmass (53%) of microaplitic plagioclase, quartz, and orthoclase; occurs at Marmato, Cauca River, Antioquia, Colombia. (Scheibe in 1926)

cordierite-anthophyllite rock   *meta.* Metamorphic rock consisting of anthophyllite (as prisms or radiating crystals) and cordierite, with varying amounts of biotite, garnet, quartz, plagioclase, and magnetite; with increasing plagioclase the rock becomes a plagioclase gneiss. (Eskola in 1914)

cornéenne   *meta.* Term used for hornfels in France. [F. *corné*, horny]

cornéite   *meta.* Biotitic hornfels produced by heat developed by crushing in sandy beds; localized in shaly beds along crests of anticlines and troughs of synclines through sharp flexion. [F. *corné*, horny]

Cornish stone   *sed.* China stone composed of feldspar, mica, and quartz, used in the manufacture of pottery.

cornstone, concretionary   *sed.* Calcareous concretion embedded in marl and grading into a concretionary limestone; term used in England. [corn, alluding to the agricultural suitability of the material]

cornstone, conglomeratic   *sed.* Fragments of marl and limestone embedded in a sandy or calcareous matrix; an English term. [corn, alluding to the agricultural suitability of the material]

cornubianite   *meta.* Contact metamorphic hornfels, consisting of mica, andalusite, and quartz; generally applied to a micaceous hornfels, although it has been suggested that the term be restricted to a tourmaline hornfels. [Cornubia, L. name for Cornwall, England] (Boase in 1832)

corsite   *ig. pluton.* Orbicular gabbro consisting of 50% concentric oval masses (1-3 in in diameter), showing alternating rings of feldspar and mafics, in an abundant groundmass; the feldspar is bytownite (consequently the rock is not a diorite, as earlier described), and the mafics include hornblende and minor hypersthene. The rock also has been named napoleonite, and incorrectly, Corsican granite, orbicular granite, and kugeldiorite. [Santa Lucia di Tallano, Corsica] (Zirkel in 1866)

cortlandtite   *ig. pluton.* Peridotite containing hornblende and olivine. [Cortlandt series in the township of Cortlandt, New York] (Williams in 1886)

Coquinite. Virginia. 10 cm.

corundolite *meta.* Any rock composed of corundum, e.g., emery. |corundum| (Wadsworth in 1891)

cosmoclastic rock *ig.?* Term applied to one of the original or primordial rocks of the Earth. |Gk. *kosmos,* order, the universe + *klastos,* broken| (Fairchild in 1904)

cosmolite *meteor.* = meteorite. |Gk. *kosmos,* order, the universe|

cotterite *sed., ig.?* Local term for quartz with inclusions of white clay; term originally used for a variety of quartz having a metallic pearly luster and forming a coating on ordinary quartz crystals at Rockforest, Ireland. |Miss Cotter of Rockforest, Ireland, its discoverer| (Harkness in 1878)

cotton rock *sed.* General descriptive name for soft, white, porous, chalky rock materials resembling cotton; used for the light-colored crusts on black flint nodules and for siliceous magnesian limestone in Missouri.

country rock *ig., sed., meta.* General term applied to the rock surrounding or enclosing an igneous intrusion or other specific rock or mineral body.

courtzilite *sed.* Variety of asphalt related to uintahite (gilsonite). (pub. by Parker in 1895/96)

covdorite *ig. pluton.* Olivine turjaite; apparently resulting from the calcium metasomatism of a peridotite or olivinite. |Kovdorozero massif, Kola, Soviet Union| (Zlatkind in 1945)

covite *ig. pluton.* Moderately light-colored nepheline syenite with hornblende and aegirine-augite as mafics. |Magnet Cove region, Arkansas| (Washington in 1901)

craigmontite *ig. pluton.* Nepheline diorite composed of nepheline, oligoclase, muscovite, and corundum; differs from congressite and raglanite in containing more nepheline and less plagioclase and corundum. |Craigmont Mountain, Ontario, Canada| (Adams and Barlow in 1910)

craignurite *ig. extru.* Glassy extrusive rock with microlites of pyroxene, plagioclase (oligoclase-andesine), and iron oxide, in whose norm there are about 25% each of quartz, orthoclase, and albite. |Craignure, Island of Mull, Scotland| (Bailey and Thomas in 1924)

crenite *sed.* Stalactitic calcite colored yellow with crenic acid or crenate of calcium (both organic

Corsite (orbicular gabbro). Island of Corsica. 13.5 cm.

compounds). [Gk. *krene,* spring; crenic acid occurs in ferruginous springs] (Wells in 1852)

creolin *sed.* Variety of jasper that is brecciated. [Possibly creole, of mixed descent]

creoline *ig., altered.* Purple epidotized basalt; originally applied to specimens from Massachusetts. [Possibly creole, of mixed descent]

creolite *sed.* Red and white banded jasper from Shasta and San Bernardino Counties, California. [Possibly creole, of mixed descent]

crinanite *ig. dike.* Term applied to olivine-analcime diabases or ophitic olivine teschenites that occur as dikes in Argyllshire, Scotland. [Loch Crinan, Argyllshire] (Flett in 1911)

criquina *sed.* Unconsolidated or poorly cemented biocalcirudite, like coquina but composed of crinoid fragments. [crinoid + coquina]

criquinite *sed.* Firmly cemented criquina composed of crinoid fragments. [criquina]

cristo-grahamite *sed.* Hydrocarbon similar to grahamite from the Cristo mine, Huasteca, Mexico. [Cristo mine + grahamite] (Kimball in 1876)

crocydite *migmat.* Migmatite with a flakelike or flufflike structure. [Gk. *krokids,* nap on cloth] (DeWaard in 1950)

cromaltite *ig. pluton.* Pyroxenite containing predominant aegirine-augite with lesser amounts of melanite, biotite, and iron oxides. [Cromalt Hills, Assynt, Scotland] (Shand in 1906)

crowstone *sed.* Very hard, siliceous, white sandstone representing the floor of a coal seam; term used in the Yorkshire and Derbyshire, England, coalfields. [crow, poor or impure bed of rock]

cryoconite *uncertain.* Gray powder covering the surface of land ice; occurring in Greenland 30 mi from the coast; may be several millimeters in thickness, often agglomerated into small round balls of loose consistency; of debated cosmic origin. Also spelled kryokonite. [Gk. *kruos,* ice + *konis,* dust] (Nordenskiöld in 1871)

cryptoclastic rock *sed.* Extremely fine-grained, clastic rock whose constituents can be seen only under a microscope; term applied to clay-rich rocks as well as aphanitic clastic carbonate rocks (e.g., finely comminuted carbonate dust). [Gk. *kruptos,* hidden + *klastos,* broken]

cryptomere *ig., sed., meta.* General term for an aphanitic rock; generally a synonym for aphanite if the rock is igneous. [Gk. *kruptos,* hidden + *meros,* part]

cryptoperthite *ig.* Variety of perthite with a structure so fine it cannot be discerned by an optical microscope; usually the submicroscopic lamellae are about 1-5 microns wide. [Gk. *kruptos,* hidden + perthite]

cryptosiderite *meteor.* Stony meteorite very poor in nickel-iron. [Gk. *kruptos,* hidden + *sideros,* iron]

crystalline carbonate *sed.* Carbonate sedimentary rock in which the original depositional textural features have been made obscure by recrystallization or replacement, e.g., dolomatized limestone.

cucalite *meta.* Metadiabase characterized by abundant chlorite; may grade into a chlorite schist as in the Rhaetian Alps. [Possibly Cuca, in modern Rumania] (Rolle in 1879)

culm *sed.* 1. Term, variously applied according to the locality, for carbonaceous shale, fissile anthracite, or fine particles of dust from anthracite mines. 2. = kolm. [M. Eng. *culme,* soot]

cumberlandite *ig. diaschist.* Greenish black, coarse-grained rock related to peridotite, containing olivine (40%), a mixture of magnetite and ilmenite (40%), and labradorite (10%) with serpentine, spinel, and apatite. [Cumberland, Rhode Island] (Wadsworth in 1884)

cumbraite *ig. extru.* Glassy andesite containing phenocrysts of bytownite-anorthite in a groundmass of labradorite, enstatite-augite, and abundant glass; in spite of its mineralogy it has a chemical composition corresponding to andesite rather than basalt. [Great Cumbrai, Firth of Clyde, Scotland]

cumulate *ig.* Any igneous rock formed by the accumulation of crystals settling out from a magma by the action of gravity. Also named accumulative rock. [L. *cumulus,* heap]

cumulophyre *ig.* Porphyritic igneous rock characterized by clustered phenocrysts, not necessarily of a single mineral. Glomeroporphyry is an example consisting of clustered feldspar phenocrysts. [L. *cumulus,* heap]

cuselite *ig. dike.* Monzonitic lamprophyre containing both plagioclase (albite) and orthoclase, with biotite, augite, hornblende, and quartz; intermediate in composition between minette and vogesite and between kersantite and spessartite. [Cusel, Saar-Nahe region, Germany] (Rosenbusch in 1887)

cuyamite *ig. hypab.* Dark teschenite dike rock composed of labradorite, analcime, hauyne, hornblende, augite and magnetite. [Cuyamas Valley, San Luis Obispo County, California] (Johannsen in 1938)

cuyunite *sed.* = binghamite. [Cuyuna range, central Minnesota]

D

dacite   *ig. extru.* Aphanitic rock containing essential quartz and sodic plagioclase, with minor biotite, amphibole, or pyroxene; like an andesite but with more quartz and plagioclase more sodic. [Dacia, Roman name for Siebenbürgen, Transylvania] (Stache in 1863)

dacitoid   *ig. extru.* Volcanic rock with the chemical composition of dacite but free of modal quartz. [dacite] (Lacroix in 1919)

dactylite   *ig.* General term for an igneous rock characterized by a dactylitic texture, where one mineral is penetrated by fingerlike projections from another. [Gk. *daktylos,* finger] (Sederholm in 1916)

dahamite   *ig. dike.* Brown variety of paisanite or microgranite characterized by the presence of abundant albite. [Dahamis, on the island of Socotra, Yemen] (Pelikan in 1902)

dallasite   *ig. extru., altered.* Rock composed of angular fragments of altered basalt cemented together with white chalcedony and quartz; occurs in open spaces in pillow basalt, esp. on the eastern coast of Vancouver Island, Canada. [Possibly Alexander Grant Dallas, 1816-1882, in charge of Hudson Bay Company in British Columbia, ca. 1858]

damkjernite   *ig. hypab.* Porphyritic rock with phenocrysts of biotite and pyroxene (titanaugite) in a fine-grained groundmass of pyroxene, biotite, magnetite, nepheline, and microcline. [Damkjern, Fen region, southern Norway] (Brögger in 1921)

dancalite   *ig. extru.* Porphyritic trachyandesite or feldspar-bearing tephrite with oligoclase, green augite rimmed with aegirine-augite, and rare amphibole, in a matrix composed of analcime, platy plagioclase, apatite, and others. [Dancala, Ethiopia] (De-Angelis in 1925)

danubite   *ig. extru.* Andesite containing amphibole and hypersthene; in the original description the labradorite was incorrectly identified as nepheline. [Possibly Danube River] (Krenner in 1910)

darlingite   *sed.* Variety of Lydian stone (basanite) from Victoria, Australia. [Possibly Charles Henry Darling, 1809-1870, governor of Victoria; other prominent Darlings and geographic features

named for them might be the source of the name] (Bleasdale in 1865)

daunialite   *sed.* Siliceous montmorillonitic clay of sedimentary origin, containing organic silica (25%) and small amounts of kaolinite, sericite, and chlorite; distinct from bentonite, which is of volcanic origin. [Dauni Mountains, Italy] (Andreatta in 1943)

davainite   *ig., meta.?* Rock consisting essentially of brown hornblende that is pseudormorphic after pyroxene, with very small amounts of other minerals such as hypersthene and feldspar; variety of hornblendite, or, if metamorphic, amphibolite. [Loch Beinn Bamhain (Loch Ben Davain), Scotland] (Wyllie and Scott in 1913)

dedolomite   *sed.* General term applied to limestones formed by the replacement of dolomite; relatively rare rock. [L. *de,* from + dolomite]

delatynite   *sed.* Variety of amber from Rumania. [Delatyn, Carpathian Mountains, Galicia] (Niedzwiedzki in 1908)

deldoradite   *ig. pluton.* Light-colored cancrinite syenite composed of microperthite, cancrinite, and accessory biotite, aegirine, apatite, titanite, and iron oxides. [Deldorado Creek, Uncompahgre quadrangle, Colorado] (Johannsen in 1939)

dellenite   *ig. extru.* Extrusive igneous rock intermediate in composition between dacite and rhyolite; equivalent to rhyodacite (also named quartz latite). [Dellen, Helsingland, Sweden] (Brögger in 1895)

deluvium   *sed.* = diluvium

denhardtite   *sed.* Pale yellow, waxy hydrocarbon similar to pyropissite; derived from plants; occurring in loam in British East Africa. [Clemens Andreas Denhardt, 1852-1928, and Gustav Denhardt, 1856-1917, brothers, German colonial explorers, who collected the material in 1878] (Potonié in 1905)

depalite   *sed.* Local name for a very fine-grained, olive- to khaki-colored chert occurring in Estill County, Kentucky; takes a good polish and is used by lapidaries. [Joseph Daugherty, Lewis Palmer, and Paul Cress, American amateur mineralogists of Kentucky, who discovered it] (pub. by Coskren in 1981) derivative rock   *sed.* General term for a rock composed of materials derived from the weathering of older rocks; sedimentary rock formed of material that has not been in a state of fusion immediately before its accumulation.

dermolith   *ig. extru.* Extrusive lava (usually basalt) manifesting subaerially a surface skin or crust that is folded and wrinkled; essentially the same as pahoehoe. [Gk. *derma,* skin] (Jaggar in 1917)

desmosite   *meta.* Adinole having a banded structure. [Gk. *desmos,* band] (Zincken in 1841)

detrital rock   *sed.* Sedimentary rock composed primarily (more than 50%) of particles or fragments detached from earlier rocks either by erosion or by weathering; term used in contrast to chemical rock. [L. *detritus* worn down] (Krynine in 1948)

detrital sediment   *sed.* Any accumulation of detritus.

detritus   *sed.* Loose fragments or particles of rocks and minerals worn off or removed by mechanical means, as by disintegration or abrasion; materials such as silt, sand, gravel derived from older rocks and moved from place of origin. [L. *detritus,* worn down]

devonite   *ig. aschist.* Diabase porphyry containing large flesh-colored phenocrysts of potassium-rich labradorite in a very dark-green ophitic groundmass of plagioclase and augite. [Mt. Devon, Madison County, Missouri] (Johannsen in 1910)

deweylite   *meta.* Greasy, resinlike variety of serpentinite (composed of clinochrysotile or lizardite with a talclike mineral), occurring as white, yellow, green, or reddish masses, and used an an ornamental stone; occurs in Pennsylvania, Maryland, Massachusetts, Tyrol, and elsewhere. [Chester Dewey, 1784-1867, American educator and botanist, Rochester University] (Emmons in 1826)

diabaros   *ig. extru.?* Ancient term, used by Theophrastus in his discussion of pumice; probably volcanic tuff of some kind. [Related to Gk. *diaboros,* porous]

diabase   *ig. dike.* Medium-grained intrusive igneous rock composed of calcic-plagioclase and pyroxene, with minor magnetite and sometimes olivine (the composition of basalt), and usually having an ophitic texture; intermediate in grain size between a gabbro and basalt. [Gk. *diabasis,* crossing over; possibly for the discordant nature of a dike] (Brongniart in 1807)

diabase, Aasby   *ig. dike.* Variety of diabase in which, in addition to labradorite, augite, and olivine, there are biotite, ilmenite, and apatite. [Aasby, Sweden] (Törnebohm in 1877)

Diabase in contact with coarser dike containing diopside and chlorapatite. Centreville, Fairfax County, Virginia. 11 cm.

diabase, albite   *ig. dike, meta.* = diabase, albitized

diabase, albitized   *ig. dike, meta.* Normal diabase altered so that the calcic-plagioclase becomes albite (or saussurite) and the ferromagnesian constituents become chloritized, epidotized, uralitized, and serpentinized; the original ophitic texture is often clearly retained. Also named albite diabase or metadiabase.

diabase, ashbed   *ig. extru., sed.* Local name for a rock on Keweenaw Point, Lake Superior, resembling a conglomerate but that was originally a very scoriaceous, amygdaloidal sheet into which much sand was washed in its early history. (Wadsworth in 1883)

diabase, bronzite   *ig. dike.* Diabase in which bronzite greatly exceeds clinopyroxene. (Rosenbusch in 1887)

diabase, enstatite   *ig. dike.* Diabase in which enstatite greatly exceeds clinopyroxene. (Rosenbusch in 1887)

diabase, glassy   *ig. dike.* Term applied to the glassy phase that sometimes develops along the contact zones of holocrystalline diabase intrusives. Also named hyalodiabase.

diabase, Hunne   *ig. dike.* Quartz diabase containing some hornblende, biotite, and a little quartz in addition to the essential plagioclase and augite; interstitial chloritic matter is present. [Hunneberg, Sweden] (Törnebohm in 1877)

diabase, hypersthene   *ig. dike.* Diabase in which hypersthene greatly exceeds clinopyroxene. (Rosenbusch in 1887)

diabase, Kinne   *ig. dike.* Swedish olivine diabase containing interstitial chloritic matter and secondary quartz. [Kinnekulle, Sweden] (Törnebohm in 1877)

diabase, Konga   *ig. dike.* Quartz diabase containing labradorite plates and intergranular augite and orthopyroxene, in a microgranitic interstitial mass of quartz and orthoclase. [Konga-Klint, Schonen, Sweden] (Törnebohm in 1877)

diabase, Oeje   *ig. dike.* Porphyritic diabase containing long plagioclase laths in an aphanitic basaltic groundmass. [Öje, Sweden] (Törnebohm in 1877)

diabase, olivine   *ig. dike.* Diabase in which olivine is an important constituent; less common than those without olivine but rather widespread.

diabase, pearl   *ig. extru.* = variolite

diabase, quartz   *ig. dike.* Diabase with quartz; composed of plagioclase (labradorite or more sodic), augite, and quartz, with accessories that may include biotite, ilmenite, magnetite, apatite,

and pyrite; relatively common variety. If the quartz is less than 5% of the total light-colored minerals, the rock is named a quartz-bearing diabase.

diabase, sodaclase   *ig. dike.* = diabase, albitized

diabase, tholeiitic   *ig. dike.* Diabase that is generally subaluminous, being saturated or slightly oversaturated with silica; generally like a tholeiitic basalt (tholeiite) but coarser in texture.

diabase, uralite   *ig. dike, meta.* = uralitite

diabasite   *ig. diaschist.* Seldom used term for a diabase aplite. [diabase] (Polenov in 1899)

diabrochite   *meta.* Metamorphic rock formed chiefly by the impregnation of ascending fluids or vapors but without the injection of a visible magma, e.g., hydrothermal metamorphism. [Possibly Gk. *diabrotikos,* corrosive]

diachyte   *migmat.* Rock formed by the mechanical or chemical contamination of anatectic magma (magma formed by melting of a preexisting rock) by mafic materials related in origin. [Gk. *diachulos,* very juicy]

diadysite   *migmat.* Migmatite with granitic veins cutting off the schistosity of the metamorphic parent rock. [Gk. *dia,* across, through + *dus,* bad]

diagenite   *sed.* General term for rock in which diagenesis has played an important role; chemical, physical, or biological changes occurred in the sediment after its initial deposition, and during and after its lithification, exclusive of weathering and metamorphism. [Gk. *dia,* through + *genes,* born]

diaglomerate   *sed.* Conglomerate in which the individual fragments are recognized as being related in their source. [Gk. *dia,* through + conglomerate]

diallagite   *ig. pluton.* Pyroxenite in which diallage (diopside or augite with excellent parting on front pinacoid) is the major component. Varieties, containing up to 25% of the additional mineral indicated, include: garnet (pyrope) d., hornblende d., hypersthene d., ilmenite d., magnetite d., olivine d. [diallage] (Cloizeaux in 1864)

diamict   *sed.* General term including diamictite and diamicton. [Gk. *dia,* through + *miktos,* mixed] (Harland and others in 1966)

diamictite   *sed.* Nongenetic comprehensive term for a nonsorted or poorly sorted, noncalcareous, terrigenous sedimentary rock containing a wide range of particle sizes; examples include tillite (glacial origin), pebbly mudstone (alluvial origin), and others. [Gk. *dia,* through + *miktos,* mixed] (Flint and others in 1960)

diamicton *sed.* Term applied to the nonlithified equivalent of diamictite, e.g., till or pebbly clay. [Gk. *dia,* through + *miktos,* mixed] (Flint and others in 1960)

diaphthorite *meta.* Crystalline metamorphic rock in which minerals characteristic of lower metamorphic grade have formed by retrograde metamorphism at the expense of minerals peculiar to a higher grade of metamorphism. [Gk. *dia,* through + *phthisis,* decay] (Becke in 1909)

diasporite *sed.* Clay rock composed of diaspore with or without boehmite; there may be subordinate kaolinite, chloritic clay, coaly material, and pyrite. [diaspore]

diatexite *migmat.* General term for rock formed by diatexis, that is, high-grade anatexis that melts even the mafic minerals. Also spelled diatectite. [Gk. *dia,* through + *tektos,* molten] (Gürich in 1905)

diatomite *sed.* Light-weight, gray to yellowish, soft, porous consolidated rock composed of the opaline tests of diatoms; the rock is often laminated with darker streaks, and diagenetic changes, including calcitization, phosphatization, and chertification, can often be observed. Common varieties include arenaceous diatomite, calcareous diatomite, and phosphatic diatomite. [diatom]

dike, breccia *sed.* Sedimentary dike formed by the injection of breccia into the country rock.

dike, clastic *sed.* Sedimentary dike formed by the injection of clastic materials, like sand, pebbles, gravel, into the country rock, e.g., sandstone dike and pebble dike.

dike, pebble *sed., ig.* Clastic dike composed largely of pebbles; term has also been applied to an igneous dike containing many sedimentary xenoliths rounded by milling or by the corrosive action of fluids.

dike, relict *granitization.* In a rock formed by granitization, relict tabular body of crystalline texture that represents a dike emplaced prior to granitization and relatively resistant to the granitization process.

dike, sedimentary *sed.* Tabular body of sedimentary material cutting across the structure or bedding of a preexisting rock in the manner of igneous dike; may form by the filling of a crack or fissure from below, above, or laterally; essentially the same as a clastic dike.

diktyonite *migmat.* Migmatite with a reticulated structure; having a network of veinlets with flexurelike appearance. [Gk. *diktuon,* net] (Sederholm in 1907)

diluvium *sed.* 1. Term applied to sorted and unsorted loose glacial deposits as contrasted to water-sorted alluvium. 2. Obsolete term for deposits produced by a great flood, e.g., the Noachian deluge. [L. *diluvium,* flood]

Dinas rock *sed.* Disintegrated high-silica sandstone, used for making refractory brick in Wales. [Craig-y-Dinas, south Wales]

diogenite *meteor.* = achondrite, hypersthene. [Gk. *diogenes,* heaven-born] (Tschermak in 1883)

diopsidite *ig. pluton.* Perknite (clinopyroxenite) composed almost entirely of diopside; minor magnetite, chromite, ferran spinel (pleonaste), garnet, and others, may be present. [diopside] (Lacroix in 1895)

diorite *ig. pluton.* 1. Phaneritic granular rock composed of essential sodic-plagioclase (usually oligoclase or andesine) and usually a mafic such as hornblende, or more rarely, biotite and/or pyroxene. [Gk. *diorizein,* to distinguish; because essential minerals are megascopic] (Haüy in 1822) 2. *IUGS,* plutonic rock with 0-5 Q, P/(A + P) greater than 90, and plagioclase more sodic than An_{50}.

diorite, albite *ig. pluton.* Diorite in which albite is the plagioclase; some mafics may be present. (Becker in 1898)

diorite, foid *ig. pluton. IUGS,* plutonic rock with 10-60 F, P/(A + P) greater than 90, and plagioclase more sodic than An_{50}. [feldspathoid]

diorite, hornblende *ig. pluton.* Normal diorite containing sodic plagioclase and hornblende; usually accompanied by minor quartz.

diorite, hypersthene *ig. pluton.* Pyroxene diorite in which hypersthene is the major pyroxene; often contains minor quartz.

diorite, leucite *ig. pluton.* Diorite containing leucite and sodic-plagioclase but no potassic feldspar; mafics may be present. (Johannsen in 1919)

diorite, mica *ig. pluton.* Diorites with biotite as the major characterizing accessory, composed of andesine, some hornblende, and a little orthoclase and quartz.

diorite, nadel *ig. pluton.* = diorite, needle. [G. *Nadel,* needle] (Gümbel in 1868)

diorite, needle *ig. pluton.* Normal diorite in which the hornblende is acicular (needlelike).

diorite, nepheline *ig. pluton.* Foid diorite containing nepheline and sodic-plagioclase but no potassic feldspar; mafics may be present. (Johannsen in 1919)

diorite, nepheline-bearing *ig. pluton.* = dungannonite

diorite, orbicular *ig. pluton.* Variety of diorite with an orbicular structure; an example occurs near Telemark, Norway. Esboite is a variety. The so-called orbicular diorites from Santa Lucia di Tallano in Corsica (corsite or napoleonite) are actually gabbro. |L. *orbis,* circle|

diorite, pyroxene *ig. pluton.* Diorite with pyroxene as the major characterizing accessory; pyroxenes may be diopside, titanaugite, or hypersthene, and quartz is also often present.

diorite, quartz *ig. pluton.* 1. Phaneritic granular rock composed of essential sodic-plagioclase (oligoclase or andesine), quartz (more than 5%), and usually a mafic such as biotite or hornblende, or more rarely pyroxene. 2. *IUGS,* plutonic rock with 5-20 Q, P/(A + P) greater than 90, and plagioclase more sodic than An$_{50}$.

diorite, sodaclase *ig. pluton.* = diorite, albite. (Johannsen in 1937)

diorite, uralite *ig. pluton, altered.* Diorite whose amphibole is secondary after an earlier pyroxene; the secondary amphibole is named uralite.

dioritite *ig. diaschist.* Seldom-used term for diorite aplite. |diorite| (Polenov in 1899)

dioritoid *ig. pluton. IUGS,* preliminary field term for a plutonic rock with Q less than 20 or F less than 10, and P/(A + P) greater than 65. |diorite|

dismicrite *sed.* Limestone consisting mainly of lithified carbonate mud (micrite) and containing irregular patches of sparry calcite (bird's-eyes). Also named bird's-eye limestone. |disturbed microcrystalline| (Folk in 1959)

dissogenite *ig., meta.* Collective name for a pegmatitic zone composed of microcline and diopside (with or without quartz and oligoclase) formed from the contact of a magma with limestone or lime-silicate rock; also called diopside pegmatite. |Possibly F. *dissoudre,* to dissolve + Gk. *genes,* born| (Lacroix in 1922)

disthenite *meta.* Rock composed almost entirely of kyanite (disthene) with some quartz. |disthene, old name for kyanite| (Lacroix in 1922)

ditroite *ig. pluton.* Nepheline syenite containing sodalite, biotite, and cancrinite, esp. those with a granular texture (foyaite is essentially the same but has a trachytoid texture). |Ditró, eastern Siebenbürgen, Transylvania| (Zirkel in 1866)

dolarenite *sed.* Dolomite rock consisting predominantly of clastic dolomite particles of sand size; consolidated dolomitic sand. |dolomite + arenite| (Folk in 1959)

dolerine *meta.* Talc schist containing feldspar and chlorite, from the Pennine Alps, Switzerland; also named feldspathic steatite. |Gk. *doleros,* deceptive| (Jurine in 1849)

dolerite *ig. dike, extru.* Term with various meanings including coarse basalt, diabase, or any dark igneous rock; in England and much of Europe, the preferred synonym for diabase (the term used in the United States). |Gk. *doleros,* deceptive; alluding to difficulty of identification because of its fine-grained nature| (Haüy, before 1822)

dolerite, nepheline *ig. extru.* Fine-grained variety of nepheline basalt. (Zirkel in 1894)

dolerite, quartz *ig. dike.* Local name for quartz diabase from the Great Whin sill, in the north of England.

dololithite *sed.* Clastic dolomite rock containing 50% or more of fragments of older dolomite rocks that have been eroded and redeposited. |dolomite|

dololutite *sed.* Dolomite rock consisting predominantly of detrital dolomite particles of silt and/or clay size; indurated dolomitic mud. |dolomite + lutite| (Folk in 1959)

dolomicrite *sed.* Consolidated or unconsolidated clay-sized dolomite of either chemical or mechanical origin (analogous to micrite, which is composed of calcite); lithified dolomite mud, like dololutite. |dolomite + micrite| (Folk in 1959)

dolomilith *sed.* = dolomite. |dolomite|

dolomite *sed.* Sedimentary carbonate rock resembling limestone in colors, grain sizes, and textures but containing more than 90% dolomite mineral and less than 10% calcite; since most form by replacement of limestone (by dolomitization), other minerals originally present in limestone may be preserved, including detrital quartz, chert, feldspars, clays, and more; some very fine-grained dolomites, formed as evaporites, contain gypsum and anhydrite. To avoid confusion with the mineral dolomite, the following synonyms have been used: dolomite rock, dolostone, dolomilith, dolomith, dolomitite. |Déodat Guy Silvain Tancrède Gratet de Dolomieu, 1750-1801, French geologist; the mineral name was later derived from the rock name| *(Saussure in 1796)*

dolomite, calcareous *sed.* = dolomite, calcitic

dolomite, calcitic *sed.* Dolomite containing 50-90%

dolomite mineral and 10-50% calcite; also named calcareous dolomite and calc-dolomite (calcdolomite).

dolomite, clastic *sed.* Rock composed of fragments derived from earlier dolomite rocks; relatively rare rock; essentially identical to dololithite.

dolomite, fetid *sed.* Some dolomite rocks that, upon breaking, give off a fetid odor from the presence of an organo-phosphoric compound belonging to the phosphenate series.

dolomite, Gurhofian *sed.* = gurhofite. [Gurhof, Austria]

dolomite, magnesian *sed.* Dolomite rock with an excess of magnesium usually due to the presence of magnesite; the rock contains 50-75% dolomite and 25-50% magnesite.

dolomite, mottled *sed.* Dolomite with a mottled appearance indicating a patchy distribution as a result of the incomplete dolomitization of a limestone, esp. seen on weathered or etched surfaces.

dolomite, primary *sed.* Dolomite rock not formed by the dolomitization of an earlier limestone but formed either by direct chemical or biochemical precipitation from sea or lake water (e.g., in evaporite deposits) or by the direct accumulation of clastic particles. Also named orthodolomite.

dolomite, zebra *sed.* Local term for dolomite rock that shows conspicuous banding (generally parallel to bedding) consisting of light gray, coarsely textured layers alternating with darker finely textured layers; occurs in Leadville district, Colorado. See zebra rock.

dolomith *sed.* = dolomite. (Grabau in 1924)

dolomitite *sed.* = dolomite. (Kay in 1951)

dolorudite *sed.* Dolomite rock consisting predominantly of detrital dolomite particles larger than sand size; consolidated dolomitic gravel. [dolomite + rudite] (Folk in 1959)

dolosiltite *sed.* Dolomite rock consisting predominantly of detrital dolomite particles of silt size; consolidated dolomitic silt. [dolomite + siltite]

dolostone *sed.* = dolomite. Term preferred by many petrologists to dolomite in order to avoid confusion with the mineral dolomite. [dolomite stone] (Shrock in 1948)

domite *ig. extru.* Light-colored, porous, altered oligoclase-biotite trachyte, containing tridymite in the pores. [Puy de Dome, Auvergne, France] (von Buch in 1802)

dopplerite *sed.* Amorphous, brownish-black,

gelatinous calcium salt of a humic acid found at depth in marsh and bog deposits. [Christian Johann Doppler, 1803-1853, Austrian physicist who first noticed the substance] (Haidinger in 1849)

doreite *ig. extru.* Andesitic lava containing about equal amounts of andesine (35%) and sodium sanidine (33%) accompanied by augite. [Mt. Doré, Auvergne, France] (Lacroix in 1923)

dorgalite *ig. extru.* Olivine basalt in which the phenocrysts are exclusively olivine. [Dorgali, Sardinia] (Amstutz in 1925)

drakonite *ig. extru.* Light-colored rock consisting of phenocrysts of sanidine, plagioclase (oligoclase to labradorite), and biotite or hornblende, in a groundmass of alkali feldspar, diopside, alkali amphibole, and minor sodalite, aegirine-augite, titanite, and others. Also named Drachenfels trachyte. [L. *draco,* dragon; for which locality Drachenfels, Siebengebirge, Germany, was named] (Reinisch in 1912)

drapery *sed.* In caves, tabular or folded speleothems that hang from ceilings or wall projections with a curtainlike appearance.

dreikanter *sed.* Three-edged, faceted wind-worn pebble (ventifact), resembling a Brazil nut; formed in a dry climate, whether hot (desert) or cold (glacial). [G. *drei-kantig,* three-cornered]

drewite *sed.* Calcareous ooze consisting of impalpable calcareous material, most of which is probably precipitated through the agency of denitrifying bacteria. [George Harold Drew, 1881-1913, British scientist who studied marine bacteria associated with this sediment in the Bahamas] (Field in 1919)

dripstone *ig. extru.* Lava shapes formed by dripping molten lava in tunnels and other open spaces.

dripstone *sed.* Any secondary cave deposit formed by dripping water.

dropstone *sed.* 1. = dripstone. 2. Rocks dropped into a silty or clayey matrix, esp. those in a pelodite.

dumalite *ig. extru.* Trachyandesite whose interstitial material is glassy and which has the composition of nepheline. [Dumala, Dych-Tau, the Caucasus] (Loewinson-Lessing in 1905)

dumortierite rock *meta.* Metamorphic rock in which dumortierite is an essential component but which may also contain andalusite, kyanite, pyrophyllite, quartz, and others; formed by the replacement of sericite schist or andalusite rock; occurs at Oreana, Nevada, and near Quartzsite, Arizona.

dumortierite quartz   *meta.* Massive, opaque quartz rock colored blue, greenish-blue, or violet-blue by intergrown dumortierite. Also named California lapis.

dune rock   *sed.* Eolianite consisting of dune sand.

dungannonite   *ig. pluton.* Diorite containing andesine (72%) and corundum (13%), with minor nepheline, biotite, muscovite, and others. [Dungannon township, Ontario] (Adams and Barlow in 1910)

dunite   *ig. pluton.* 1. Peridotite composed essentially of olivine; minor chromite, magnetite, ilmenite, pyrrhotite, pyrope, and spinel may be present. [Dun Mountain, New Zealand] (Hochstetter in 1859) 2. *IUGS,* Plutonic rock with M equal to or greater than 90 and ol/(ol+opx+cpx+hbd) greater than 90.

dunite, chromite   *ig. pluton.* Dunite containing 5-50% chromite. (Vogt in 1894)

dunite, ilmenite   *ig. pluton.* Dunite containing 5-50% ilmenite, in place of chromite.

dunite, magnetite   *ig. pluton.* Dunite containing 5-50% magnetite, in place of chromite.

dunite, pyrrhotite   *ig. pluton.* = alexoite

dunstone   *ig. extru., altered.* Local name for an amygdaloidal spilite; occurs in the Plymouth area, England. [O. Eng. *dun,* brown]

dunstone   *sed.* 1. Local name for a hard granular, yellowish or cream-colored magnesian limestone; occurs in Matlock, England.   2. In Wales, hard fireclay or underclay; in England, shale. [O. Eng. *dun,* brown]

durain   *sed.* Dull coal; hard, black to lead-gray, lacking luster and having a matte or earthy appearance, consisting of cuticles, spores, and others. Also spelled durite. [L. *durus,* hard] (Stopes in 1919)

durbachite   *ig. hypab.* Syenite consisting of orthoclase (Carlsbad twins) phenocrysts in an aggregate of coarse biotite flakes accompanied by a little orthoclase; minor plagioclase, hornblende, titanite, zircon, and quartz also are present. [Durbach, Black Forest, Germany] (Sauer in 1891)

duricrust   *sed.* General term for a hard crust on the surface of, or a layer in the upper horizons of, soil in a semiarid region; formed from minerals deposited by mineral-bearing waters brought to the surface by capillary action and evaporation during the dry season. [L. *durus,* hard + crust]

durinode   *sed.* In soil, concretion cemented or indurated with silica. [L. *durus,* hard + *nodus,* knot]

dust, volcanic   *ig. extru.* 1. Unconsolidated pyroclastic material originating from a volcanic eruption and consisting of clay and silt-size particles; because of its fine size it can be readily lifted and carried considerable distances in suspension before falling back to the earth's surface.   2. *IUGS,* dust grains that are pyroclasts with mean diameters smaller than 1/16 mm. Also named fine ash.

duxite   *sed.* Dark brown, opaque variety of retinite containing about 0.42% sulfur; related to walchowite. [Dux, Bohemia, Czechoslovakia] (Doelter in 1874)

dy   *sed.* Dark, jellylike, freshwater mud, consisting largely of unhumified or peaty organic matter, such as that derived from an acidic peat bog, transported as a colloid, and precipitated in a nutrient-deficient lake. [Sw. *dy,* mud, sludge] (Post in 1861)

dynamogranite   *meta.* Augen gneiss containing much potassium feldspar (microcline and/or orthoclase). [Gk. *dunamis,* force, power (alluding to its formation by metamorphism) + granite (which it resembles)] (Krivenko and Lapchik in 1934)

dysodile   *sed.* 1. Yellow to greenish-gray, elastic hydrocarbon that burns with a highly fetid odor, from Melili, Sicily, and certain German lignite deposits. (Boccone in 1674) 2. Lignitic coal, derived from diatomaceous sediments under anaerobic conditions; burns with a bad odor. [Gk. *dusodes,* ill-smelling]

dzhetymite   *sed.* Nonsorted clastic sedimentary rock composed of approximately equal proportions of large angular clasts (1-10 mm diameter), sand (0.1-1 mm), and mud (under 0.1 mm). [Possibly Dzhetym-Too Range, Soviet Union] (Dzholdoshev, date uncertain)

E

eaglestone   *sed.* Walnut-sized concretion (generally clay-rich ironstone), often containing loose stones in a hollow interior (like a klapperstein); the ancients believed that the eagle transported these stones to the nest to facilitate the laying of eggs. Also named aetite.

earth   *sed.* 1. The softer loose part of land, as distinguished from rock; soil.   2. Very fine-grained substance, such as clay or any material resembling clay. [O. Eng. *eorthe*]

earth, Barbados   *sed.* Fine-grained siliceous deposit containing abundant radiolarian remains; occurs in Barbados, West Indies.

earth, black   *sed.* 1. Any black soil, usually rich in carbonaceous materials.   2. Finely ground brown coal used as a pigment. Also named Cassel brown and Cassel earth.

earth, brick   *sed.* Any earth, clay, or loam suitable for making bricks; usually consisting of ferruginous clay mixed with quartz.

earth, diatomaceous   *sed.* Very fine, light-colored earthy material, consisting chiefly of the minute opaline skeletons of diatoms; clay minerals, quartz, volcanic shards, sponge spicules, radiolaria, and other components may be present.

earth, fuller's   *sed.* Any clay with an adequate decolorizing and purifying capacity to be used commercially in oil refining without chemical treatment; attapulgite and some montmorillonites possess superior decolorizing properties; originally, the earths were used in cleaning or fulling wool with a water slurry of the earth, thereby removing oil and dirt particles from the fiber. [O. Eng. *fullere,* person who fulls cloth]

earth, infusorial   *sed.* Very fine earthy material, consisting chiefly of the minute opaline skeletons of diatoms. Nearly equivalent to diatomaceous earth. [*Infusoria,* class of the animal phylum *Protozoa;* poorly named, since diatoms are plants]

earth, radiolarian   *sed.* Very fine, light-colored earthy material, consisting chiefly of the opaline latticelike skeletal frameworks of radiolarians.

earth, saltpeter   *sed.* Earthy cave fill containing saltpeter (the mineral usually being nitrocalcite).

earth, siliceous   *sed.* General term for very fine, white or light-colored, earthy materials consisting chiefly of siliceous particles, for the most part derived from the remains of organisms such as diatoms, radiolarians, and others.

earth, Verona   *hydrotherm.* Earthy celadonite filling amygdaloidal cavities in lava at Mt. Baldo near Verona, Italy. Also named veronite. (de Lisle in 1783)

earth, walker's   *sed.* = earth, fuller's. [G. *Walkerde* from *Walker,* fuller + *Erde,* earth]

earth stone   *sed.* = amber, earth

echodolite   *ig. extru.* = phonolite. [Gk. *echo,* echo + *dosis,* related to *didonai,* to give]

eclogite   *meta.* Granular metamorphic rock composed of pink garnet (almandine-pyrope) and grass green sodic pyroxene (omphacite) with minor accessories like mica, chlorite, magnetite, and others; probably formed at very high pressures in an anhydrous environment. [Gk. *eklogos,* picked out, choice selection; presumably alluding to its beautiful appearance] (Haüy in 1822)

eclogite, hornblende   *meta.* Probably garnet amphibolite; originally discovered near Loch Maree and in the district of Scourie and Loch Laxford in Sutherland, Scotland. (Bonney in 1880)

ectectite   *migmat.* = ectexite

ectexite   *migmat.* Migmatite formed by ectexis (migmatization in which the mobile component is formed in situ) [Gk. *ek,* out + *tektos,* molten]

ectinite   *meta.* Rock essentially formed by isochemical regional metamorphism, that is, without notable associated metasomatism. [Possibly Gk. *ek,* out + *teinein,* to stretch]

edolite   *meta.* Variety of hornfels containing essential feldspar and mica; varieties include cordierite edolite and andalusite edolite. [Edelo, Italian Alps] (Salomon in 1898)

eggstone   *sed.* = oolite. [From resemblance to fish eggs]

ehrwaldite   *ig. dike, extru.* 1. Mafic dike rock consisting of phenocrysts of altered olivine, biotite, barkevikite, and titanaugite, in a groundmass of augite microlites and glass. [Ehrwald, Leermoos,

Tyrol] (Pichler in 1875) 2. Augitite containing both orthopyroxene and clinopyroxene. (Cathrein in 1890)

Eilat stone *altered ore.* Rock composed of malachite, chrysocolla, and related copper minerals; occurs in the area of King Solomon's ancient copper mine near Eilat, Israel.

einkanter *sed.* Wind-worn pebble (ventifact) having only one face or a single sharp edge. [G. *ein,* one + *kantig,* edged]

ejecta *ig. extru., impact.* 1. Material thrown out by a volcano; can consist of lava or solid rock materials from the throat of the volcano. 2. Material thrown from an impact crater during formation that can consist of glass, shock-metamorphosed rock fragments, and other rocks. [L. *ejectus,* thrown out]

ejecta, capillary *ig. extru.* = Pele's hair

ekerite *ig. pluton.* Syenite or quartz syenite containing sodium microcline and microperthite with arfvedsonite and aegirine; in part porphyritic. [Eker, Christiana, Norway] (Brögger in 1906)

elaterite *sed.* Soft, elastic asphaltite used for waterproof and insulating paints; occurs in veins in Derbyshire and elsewhere. [Gk. *elater,* driver, hurler] (pub. by Emmons in 1826)

eldoradoite *ig.?* Blue variety of quartz used as a gemstone; term also applied to blue chalcedony and to iridescent quartz. Also spelled El Doradoite. [El Dorado County, California] (Watkins in 1912).

elixirite *ig. extru.* Local name for a kind of banded rhyolite (wonderstone) found near Truth or Consequences, Sierra County, New Mexico. [elixir]

elkerite *sed.* Variety of bitumen formed through the slow oxidation of petroleum. [Elk Hills, California] (Hackford in 1932)

elkhornite *ig. hypab.* Syenite dike rock composed of microcline and augite with minor labradorite and titanite. [Elkhorn district, Montana] (Johannsen in 1937)

eluvium *sed.* Clay or sand moved and deposited by the wind, as in a sand dune or loess deposit. [L. *eluere,* to wash out]

elvanite *ig. dike.* Local term for granitic rocks, essentially fine-grained granite porphyry or rhyolite porphyry; phenocrysts of quartz and orthoclase, both rounded or corroded, occur in a microcrystalline groundmass; term used in Cornwall. Also spelled elvan. [Celt. *el,* rock + *van,* white]

embrechite *migmat.* Granitic migmatite gneiss with a regularly parallel stratification or schistosity; often with phenoblasts of feldspar (augen structure) or lenses and small layers of granite; some features of pre-existing rocks are preserved. [Gk. *en,* into + *brechein,* to wet]

emery *meta.* Massive, lenticularly foliated, or layered dark-colored rock composed of major amounts of corundum and lesser amounts of spinel and magnetite, with additional minerals such as hematite, margarite, chloritoid, quartz, hoegbomite, and anorthite. [Traceable to F. *emeri,* Gk. *smiris, smeris*]

emery, spinel *meta.* Rare variety of emery in which spinel exceeds corundum; hematite is usually present.

emery rock *meta.* = emery

empirite *tekt.?* Glass found in Georgia similar to tektites. [Empire, Georgia, U.S.]

encrinite *sed.* Crinoidal limestone (calcarenite), relatively coarse-grained and in which the crinoid fragments appear usually well-worn and sorted into layers that differ according to the grain size of their fragments. [genus *Encrinus,* in which fossil crinoids belong]

enderbite *ig., meta.?* Charnockite containing quartz, much plagioclase (commonly antiperthitic), hypersthene, and minor magnetite. [Enderby Land, Antarctica] (Tilley in 1936)

enderbite, m- *ig., meta.?* Name proposed for a charnockite in which mesoperthite (an intimate mixture of about equal amounts of potassic feldspar and plagioclase) and free plagioclase are both present. (Tobi in 1971)

endoskarn *meta.* Skarn formed by reactions within the intruded igneous rock produced by the assimilation of the older country rock. [Gk. *endo,* within + skarn]

engadinite *ig. pluton.* Aplite granite poor in quartz; orthoclase microperthite (47%), plagioclase (25%), quartz (23%), mafics and accessories (about 5%) are present. [Engadin, Switzerland] (Niggli in 1923)

engelburgite *ig. pluton.* Granodiorite with titanite spots; plagioclase (42%), potassic feldspar (17%), quartz (25%), with biotite, titanite (some as leucoxene), and others. [Engelburg (Englburg), Fürstenstein bei Passau, Germany] (Frentzel in 1911)

enhydrite *ig., sed.* General term for a rock or mineral having cavities containing water, e.g., enhydros. [Gk. *en,* in + *hudor,* water]

enhydros *sed., ig.* 1. Hollow, thin-walled nodule or geode of chalcedony carrying water sealed within it; may be found wherever chalcedony occurs but relatively rare. 2. Term also applied to crystals of quartz, and more rarely other minerals, containing water or other fluid enclosed, esp. within polyhedral openings; the term enhydrous quartz is often used. [Gk. *en*, in + *hudor*, water]

enstatitfels   *ig.* = enstatolite. (Streng in 1864)

enstatitite   *ig.* = enstatolite

enstatolite   *ig.* Coarse, yellowish to greenish-gray perknite (pyroxenite) composed of enstatite, with accessory chromite and magnetite. [enstatite] (Pratt and Lewis in 1905)

entectite   *migmat.* = entexite

entexite   *migmat.* Migmatite formed by entexis, where the more mobile part in the genesis of the rock was introduced from without. [Gk. *en*, in + *tektos*, molten]

eobasalt   *ig., meta.* Term applied to ancient basalt, which is often metamorphosed. [Gk. *eos*, dawn (early) + basalt] (Nordenskiöld in 1893)

eolianite   *sed.* Indurated sedimentary rock consisting of clastic materials deposited by the wind. [L. Aolus, god of the winds]

eorhyolite   *ig., meta.* = aporhyolite. [Gk. *eos*, dawn + rhyolite] (Nordenskiöld in 1893)

eosite   *meta.?* Trade name for a rose-colored Tibet stone. [Gk. Eos, goddess of dawn, alluding to rose color]

epibolite   *migmat.* Migmatite with granitic layers concordant to a nongranitic parent rock. [Gk.

epi, upon + *ballein*, to throw] (Jung and Roques in 1952)

epibugite   *ig., meta.?* Hypersthene quartz diorite; composed of andesine (64%), quartz (25%), hypersthene (8%), with minor mafics, apatite, and zircon. [Gk. *epi*, upon, over + bugite] (Bezborodko in 1931)

epidiabase   *meta.* Diabase or basaltic rock in which the augite has suffered alteration to hornblende (uralite), so that the rock approaches the composition of a diorite; the rock was originally called an epidiorite, but this new term was proposed in order to avoid confusion with the rock diorite. [Gk. *epi*, upon, beside + diabase] (Issel in 1892)

epidiorite   *meta.* = epidiabase. (Gümbel in 1879)

epidosite   *meta.* Metamorphic rock composed largely of epidote and quartz, with minor albite, chlorite, and actinolite. [epidote] (Reichenbach in 1834)

epidotite   *meta.* Term sometimes applied to massive or poorly foliated metamorphic rocks composed mainly of epidote, often with minor hornblende, plagioclase, chlorite, quartz, and others; term variably used. [epidote]

epixenolith   *ig.* Xenolith derived from adjacent wall rock. [Gk. *epi*, upon + xenolith] (Goodspeed in 1947)

erlanfels   *meta.* Metamorphic foliated pyroxene-feldspar rock with accessory vesuvianite, titanite, zoisite, and others; augite schist. Also named erlan. [Erla, near Crandorf, Saxony + G. *Fels* rock] (Breithaupt in 1836)

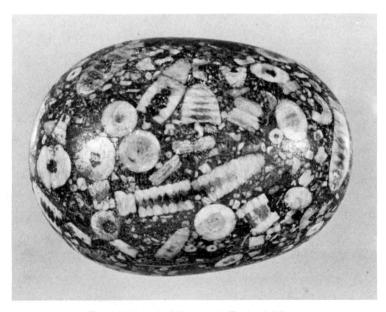

Encrinite (encrinal limestone). England. 6.5 cm.

erzbergite *sed.* Calcareous deposit consisting of alternate layers of calcite and aragonite. [Erzberg, Eisenerz, Styria] (Hatle in 1892)

esboite *ig. pluton.* Oligoclase orbicular diorite; orbicules are composed of oligoclase with biotite-rich layers and make up about 60% of the rock. [Esbo, Helsinki, Finland] (Sederholm in 1928)

esmeraldite *ig.* White to gray, fine to coarse rock with essential quartz and muscovite, occurring as dikes or border facies of other igneous rocks, esp. granite. Not to be confused with esmeraldaite, a hydrous ferric oxide mineral from the same area. [Esmeralda County, Nevada] (Spurr in 1906)

espichellite *ig. hypab.* Lamprophyre containing nearly equal amounts of zoned plagioclase (labradorite) and basaltic hornblende, with minor titanaugite, analcime, and others. [Cape Espichel, south of Lisbon, Portugal] (Souza-Brandão in 1907)

essexibasalt *ig. extru.* Nepheline basanite containing very calcic plagioclase (bytownite); in France, a synonym for alkaline basalt. [essexite + basalt] (Lehmann in 1924)

essexite *ig. pluton.* 1. Alkali gabbro composed of calcic plagioclase, greenish hornblende, potassic feldspar, nepheline, and others. [Essex County, Massachusetts] (Sears, 1891) 2. *IUGS,* plutonic rock with 10-60 F, and 50-90 P/(A + P); synonyms are foid monzodiorite and foid monzogabbro.

essexite, Oslo *ig. pluton.* Local term for biotite-rich gabbro that also consists of augite (violet to greenish), and andesine-labradorite, with minor potassic feldspar, iron-titanium oxides, and apatite. [Oslo, Norway, and alluding to its possible transition toward the alkali gabbro, essexite] (Barth in 1944)

essexite, pallio- *ig. pluton.* Term applied to a border-zone rock associated with a nepheline-bearing diorite (incorrectly, essexite); this border rock contains biotite, plagioclase, olivine, and others; occurs at Prospect Hill near Sydney, New South Wales, Australia. [L. *pallium,* cloak; alluding to the compact envelope of rapidly cooled rock] (Jevons and others in 1912)

esterellite *ig. dike.* Quartz diorite porphyry (blue porphyry), containing phenocrysts of quartz, zoned andesine, and hornblende. [Esterel, France] (Michel-Lévy in 1897)

etindite *ig. extru.* Dark-colored igneous rock intermediate between leucitite and nephelinite; phenocrysts of augite occur in a dense groundmass of leucite, nepheline, and augite. [volcano Etinde, the Cameroons] (Lacroix in 1923)

etnaite *ig. extru.* Alkali olivine basalt. [Possibly Etna, eastern Sicily]

eucrite *ig. hypab.* Gabbro composed of calcic plagioclase (bytownite, anorthite) and clinopyroxene (augite), with accessory olivine. [From similarity to eucrite meteorite]

eucrite *meteor.* = achondrite, pyroxene-plagioclase. [Gk. *eukratos,* well-tempered] (Rose in 1863)

euktolite *ig. extru.* = venanzite. Also spelled euktolith. [Gk. *euktos,* desired] (Rosenbusch in 1899)

eulysite *meta.* Massive to crudely foliated metamorphic rock characterized by the presence of iron-rich olivine (fayalite), which is usually associated with iron-rich pyroxene (diopside-hedenbergite), amphibole, and garnet; often in close association with grunerite schist, hedenbergite-garnet-magnetite rock, almandine rock, and other rocks rich in iron or manganese. The major varieties, based upon characterizing minerals, are: anthophyllite e., cordierite e., grunerite e. (collobriérite), hypersthene e. [Gk. *eu,* easy + *lutikos,* dissolved; olivine gelatinizes easily in acids] (Erdmann in 1849)

euosmite *sed.* Brownish yellow resin, found in brown coal; has a low oxygen content and a pleasant odor. [Gk. *eu,* good + *osme,* odor; alluding to aromatic odor when burned] (Gümbel in 1864)

eupelagite *sed.* Sediments formed on the sea floor far from the continents by the slow accumulation of biogenic materials, pelagic clays, and oozes; less than 25% of the fraction coarser than 5 microns is of terrigenous or volcanogenic origin. [Gk. *eu,* good + *pelagos,* sea]

eupholite *ig., altered.* Variety of euphotide characterized by the presence of talc. [Gk. *eu,* well + *phos,* light; alluding to the reflection of light by the rock] (Cordier in 1868)

euphotide *ig., altered.* Term sometimes applied to gabbros whose feldspars have been saussuritized; originally applied to gabbro, referring to reflection of light by the green diallage (clinopyroxene with excellent parting planes) present. [Gk. *eu,* well + *phos,* light; alluding to reflections from diallage] (Haüy in 1822)

eurite *ig. extru.* Broad term for compact, felsitic igneous rocks without phenocrysts, having the composition of rhyolite or trachyte; term also applied to all aphanitic rocks of granitic composition whether porphyritic or not. [Gk. *eurys,* wide, broad]

eustratite *ig. hypab.* Compact lamprophyric rock with rare phenocrysts of olivine and corroded hornblende and with still fewer green augite and oligoclase crystals in a groundmass of augite, magnetite, mica, glass, and possibly feldspathoid. |Island of Haghios Eustratios, Egyptian Islands| (Kténas in 1928)

eutaxite *ig. extru.* Taxite whose components have aggregated into separate bands. |Gk. *eu,* good + taxite|

eutectofelsite *ig. extru.* = eutectophyre. |Gk. *eutektos,* well-fused + felsite|

eutectoperthite *ig. pluton.* = mesoperthite. |Gk. *eutektos,* well-fused + perthite|

eutectophyre *ig. extru.* Light-colored tuffaceous rhyolite composed of interlocking quartz and orthoclase crystals. |Gk. *eutektos,* well-fused|

euvitrain *sed.* Very common, structureless amorphous vitrain. |Gk. *eu,* good + vitrain|

evaporite *sed.* General term applied to nonclastic sedimentary rocks composed of minerals precipitated from saline solution by extensive or total evaporation of the solvent; these include rocks composed of halite, sylvite, carnallite, gypsum, anhydrite, as well as some dolomite, barite, celestite, and borate rocks. |L. *evaporare,* to go out in vapor|

evergreenite *ig. dike.* Variety of nordmarkite containing sulfide minerals such as chalcopyrite and bornite; principal minerals are microcline perthite (43%), quartz (24%), and wollastonite (22%). Also named wollastonite-nordmarkite. |Evergreen mine, Gilpin County, Colorado| (Ritter in 1908)

evisite *ig. pluton.* Collective name for aegirine- and riebeckite-granite and syenite from Evisa, Corsica. |Evisa| (Niggli in 1923)

excretion *sed.* Concretion growing progressively inward from the exterior; illustrated by a shell of sand cemented by iron oxide and filled by unconsolidated sand or other shells of cemented sand. |L. *ex,* from + *crescere,* to grow| (Todd in 1903)

exhalite *sed., hydrothermal.* Chemical sediment containing iron, manganese, base metals, and gold as cations, and oxide, carbonate, or sulfide as anions; formed by the issuance of volcanically derived fluids onto the sea floor or into the sea. |exhalation|

exinonigritite *sed.* Type of nigritite derived from spore exines (the outer layer forming the wall of spores and pollen). |exine + nigritite|

exoskarn *meta.* Skarn formed by reactions outside the intruded igneous rock, by replacement of limestone or dolomite; contrast endoskarn. |Gk. *ex,* from, out of + skarn|

eyestone *ig., altered.* Orbicular variety of thomsonite, sometimes used as a gem.

F

fahlband   *meta.* Sulfide-rich bands in metamorphic rocks; although too sparse to form an ore lens, the sulfides are too abundant to be classed as accessory minerals. [G. *fahl,* fallow + *Band,* band]

fairburnite   *sed.* Colorful variety of fortification agate occurring in southwestern South Dakota and northwestern Nebraska. Usually named Fairburn agate. [Fairburn, South Dakota]

fairy stone   *sed.* Strange or fancifully shaped calcareous or ferruginous concretion formed in alluvial clays; not to be confused with twinned staurolite crystals which have the same name. [Name alluding to fanciful shapes]

fallback   *ig. extru., impact.* Material formed from an explosion or impact crater during formation, redeposited within (falls back) and partly filling the true crater almost immediately after its formation.

fanglomerate   *sed.* Conglomerate composed of heterogeneous, slightly water-worn, fragments of all sizes, deposited in an alluvial fan and, after deposition, cemented into solid rock. [fan, alluding to alluvial fan + conglomerate] (Lawson in 1913)

farrisite   *ig. dike.* Fine-grained, chocolate-brown rock composed of a melilitelike mineral (forming about one-third of the rock), brown barkevikite, light green pyroxene, and biotite; there is no feldspar or feldspathoid. [Farris Lake, Oslo district, Norway] (Brögger in 1898)

farsundite   *ig. pluton.* Granodiorite with plagioclase (40%), microcline (20%), quartz (25%), and mafics (hornblende and hypersthene) (11%). [Farsund, southern Norway] (Kolderup in 1903)

faserkiesel   *meta.* = quartz sillimanitisé. [G. *Faser,* fiber + *Kiesel,* silica] (Lindacker in 1792)

fasernephrite   *meta.* Variety of nephrite (jade) showing, in part, a fibrous structure owing to the parallel rather than matted aggregation of the actinolite fibers; occurring at Radauthal, Harz, Germany. Also named nephritoid. [G. *Faser,* fiber + nephrite] (Uhlig in 1910)

fasibitikite   *ig. pluton.* Sodium-rich granite containing orthoclase microperthite, albite, quartz, riebeckite, aegirine, and zircon. [Ampasibitika, Madagascar] (Lacroix in 1915)

fasinite   *ig. pluton.* Melteigite composed of violet titanaugite (75%), nepheline, and lesser amounts of biotite, olivine, and sodium microcline; found in boulders in a stream near Ambaliha, Madagascar. [Ampasindava, Madagascar] (Lacroix in 1916)

favas   *sed.* Collective term given to the various minerals occurring as rolled pebbles in the diamond sands of Brazil; consist of the oxides of zirconium and titanium, of goyazite, and others. [Port. *fava,* bean] (Hussak in ca. 1899)

feldspathoidite   *ig. extru.* Group name for the single-mineral end-members of the extrusive feldspathoidal rocks, including rock such as nephelinolith, analcimolith, and others. [feldspathoid] (Johannsen in 1939)

fels   *meta.* General term applied to granoblastic metamorphic rocks; those rocks without foliation. [G. *Fels,* rock]

felside   *ig. extru.* Informal field term applied to any fine-grained, light-colored, aphanitic igneous rock, e.g., nonporphyritic rhyolite, trachyte, and latite. [Possibly G. *Fels,* rock] (Johannsen in 1938)

felsite   *ig. extru.* General field or hand-specimen name for megascopically aphanitic, light-colored, extrusive or hypabyssal igneous rocks, including rhyolite, trachyte, quartz latite, latite, dacite, and others. [G. *Fels,* rock] (Gerhard in 1814)

felsite, minette   *ig. dike.* Minettelike rock having a micro- to cryptocrystalline groundmass; minette proper has a fine-grained, panerocrystalline groundmass. (Bonney and Haughton in 1879)

felsite, quartz   *ig. extru., dike.* Felsite with quartz phenocrysts; term can apply to a rhyolite porphyry as well as a dacite porphyry.

felsitoid   *ig. extru.* Informal term nearly equal to felsite. [felsite]

felsoandesite   *ig. extru.* Andesite whose groundmass is too fine-grained for the individual constituents to be determined but not glassy. Also named felsitic andesite. [felsite + andesite] felsophyre   *ig. extru.* General term for a porphyritic felsite; usually applied to rocks of rhyolitic composition. [felsite + porphyry] (Vogelsang in 1872)

fenite   *ig. pluton., hypab.* Light-colored rock consisting mostly of alkali feldspar, some aegirine, subordinate alkali hornblende, and accessory titanite and apatite; formed by metasomatism where carbonatite rocks come in contact with biotite granite or other quartzo-feldspathic country rock. [Fen area, Norway] (Brögger in 1921)

fenite, ijolitic   *ig. pluton., hypab.* Rock of fenitic origin derived from the metasomatism of a gabbro or gneiss and containing nepheline, biotite, pyroxene, and lesser magnetite, apatite, and calcite.

fenite, mafic   *ig. pluton., hypab.* Rock of fenitic origin rich in mafic minerals, esp. pyroxene; from the metasomatism of an amphibolite rather than granite.

fenite, pulaskitic   *ig. pluton., hypab.* Nepheline-bearing igneouslike intrusive rock that originated as a metasomatic fenite and later liquefied.

fergusite   *ig. pluton.* 1. Granular, light gray intrusive rock consisting of dominant leucite with subordinate augite, biotite, apatite, iron oxides, and others. [Fergus County, Montana] (Pirsson in 1905) 2. *IUGS,* plutonic rock in which F is 60-100, M is 30-50, and potassium exceeds sodium.

ferralite   *sed.* 1. Term used in North Africa for a soil composed of a mixture of hydrates of iron, aluminum, and sometimes manganese and titanium that originated from mafic crystalline rocks having undergone chemical change. 2. Humid, tropical soil formed by the leaching of silica and bases by mildly acidic or neutral solutions, characterized by a large amount of iron oxide. [L. *ferrum,* iron + aluminum]

ferricrete   *sed.* Conglomerate formed by the cementation of gravels into a hard mass by iron oxides, derived from the oxidation of percolating solutions of iron salts. [ferruginous + concrete] (Lamplugh in 1902)

ferricrust   *sed.* Indurated soil horizon cemented with iron oxides, mainly hematite. [ferruginous + crust]

ferrilith   *sed.* = ironstone. [L. *ferrum,* iron]

ferrite   *sed.* Cemented iron-rich sediment whose particles do not interlock. [L. *ferrum,* iron] (Tieje in 1921)

ferroandesite   *ig. extru.* Andesite characterized by a high Fe/Mg ratio. [L. *ferrum,* iron + andesite]

ferrobasalt   *ig. extru.* Basaltic lava marked by a strong enrichment in iron; total iron exceeds 12-13 %, and MgO is less than 6%; SiO_2 is about 48-50%. [L. *ferrum,* iron + basalt] (McBirney and Williams in 1969)

ferrocarbonatite   *ig. hypab.* Carbonatite composed of iron-rich carbonate minerals, e.g., siderite and ankerite. [L. *ferrum,* iron + carbonatite]

ferrodiorite   *ig. pluton.* Diorite in which the actual plagioclase is less calcic than about An_{50}, and the ferromagnesian minerals are iron-rich. [L. *ferrum,* iron + diorite] (Wager and Brown in 1967)

ferrogabbro   *ig. pluton.* Obsolete term applied to igneous rocks containing iron-rich pyroxenes and olivine (fayalite), from the upper zone of the Skaergaard intrusion, East Greenland; term withdrawn by the original author (Wager) in favor of ferrodiorite. [L. *ferrum,* iron + gabbro] (Wager and Deer in 1939)

ferrogranophyre   *ig. extru., aschist.* Granophyre with an unusually high FeO/(FeO+MgO) ratio; these consist mainly of sodic-plagioclase that passes marginally into potassic feldspar micrographically intergrown with quartz; the mafics are hedenbergite, fayalitic olivine, and iron oxides. [L. *ferrum,* iron + granophyre]

ferrolite   *sed., meta.* General term applied to rocks composed of the iron oxides. [L. *ferrum,* iron] (Wadsworth in 1891)

fiamme   *ig. extru.* Name applied to small, dark, vitric lenses in welded tuffs; perhaps formed by the collapse of pumice fragments. [It. *fiamma,* flame]

fiasconite   *ig. extru.* Dark gray, highly porphyritic, aphanitic leucitite-basanite, containing anorthite, augite, olivine, nepheline, and iron oxides. [Mt. Fiascone, Vulsinian district, Italy] (Johannsen in 1939)

figure stone   *meta.* = agalmatolite

fimmenite   *sed.* Variety of peat derived mainly from pollen. [Mr. D. Fimmen] (Früh in 1885)

finandranite   *ig. pluton.* Coarse, potassium-rich syenite, composed of microcline, amphibole, and some biotite, ilmenite, and apatite. [Madagascan *ambato-finandrahanan,* of the rock which can be cut with a chisel] (Lacroix in 1922)

fiorite   *sed.* = siliceous sinter. [Mt. Santa Fiora, Tuscany]

fireclay   *sed.* 1. General term for clays rich in hydrous aluminum silicates that resist exposure to high temperatures without deforming (either disintegrating or becoming soft and pasty), used for the manufacture of refractory ceramic products. 2. Term formerly used for underclay (seatearth); however, not all underclays are refractory, so the name, alluding to refractory properties, often does not fit.

firestone *sed.* 1. = fireclay. 2. Fine-grained, siliceous rock formerly used for striking fire, e.g., flint. 3. Fine-grained, siliceous rock that can endure high heat and can be used for lining furnaces and kilns, e.g., certain sandstones.

firn *glacial.* Granular, loose or consolidated snow of high altitudes before it forms glacial ice below; transitional between snow and glacier ice. [G. *Firn,* snow of last year]

fitzroyite *ig. hypab., extru.* Lamproite consisting of phenocrysts of leucite and phlogopite in a very fine-grained groundmass. [Fitzroy Basin, Western Australia] (Wade and Prider in 1940)

flag *sed.* 1. = flagstone. 2. Bed of hard marl overlying the top stratum of a salt bed. [M. Eng. *flagge,* probably from O.N. *flaga,* slab of stone]

flagstone *sed.* Thin-bedded, usually argillaceous or micaceous, sandstone; because it splits along bedding planes into thin slabs, it can be used for walkways, floors, retaining walls, and the like. Less commonly, term may be applied to a thin-bedded, sandy limestone of similar properties. [M.Eng. *flagge,* probably from O.N. *flaga,* slab of stone]

flaxseed ore *sed.* Oolitic, hematite-rich iron ore of primary origin, making up well-stratified deposits; the ooliths are small oblate spheroids flattened in the plane of bedding and consist of concentric shells of hematite, goethite, and other minerals. Also named flaxseed stone. [Name alluding to shape of the grains]

fleckschiefer *meta.* Spotted slate characterized by small flecks or spots of minute graphite flakes or clusters of tiny magnetite grains or some indeterminate material. [G. *Fleck,* spot or speck + *Schiefer,* slate]

flint *sed.* 1. Carbonaceous chert, usually black, gray, or brown, occurring as fine-grained nodules in chalk deposits; usually with a waxy luster and excellent conchoidal fracture. 2. Term often applied to numerous hard materials with a fracture similar to flint, including various varieties of chert, jasper, chalcedony, silicified wood, silicified rhyolite, quartzite, milky quartz, and others. [O.Eng. *flint*]

flinty crush-rock *meta.* Black ultramylonite, formed by extreme cataclastic metamorphism, fritted or partly fused into a vitreous mass; under the microscope it appears to be a nonpolarizing glassy substance, but X-ray diffraction shows it is cryptocrystalline. [Name alluding to its appearance (like flint) and origin] (Clough, 1907)

floatstone *sed.* Variety of siliceous sinter consisting of a network of interlacing siliceous (opaline) fibers enclosing numerous interstitial spaces that confine sufficient air to cause the stone to float upon water.

floitite *meta.* Metamorphic rock consisting of biotite and those minerals typical of the greenschist facies. [Floitenthal, Zillerthal, Tyrol] (Becke, 1913)

floridine *sed.* Trade name for a fuller's earth, used largely for decolorizing mineral oils, from Quincy, Florida. Also spelled floridin. [Florida]

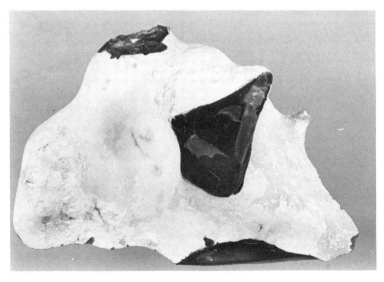

Flint (black nodule coated with chalk). Dover, England. 15 cm.

floridite *sed.* Variety of phosphorite from Florida. [Florida] (Cox in 1891)

florinite *ig. hypab.* Dark-colored biotite-containing monchiquite; major minerals are olivine and augite (about equal in amount), with biotite (8%), and a groundmass of zeolites (38%). [St. Florine, Brassac, France] (Lacroix in 1933)

flos ferri *sed.* Coralloidal form of aragonite consisting of snow-white, divergent and ramifying branches, esp. occurring in cavities associated with siderite rock. [L. *flos ferri,* flower of iron; from its association with iron deposits] (Linnaeus in 1768)

flour, glacial *sed.* = flour, rock

flour, rock *sed.* Finely pulverized, chemically unweathered material, consisting of silt- and clay-sized particles of rock-forming minerals, chiefly quartz; formed when rock fragments are pulverized while being transported or are crushed by the weight of the overlying glacier or ice sheet. [flour, alluding to powdery nature]

flowstone *sed.* Smooth, sheetlike speleothem formed by films of flowing water in caves.

flowtill *sed.* Superglacial till modified and transported by plastic mass flow. (Hartshorn in 1958)

fluobaryt *hydrotherm. ?* Compact, intimate mixture of fluorite and barite from Derbyshire, England; once thought to be a single compound. [fluorite + barite (barytes)] (Hausmann in 1847)

fluolite *ig. extru.* = pitchstone. [Possibly L. L. *fluor,* flow, flux]

fluvioclastic rock *sed.* Water-transported clastic rock containing current- or river-worn fragments. [L. *fluvius,* river + Gk. *klastos,* broken] (Grabau in 1924)

fluxoturbidite *sed.* Sediment produced by a mechanism related both to the deposition from turbidity currents and to submarine sliding or slumping; coarse-grained, thick-bedded, and with a poor development of grading and sole marks. [L. *fluxus,* flux, flow + turbidity]

fluxstone *sed.* Any rock (e.g., limestone, dolomite) used in metallurgical processes to lower the fusion temperature of the ore, to combine with impurities, and to make a slag. [L. *fluxus,* flux, flow]

flysch *sed.* Sedimentary facies, of marine origin, characterized by a thick sequence of poorly fossiliferous, thinly bedded, graded deposits composed chiefly of clastic materials; the principal material is shale or silty shale regularly interbedded with graded beds of dark sandstone (usually graywackes) varying from a few centimeters to a couple of meters in thickness; the most characteristic sedimentary suite of Alpine-type orogenic belts. [G., used in Switzerland for crumbly or fissile material that slides or flows]

foidite *ig. extru.* Volcanic rock in which feldspathoids make up 60-100% of the light-colored components; some restrict the term to those rocks with 90% or more feldspathoids. The coarser-grained plutonic equivalents are called foidolites. [feldspathoid]

foidolite *ig. pluton. IUGS,* feldspathoidal plutonic rock with F greater than 60%; examples include missourite, ijolite, and urtite. [feldspathoid]

foliate *meta.* General term for any foliated rock, e.g., gneiss, schist, phyllite, and slate. [L. *folium,* leaf] (Bastin in 1909)

foraminite *sed.* Rock composed predominantly of the remains of foraminifers, a protozoan belonging to the subclass *Sarcodina,* order *Foraminifera.* [foraminifers]

forcherite *sed.?* Orange yellow opal, colored by orpiment; from gneiss at Reittelfeld, Upper Styria. [Vinzenz Forcher, Austrian chemist and discoverer of the material] (Aichhorn in 1860)

forellenstein *ig. pluton.* = troctolite. [G. *Forelle,* trout + *Stein,* stone; alluding to its spotted appearance] (Rath in 1855)

fortunite *ig. extru.* Brownish-gray trachyte showing occasional brown mica flakes, crystals of pyroxene, and nut-sized nodules of olivine sparsely scattered through the rock; the groundmass is composed of ortho- and clinopyroxene, mica, feldspar, and some glass. [Fortuna, province Murcia, Spain] (De Yarza in 1893)

fourchite *ig. dike.* Olivine-free monchiquite, with as much as 75% pyroxene (augite). [Fourche Mountains, Arkansas] (Williams in 1891)

foyaite *ig. pluton.* Nepheline syenite containing a predominance of potassium feldspar (orthoclase); although the term is essentially a synonym for nepheline syenite, it once was restricted to varieties of that rock having a trachytic texture. [Foya, Portugal] (Blum in 1861)

foyaite, Diamond Jo-type *ig. dike.* Nepheline syenite dike composed of orthoclase, nepheline, and cancrinite, and having a trachytic texture. [Diamond Jo quarry, Magnet Cove, Arkansas] (Williams in 1891)

fragipan *sed.* Dense, subsurface layer of soil whose hardness and relatively slow permeability to water are chiefly due to extreme compactness rather than to clay (claypan) or cementation (hardpan); shows some brittleness when moist. [L. *fragilis,* fragile + pan]

fraidronite *ig. dike.* Variety of minette consisting

of a dirty green, dense, carbonate-rich ground-mass with mica; name used by early French geologists, esp. for rocks in Lozère and Cevennes. (pub. by Dumas in 1846)

framesite   *ig.* Aggregate of diamond, bort, and carbon, esp. from the Premier diamond mine, near Pretoria, Transvaal. [Percival Ross Frames, 1863-1947, South African lawyer of Johannesburg, who collected the first specimens]

frangite   *sed., meta.* Broad term for all sedimentary rocks (unconsolidated or cemented), and their dynamically metamorphosed representatives, formed from the disintegration of igneous rocks without extensive decomposition or mechanical sorting, e.g., arkose, graywacke, paragneiss. [L. *frangere,* to break] (Bastin in 1909)

freestone   *sed.* Term applied to those sandstones (and more rarely limestones, e.g., konite) that submit readily to tool treatment; quarryman's term. [Named for ability to be freely worked in any direction]

frost stone   *sed.* Translucent gray chalcedony with white patches or tufts, like snowflakes, scattered through it, esp. materials from the Mojave Desert, California. Also named frost agate.

fruchtschiefer   *meta.* Spotted slate characterized by concretionary spots (often rich in andalusite or cordierite) suggestive of grains of wheat; occurs in contact metamorphic aureoles in slates and phyllites. [G. *Frucht,* fruit or grain + *Schiefer,* slate]

fucosite   *sed.* Bitumen derived from the hydration of *fucose pentosane* (a seaweed) and found among clays and sands in California. [L. *fucus,* seaweed] (Hackford in 1932)

fulgurite   *meta.* Irregular, glassy tube or crust produced by the fusion of loose sand (or rarely, compact rock) by lightning strokes; common in sands of dunes on shores and deserts and on exposed mountain tops. Varieties include sand fulgurite (dune sands) and rock fulgurite (superficial glassy coatings formed on solid rock). [L. *fulgur,* lightning] (Arago in 1821)

fulvurite   *sed.* = coal, brown; lignite. [L. *fulvus,* tawny, brownish-yellow]

fusain   *sed.* In banded coal, the carbonized wood that resembles charcoal; it is high in ash. Also named fusite, and more rarely, mineral charcoal or mother of coal. [F. *fusain,* spindle tree, the charcoal made from it] (Grand 'Eury in 1882)

G

gabbride *ig. pluton.* General field term for any igneous rock in which pyroxene is the only dark mineral; feldspars are present but not as abundant as pyroxene. [gabbro]

gabbrite *ig. dike.* Seldom used term for aplite with the composition of gabbro. [gabbro] (Polenov in 1899)

gabbro *ig. pluton.* 1. Dark-colored, phaneritic, igneous rock composed of calcic-plagioclase (commonly labradorite or bytownite) and often clinopyroxene (usually augite), with or without olivine and orthopyroxene; the plutonic equivalent of basalt. [Possibly Gabbro, Tuscany, Italy] (Florentine term in print in 1768). 2. *IUGS,* plutonic rock with 0-5 Q, P/(A + P) greater than 90, and plagioclase more calcic than An_{50}.

gabbro, biotite *ig. pluton.* Plutonic rock containing biotite in addition to the usual minerals of a normal gabbro; quartz usually present when biotite is abundant, and plagioclase near $Ab_{50}An_{50}$.

gabbro, corundum *ig. pluton.* Gray to purplish, medium to fine-grained gabbroic rock composed of calcic-plagioclase, hornblende, and corundum; type specimen from South Sherbrooke, Ontario, Canada.

gabbro, cristobalite *lunar.* Medium to coarse-grained, vuggy, crystalline ophitic gabbro, collected during the Apollo 11 mission to the moon, commonly containing cristobalite (up to 6%).

gabbro, feldspathoidal *ig. pluton.* = gabbro, foid

gabbro, flaser *meta.* Metamorphosed gabbro in which mechanically fragmented minerals or mineral aggregates are drawn out in schlieren (bands or streaks) at right angles to the pressure; they show a fluidal texture, with drawn-out lenses between more schistose material; blastomylonite derived from a gabbro. [G. *Flaser,* vein, streak]

gabbro, flowering *ig. extru.* = chrysanthemum stone

gabbro, foid *ig. pluton.* 1. = theralite. Also named feldspathoidal gabbro and foidal gabbro. [feldspathoid] 2. *IUGS,* plutonic rock with 10-60 F, P/(A + P) greater than 90, and plagioclase more calcic than An_{50}.

gabbro, foidal *ig. pluton.* = gabbro, foid

gabbro, gneissoid *meta.* Metamorphosed gabbro resembling gneiss, in which enough of the original texture and constituents remain so that the original gabbroic character can be determined.

gabbro, hornblende *ig. pluton.* Gabbro composed of calcic-plagioclase and primary hornblende rather than the normal clinopyroxene (augite). Also named bojite. (Streng and Kloos in 1877)

gabbro, olivine *ig. pluton.* Rock differing from normal gabbro by containing olivine in addition to the usual calcic-plagioclase and clinopyroxene; generally contains somewhat less plagioclase and more mafic minerals.

gabbro, orbicular *ig. pluton.* Gabbro in which some of the minerals are aggregated into concentric zoned orbs; among these rocks is corsite (napoleonite) from Santa Lucia di Tallano, Corsica.

gabbro, quartz *ig. pluton.* Rock close to normal gabbro in mineralogy and appearance but that may contain as much as 15% quartz (often visible megascopically).

gabbro, saussurite *meta.* Gabbro in which, as a result of metamorphism, the plagioclase is altered to saussurite (a gray to greenish mixture of fine-grained epidote, zoisite, albite, white mica, quartz, and others); contemporaneous with this saussuritization, pyroxenes may alter to uralite, chlorite, and iron oxides, and olivine may alter to serpentine, talc, and chlorite.

gabbro, scapolite *meta.* Scapolite-containing rock derived from the alteration of a gabbro or hyperite; in addition to scapolite and remnants of original calcic-plagioclase and pyroxenes, the rock may contain amphiboles, zoisite, titanite, and others.

gabbro, schistose *meta.* Metamorphosed gabbro resembling a schist in which enough of the original texture and constituents remain so that the original gabbroic character can be determined.

gabbro, uralite *meta.* Gabbro in which the original pyroxene has been altered to secondary amphibole (uralite); sometimes difficult to separate from a hornblende gabbro (bojite), but the secondary amphibole may surround cores of pyroxene and penetrate them as irregular needles and fibers.

gabbro-granite *ig. pluton.* Obsolete name for an intermediate rock type; similar to gabbro, but also containing orthoclase as in granite.

gabbro-syenite *ig. pluton.* Obsolete name for a rock intermediate between gabbro and syenite; pyroxene syenite or syenogabbro.

gabbroid *ig. pluton.* 1. Any rock resembling gabbro. 2. *IUGS,* preliminary field term for a plutonic rock with Q less than 20 or F less than 10, P/(A + P) greater than 65, and 10-90 pl/(pl+px+ol).

gabbronorite *ig. pluton. IUGS,* plutonic rock satisfying the definition of gabbro, in which pl/(pl+px+ol) and pl/(pl+px+hbl) are 10-90 and ol(pl+px+ol) and hbl/(pl+px+hbl) are less than 5.

gabbronorite, olivine *ig. pluton. IUGS,* plutonic rock satisfying the definition of gabbro and in which pl/(pl+px+ol) is 10-90, px/(pl+px+ol) is greater than 5, and ol/(pl+px+ol) is greater than 5.

gabbronorite, pyroxene-hornblende *ig. pluton. IUGS,* plutonic rock satisfying the definition of gabbro, and in which pl/(pl+hbl+px) is 10-90, and px/(pl+hbl+px) and hbl/(pl+hbl+px) are greater than 5.

gabbrophyre *ig. dike.* = odinite. (Chelius, 1892)

gabbrophyrite *ig. pluton., hypab.* = gabbro porphyry (see porphyry). (Polenov in 1899)

gagat *sed.* = jet. [G. *Gagat,* jet]

gagatite *sed.* Coalified woody material resembling jet. [G. *Gagat,* jet]

galliard *sed.* Hard flinty siliceous sandstone used for road metal; ganister. Also spelled calliard. [Possibly F. *galliard,* gay, lively, hardy]

gangmylonite *meta.* Mylonite or ultramylonite showing intrusive relations with the adjacent rock without evidence of fusion. [G. *Gang,* duct, passage, vein + mylonite] (Hammer in 1914)

gangue *hydrotherm.* Nonvaluable, usually nonmetallic, minerals associated with the ore in a vein or related deposit. Common mineral examples are dolomite, calcite, quartz, barite, and others. [F. *gangue,* from G. *Gang,* duct, vein]

ganister *sed.* 1. Compact, highly siliceous sandstone, consisting of angular quartz grains cemented by secondary quartz and possessing a characteristic splintery fracture; with a more granular texture than chert. Also spelled gannister. [M.H.G. *ganster,* spark] 2. Siliceous fire clay.

garbenschiefer *meta.* Variety of contact metamorphic spotted slate (knotenschiefer), whose concretionary spots resemble caraway seeds. [G. *graben,* to sheave, to weld + *Schiefer,* schist, slate]

garéwaite *ig. dike.* Nearly feldspar-free, dark-colored end-member of the vogesite-spessartite-odinite lamprophyre series, which has phenocrysts of corroded diopside in a fine-grained, crystalline groundmass of olivine, pyroxene, chromite, and magnetite. [Garéwaia River, Tilai Range, northern Urals, Soviet Union] (Duparc and Pearce in 1904)

garganite *ig. dike.* Lamprophyre intermediate between a vogesite and spessartite, containing both amphibole and augite (malacolite), associated with olivine, orthoclase, and plagioclase; may be a hybrid rock. Also spelled garganito. [Gargano Mountain, Italy] (Rosenbusch in 1896)

garganito *ig. dike.* = garganite. [Gargano Mountain, Italy] (Viola and di Stefano in 1893)

garnerite *sed.* Local name for a variety of agate consisting of small chalcedony nodules more or less firmly cemented together with additional chalcedony; in cross section specimens display attractive circular patterns; occurs at Panoche Pass, San Benito County, California. [Ralph Garner of Fresno, California, who discovered it in 1951]

garnet rock *meta.* = garnetite

garnetite *meta.* Rock essentially composed of interlocking garnet grains, e.g., spessartine rocks related to gondite, and the rocks granatite and queluzite.

gastrolith *sed.* Rounded stone or pebble, often highly polished and with circular scratches, from the stomach of some vertebrates (esp. extinct fossil reptiles), thought to have been used in grinding up their food; often these consist of jasper, agate, chalcedony, and other quartz rocks. Also named dinosaur gizzard stone. [Gk. *gaster,* stomach]

gault *sed.* = marl, folkestone. [Eng. dial. (Northamptonshire) *gault;* related to O. Sw. *galt,* barren]

gaussbergite *ig. extru.* Leucitite consisting of phenocrysts leucite, augite, and olivine, in a glassy groundmass that is potentially sanidine; similar to orendite but with augite and olivine in place of phlogopite. [Gaussberg volcano, Kaiser Wilhelm II Land, Antarctica] (Lacroix in 1926)

gauteite *ig. hypab.* Rock related to trachyte or trachyandesite having phenocrysts of hornblende, augite, occasional biotite, and abundant plagioclase plates in a groundmass composed of feldspar (sodium sanidine) with minor mafics. |Gaute (Kout), Czechoslovakia| (Hibsch in 1897)

geburite-dacite   *ig. dike.* Holocrystalline, fine-grained, dark green to black dacite, characterized by the presence of hypersthene, from Mt. Macedon, Victoria, Australia. |Gebur (Geboor), aboriginal name for Mt. Macedon| (Gregory in 1902)

gedanite   *sed.* Clear, wine yellow resin resembling amber but not containing succinic acid and less rich in oxygen; found with amber on the shores of the Baltic. |L. *Gedanum,* Danzig, now Gdansk, Poland| (Helm in 1878)

geest   *sed.* 1. = saprolite. 2. Provincial name for sandy soil in northern Germany and Holland. |G. *Geest,* sandy or high and dry soil| (DeLuc in 1816)

gelite   *sed.* Opal or chalcedony deposited as a secondary accessory mineral, usually as a cement, or in fractures, in sandstone. |gel, shortened from gelatin| (Walther in 1935)

geocerite   *sed.* Natural, white, waxy substance occurring in brown coal at Gesterwitz near Weissenfels, eastern Germany. Also spelled geocerain and geocerin. |Gk. *gē,* earth + *keros,* wax| (Brückner in 1852)

geode   *sed.* Hollow, spherical or subspherical body, usually lined with crystals projecting inward from the walls; commonly separable as a discrete nodule from the country rock (usually calcareous or argillaceous but may be extrusive igneous) in which it occurs; many geodes are characterized by an outermost layer of banded chalcedony, and have quartz crystals projecting inward, but numerous other minerals (calcite, barite, celestite, gypsum) may occur with or without quartz. |Gk. *gē,* earth + *eidos,* form| (In use as early as 1619)

geolyte   *sed.* Those constituents of soils that are readily soluble and of indefinite mineralogical composition but have little in common with zeolites. |Gk. *gē,* earth + *lutos,* dissolved| (Wülfing in 1900)

georgiaite   *tekt.?* Greenish tektitelike body from Georgia, U.S.

geröllton   *sed.* Conglomeratic mudstone or pebble-bearing clay in which the pebbles and clay were deposited simultaneously; tilloid, since its origin is nonglacial. |G. *Geröll,* pebbles, gravel + *Ton,* clay| (Ackerman in 1951)

gestellstein   *ig. diaschist.* Old German term for igneous rock consisting only of a mixture of quartz and mica; esmeraldite. |G. *Gestell,* framework + *Stein,* stone| (Brückmann in 1773)

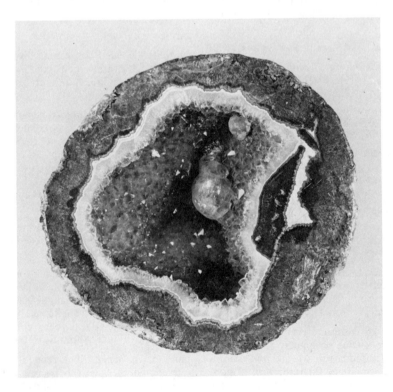

Geode lined with smoky quartz and containing calcite. Mexico. 11 cm.

geyserite *sed.* Lightweight, soft, porous, light-colored aggregate composed mainly of a delicate network of minute fibers or concretions of opal and deposited by a geyser; like a siliceous sinter except for its mode of occurrence. [geyser] (Delamétherie in 1812)

ghizite *ig. extru.* Analcime andesite also containing olivine and biotite. [Ghizo, west of Monte Urtigu, Sardinia] (Washington in 1914)

gibber *sed.* Pebble or boulder, esp. one of the wind-polished or wind-sculptured stones (ventifacts) that make up a desert pavement or the lag gravels of an arid region. [Australian aboriginal *gibber,* stone]

gibbsite rock *sed.* Although rare, nearly pure gibbsite nodules and cylindrical tubes occurring esp. in bauxite deposits.

gibelite *ig. extru.* Trachyte characterized by large and abundant phenocrysts of sodium-rich microcline, sometimes microperthite, and a few small prisms of colorless augite, in a groundmass of microcline exhibiting a flow texture. [Monte Gibelé, Pantelleria Island, western Mediterranean] (Washington in 1913)

Gibraltar stone *sed.* Light-colored, onyx marble, esp. that found in a cavern at Gibraltar.

gibsonite *ig., altered.* Fibrous pink variety of thomsonite from Renfrewshire and Dumbartonshire, Scotland. [Possibly Alexander Gibson, 1800-1867, Scottish student of Indian organic natural resources] (Haidinger, pub. by Dufrenoy in 1847)

gidderite *meta.* Aplitic gneiss occurring in Cameroon. [Possibly Guider (Guidder) in northern Cameroon] (Passarge in 1895)

gieseckite-porphyry *ig., altered.* Porphyry from Greenland whose nepheline phenocrysts are altered to a fine-grained aggregate of muscovite scales (named gieseckite because it was thought to be a new mineral). [Karl Ludwig Giesecke, 1761-1833, German geologist who brought the rock from Greenland]

gilsonite *sed.* Natural asphaltite of high purity used in varnishes, japans, inks, roofing, and molded articles; occurs in veins in Utah; essentially identical to uintahite. [S. H. Gilson, Salt Lake City, Utah, 19th century]

girasol *sed.* Relatively transparent, precious opal with a rather uniform reddish or bluish wavy type of internal light. Also named girasol opal, girasole, and gyrasol. Not to be confused with cat's-eye corundum, which rarely is given the same name. [L. *gyrare,* to turn + *sol,* sun]

giumarrite *ig. hypab.* Olivine-free amphibole monchiquite [Giumarra, Ramacca, Sicily] (Viola in 1901)

gizzard stone *sed.* = gastrolith

glacialite *sed.* Trade name for a white clay used as a fuller's earth, from Enid, Oklahoma. [Possibly glacial, alluding to its icy or slippery properties] (Term in use by 1897)

gladkaite *ig. dike.* Fine-grained, gray quartz lamprophyre, consisting of a fine granular aggregate of oligoclase to andesine, abundant quartz,

Geode (quartz). Illinois. 18 cm.

hornblende, and a little biotite. [Gladkaia Sopka, northern Urals, Soviet Union] (Duparc and Pearce in 1905)

glamaigite   *sed.* Intrusive breccia containing dark patches of marscoite in a lighter groundmass; term restricted to original locality in Scotland. [Glamaig, Isle of Skye, Scotland] (Harker in 1904)

glaskogel   *tekt.* = billitonite. [Du. *glas,* glass + *kogel,* ball] (pub. by van Dijk in 1878)

glass, andesitic   *ig. extru.* Glass of andesite composition that may be colorless, gray, pale green, yellow brown, and brown, and contain minute bubbles, iron oxide specks, and various crystallites; its index of refraction is in the range 1.49-1.54, although values as low as 1.48 and as high as 1.60 are known.

glass, basaltic   *ig. extru.* = hyalobasalt

glass, dacitic   *ig. extru.* Natural volcanic glass of dacite composition; resembling obsidian but tends to be perlitic.

glass, Darwin   *impact., uncertain.* Naturally occurring glass having a slaggy appearance, vesicularity, and a high silica content; found in the Jukes-Darwin mining field of western Tasmania; probably impactite glass or fused ash from a prehistoric peat bog fire, although once considered to be a tektite. Also named queenstownite.

glass, Libyan Desert   *impact., uncertain.* Silica-rich (97.58% SiO_2) natural glass found scattered over the Libyan Desert (North Africa); considered to be akin to tektites by some investigators and to impactite by others, its origin is uncertain.

glass, rhyolitic   *ig. extru.* General term for glassy rocks having the composition of rhyolite (and granite); includes obsidian, pitchstone, perlite, and pumice.

glass, Sakado   *ig.?* Natural glass from Japan once considered to be a tektite. [Sakado, Japan]

glass, silica   *impact., uncertain.* Natural glass very rich in silica, e.g., Libyan Desert glass (over 97% SiO_2) and glass from the meteorite craters at Henbury, central Australia, and Wabar in Arabia.

glass, trachytic   *ig. extru.* = obsidian, trachyte

glass, volcanic   *ig. extru.* Any natural glass formed by the rapid cooling of lava, e.g., obsidian, trachyte obsidian, hyalobasalt, pitchstone, pumice, and shards.

glass, Wabar   *impact.* = impactite, Wabar

glass agate   *ig. extru.* Misnomer for obsidian

glass lava   *ig. extru.* Obsidian or other volcanic glass.

glass meteorite   *tekt.* Obsolete name for tektite, esp. moldavite.

glass rock   *sed.* Regional name for pure, cryptocrystalline limestone of Trentonian age in northern Illinois and southern Wisconsin.

glass sand   *sed.* Sand used for the manufacture of glass, because of its high silica content (more than 93%) and its low content of iron, chromium, cobalt, and other colorants.

glauconarenite   *sed.* = sandstone, glauconitic. [glauconite + arenite]

glauconite rock   *sed.* Glauconite-rich rocks including greensands and glauconitic sandstone; composed of glauconite granules (pellets, thin tablets, vermicular crystals, casts of foraminiferal shells) mixed with detrital quartz, clays, calcite, shell pieces, pyrite, and, if weathered, gypsum and jarosite.

glaucophanite   *meta.* Metamorphic rock related to a glaucophane amphibolite in which the major mineral is glaucophane; two varieties, epidote glaucophanite and garnet glaucophanite, have been recognized. [glaucophane]

glendonite   *sed.* Large, euhedral pseudomorphs of granular calcite after an original mineral of uncertain composition; the original mineral may have been glauberite or ikaite (calcium carbonate hexahydrate); found esp. in high latitude environments. [Glendon on the Hunter River, New South Wales, Australia] (David and Taylor in 1905)

glenmuirite   *ig. pluton.* Rock similar to essexite but having analcime rather than nepheline as the principal feldspathoid; composed of labradorite (35%), augite (27%), analcime (15%), olivine (13%), orthoclase, biotite, and others. [Glenmuir Water, Ayrshire, Scotland] (Johannsen in 1938)

glessite   *sed.* Red brown to brown, nearly opaque, fossil resin from the shores of the Baltic; contains minute spherical cell-like forms. [L. *glessum,* amber, mentioned by Tacitus] (Helm in 1881)

glimmerbasalt   *ig. extru.* = basalt, mica. [G. *Glimmer,* mica + basalt] (Mohl in 1874)

glimmergabbro   *ig. pluton.* = gabbro, biotite. [G. *Glimmer,* mica + gabbro] (Eichstädt in 1887)

glimmerite   *ig. pluton.* = biotitite. [G. *Glimmer,* mica] (Larsen and Pardee in 1929)

glimmertrachyte   *ig. extru.* = selagite. [G. *Glimmer,* mica + trachyte] (Rosenbusch in 1923)

glomeroporphyry   *ig. extru., hypab.* Porphyritic rock containing clustered phenocrysts (like rosettes) of a single mineral, e.g., basalt glomeroporphyry containing clusters of calcic-plagioclase in an aphanitic groundmass. [L. *glomerare,* to wind into a ball + porphyry] (Judd in 1886)

glyptolith   *sed.* Wind-cut stone; ventifact. [Gk. *glyptos,* carved] (Woodworth in 1894)

gneiss   *meta.* Foliated rock formed by regional metamorphism and consisting of alternating schistose bands and granular bands (or sometimes, granular lens-shaped zones); usually less than 50% of the minerals show preferred parallel orientation; most gneisses have feldspar and quartz, but they are not essential components. Many varieties have been recognized and named. Varieties based on mineralogy include: anthophyllite g., arfvedsonite g., biotite g., chlorite g., cordierite g., diopside g., epidote g., garnet g., garnet-zoisite g., gedrite-cordierite g., hornblende g., hornblende-biotite g., hypersthene g., kyanite g., muscovite g., pyroxene g., quartz-feldspar g., quartz-oligoclase g., riebeckite g., sapphirine g., sillimanite g., staurolite g., and wollastonite g. Varieties based on chemical considerations include: alkali g., calc-silicate g., and lime-silicate g. Varieties with compositions similar to igneous rocks include: diorite g., gabbro g., granite g., monzonite g., quartz diorite g., and syenite g. Varieties derived from sediments (paragneiss) include: conglomerate g., psammitic g., psephitic g., quartzite g., and stretched-pebble g. Varieties based on unusual modes of origin or structural features are listed below. [G. *Gneis,* old

miners' term for rock traceable to the 16th century in the Erzgebirge mining district of Saxony; may be related to M.H.G. *gneiste,* spark]

gneiss, aplite   *meta.* Gneiss in which the granular components resemble aplite in that they have a fine-grained, saccharoidal appearance.

gneiss, augen   *meta.* Gneiss containing large lenticular mineral grains or mineral aggregates, having the shape of eyes in cross section; these lenses may represent granulated pebbles or phenocrysts of the parent rock. [G. *Augen,* eyes]

gneiss, bacillar   *meta.* Gneiss with a parallel, rod-shaped structure. [L. *baculum,* rod, stick]

gneiss, banded   *meta.* Gneiss consisting of alternating layers of different compositions or textures; bands may be straight or sinuous.

gneiss, calc-silicate   *meta.* Gneiss composed primarily of calcium and/or magnesium silicates, usually with some calcite; a variety of calc-silicate rock.

gneiss, composite   *migmat.* = gneiss, injection

gneiss, flaser   *meta.* = gneiss, mylonite. [G. *Flaser,* streak, vein]

gneiss, flow   *ig. pluton.* = gneiss, fluxion

gneiss, fluxion   *ig. pluton.* Banded (gneisslike) igneous rock with a flowage structure due to the movements of a viscous magma during the later stage of crystallization. Also named flow gneiss, protoclastic gneiss, primary gneiss, and gneiss-

Gneiss (bacillar structure). Doubravcan, Bohemia, Czechoslovakia. 11.5 cm.

oid granite (if granite in composition). [L. *fluxus*, flux, flow] (Gregory in 1894)

gneiss, granite   *meta.* 1. Gneiss having the mineral composition of a granite; may be derived from a sedimentary or igneous parent rock. 2. Metamorphosed granite.

gneiss, ice   *sed.* Frozen ground with ice segregated in laminae so as to resemble a gneissic rock. (Taber in 1943)

gneiss, injection   *migmat.* Composite rock resembling a gneiss but whose banding is wholly or partly due to the interlaminar injection of igneous magma (usually granitic) into schistose, fissile, or otherwise penetrable rocks.

gneiss, lenticular   *meta.* Gneiss containing large lenticular mineral grains or mineral aggregates; the lenses may look like eyes (augen gneiss) or they may be more elongated.

gneiss, lit-par-lit   *migmat.* = gneiss, injection. [F. *lit-par-lit*, bed by bed]

gneiss, mixed   *migmat.* = gneiss, injection

gneiss, mylonite   *meta.* Partly granulated and partly recrystallized rock, intermediate in its characteristics between mylonite and schist; the felsic minerals are shattered without much recrystallization and often occur in lenticular aggregates surrounded by and alternating with schistose streaks

of recrystallized mafic minerals. (Quensel in 1916)

gneiss, pearl   *meta.* Plagioclase blastite (see blastite).

gneiss, pelite   *meta.* Paragneiss whose parent material was argillaceous sediment. (Rosenbusch in ca. 1910)

gneiss, pencil   *meta.* = gneiss, stengel

gneiss, primary   *ig. pluton.* gneiss, fluxion

gneiss, protoclastic   *ig. pluton.* Gneissoid granite (primary or fluxion gneiss) formed as a marginal facies of some plutons by shearing during emplacement. [Gk. *protos*, first + *klastos*, broken]

gneiss, stengel   *meta.* Gneiss that breaks into roughly cylindrical, pencil-like quartz-feldspar aggregates, often mantled by mica flakes; a linear arrangement of its components rather than the usual lamellar arrangement; related to a bacillar gneiss in structure. Also named pencil gneiss. [G. *Stengel*, stem, stalk]

gneiss, trapshotten   *meta.* Gneiss impregnated with a very black indurated material, originally supposed to be injections of trap rock but now identified as flinty crush-rock. [In allusion to having been shot through with trap] (King and Foote in 1864)

gneiss, veined   *migmat.* Injection gneiss with irregular layering.

Gneiss (granite composition). Locality unknown. 9 cm.

gondite *meta.* Manganiferous metamorphic rock composed of spessartine and quartz, with various amounts of magnetite, rhodonite, and manganophyllite (manganoan biotite). Major varieties include amphibole gondite and rhodonite gondite. [Gonds, of the central provinces of India] (Fermore in 1909)

gooderite *ig. pluton.* Feldspathoidal rock differing from ordinary nepheline syenite by containing more albite than potassic feldspar. [Gooderham, Glamorgan Township, Ontario, Canada] (Johannsen in 1938)

goodletite *meta.* 1. Green-pyroxene or green-amphibole rock containing ruby corundum crystals; term used esp. in Australia and New Zealand, although a similar amphibole rock occurs in Clay County, North Carolina. Not to be confused with the ruby-containing green-zoisite rock from Longido, northeastern Tanzania. [Possibly Mr. Goodlet] 2. Rarely, term applied to a white crystalline marble at Mogok, Burma, containing ruby and spinel.

gordunite *ig. pluton.* Garnet-bearing wehrlite composed of olivine (62%), diopside (26%), and pyrope (10%), with picotite (chromian spinel), iron oxides, serpentine, and others. [Gordunotal, Tessin, Switzerland] (Grubenmann in 1908)

gossan *sed.* Weathered, or otherwise decomposed, upper zone of a mineral lode, characterized by an abundance of oxidized and hydrated alteration products, esp. goethite; principally derived from the weathering of sulfides of iron and copper or of iron-containing carbonates. [Cornish dial.]

gouge *meta.* Fine-grained, weakly cohesive material formed by extreme cataclastic metamorphism along fault zones under weak confining pressure; materials are commonly quartz, clay minerals, sericite, chlorite, and others. Also named fault gouge. [M. Eng. *gowge*, related to L.L. *guvia*, a kind of chisel]

grahamite *sed.* Jet-black, lustrous hydrocarbon with a conchoidal fracture and soluble in carbon disulfide and chloroform but not in alcohol; occurs in veinlike masses in West Virginia, Colorado, and Oklahoma. [J. Lorimer Graham, of New York, and Colonel J. A. Graham, of Baltimore, early owners of the mine where it occurred in West Virginia] (Wurtz in 1865)

grahamite *meteor.* Obsolete name for a mesosiderite (stony-iron meteorite); originally thought to contain more plagioclase than mesosiderites. [Possibly Thomas Graham, 1805-1869, Scottish chemist] (Tscharmak in 1883)

grainstone *sed.* Mud-free, grain-supported carbonate sedimentary rock; the sand-size particles are so abundant they are in three-dimensional contact and are able to support one another. (Dunham in 1962)

granat *sed.* Term used in Ireland for a quartzose grit. [Possibly corruption of granite]

granatite *meta.?* Rock composed of garnet, occurring in Ala Valley, Piemonte, Italy. [It. *granato*, garnet]

granide *ig. pluton.* General field term for a holocrystalline, medium- to coarse-grained igneous rock, consisting of quartz and any kind of feldspar, and with one or more mafics, generally biotite or hornblende; all granites and light-colored quartz diorites are included under this term. [granite] (Johannsen in 1931)

granilite *ig.* Obsolete term applied to any crystalline (granular) igneous rock composed of more than three ingredients. [L. *granum*, grain] (Page in 1859)

granite *ig. pluton.* 1. Light-colored, granular, plutonic rock composed of essential quartz and potassic feldspar (orthoclase and/or microcline), with nonessential sodic-plagioclase (oligoclase or andesine) and one or more mafic minerals (usually biotite or hornblende). Numerous varieties are listed below. [Possibly L. *granum*, grain] (In print in Italy as early as 1596). 2. *IUGS,* plutonic rock with 20-60 Q and 10-65 P/(A + P).

granite, albite *ig. pluton.* Granite in which quartz and potassic feldspar are accompanied by accessory albite rather than the usual oligoclase or andesine. Also named sodaclase granite.

granite, alkali *ig. pluton.* Granite typically rich in sodium, in which potassic feldspars are strongly perthitic or anorthoclase, plagioclase is albite or sodic oligoclase, biotite is iron-rich, amphiboles are hastingsite, arfvedsonite, or riebeckite, and pyroxenes may include aegirine-augite and aegirine. (Rosenbusch in 1907)

granite, alkali-feldspar *ig. pluton. IUGS,* plutonic rock with 20-60 Q, and P/(A + P) less than 10.

granite, amphibole *ig. pluton.* = granite, hornblende

granite, andalusite *ig. pluton.* Granite containing andalusite; usually formed by the contamination of the granitic magma by adjacent pelitic sediments.

granite, binary *ig. pluton.* 1. Relatively uncommon variety of granite containing two micas, biotite and muscovite; also named two-mica granite, biotite-muscovite granite (if biotite dominates), or muscovite-biotite granite (if muscovite dominates). 2. Term originally applied to granite that contained only the two essential minerals, quartz and potassic feldspar. [L. *binarius*, two] (Keyes in 1895)

granite, biotite   *ig. pluton.* Granite whose only important mafic mineral is biotite; the most abundant of all granites. Also named granitite.

granite, biotite-augite   *ig. pluton.* Variety of pyroxene-biotite granite in which original augite was partially resorbed and recrystallized as hornblende.

granite, biotite-hornblende   *ig. pluton.* Granite intermediate between biotite granite and hornblende granite; usually with increasing amphibole (hornblende) the plagioclase increases and becomes more calcic.

granite, black   *ig. pluton.* 1. Popular term applied to dark-colored rocks, such as gabbro, diabase, and diorite, that can be used as attractive building stones.   2. Dark-colored true granite containing large amounts of mafics (biotite and/or hornblende).

granite, calc-alkali   *ig. pluton.* General term for the more widespread granite composed of perthitic orthoclase or microcline, quartz, biotite, hornblende, oligoclase (rarely andesine), diopside, hypersthene, and others; used in contrast with alkali granite, which is more sodic. (Rosenbusch in 1907)

granite, calcite   *ig. pluton.* Granite with somewhat limited amounts of primary calcite whose origin may be magmatic or from contamination. (Trinker in 1853)

granite, cordierite   *ig. pluton.* Rare variety of granite containing cordierite; probably a result of contamination of the magma from adjacent rocks.

granite, diallage   *ig. pluton.* Pyroxene granite whose pyroxene is lamellar diallage (augite or diopside).

granite, epidote   *ig. pluton., altered.* Granite that usually contains enough epidote to have a green color; in most of the reported localities, the epidote is regarded as secondary. Unakite is a variety.

granite, flaser   *meta.* Granite subjected to cataclastic metamorphism and showing streaky layers of parallel scaly mica and other components surrounding lenticular bodies of quartz and feldspar. [G. *Flaser*, streak, vein]

granite, gneissoid   *ig. pluton.* Variety of igneous granite with a gneissoid structure. Also named fluxion gneiss, flow gneiss, protoclastic gneiss, and primary gneiss.

granite, graphic   *ig. pluton, dike.* Coarse- to very coarse-grained, quartz-potassic feldspar rock in which large crystals of the two minerals are intergrown in such a way that in cross section the intercalates of quartz have the appearance of cuneiform, semitic, or runic characters. Also called Hebraic granite and runite. [G. *graphein*, to write]

granite, graphite   *ig. pluton.* Granite in which graphite is a characteristic accessory mineral. (Zirkel in 1866)

granite, Hebraic   *ig. pluton., dike.* = granite, graphic

granite, hornblende   *ig. pluton.* Granite in which the major characterizing mafic is hornblende; most of these rocks also contain some accessory biotite or pyroxene (a nucleus for hornblende). (Naumann in 1849)

granite, laneite   *ig. pluton.* Quartz-poor, alkali granite composed of sodium-orthoclase perthite, albite, riebeckite, laneite (a variety of barkevikite), arfvedsonite, and quartz (18%).

granite, lithionite   *ig. pluton.* Granite in which lithian mica (zinnwaldite, lepidolite) takes the place of biotite. [lithium]

granite, miarolitic   *ig. pluton.* Term applied to those granites with small cavities (called miarolitic cavities) into which well-terminated crystals project.

granite, muscovite   *ig. pluton.* Granite carrying only muscovite in addition to the essential quartz and potassic feldspar; not to be confused with quartz-mica rocks like esmeraldite or greisen.

granite, orbicular   *ig. pluton.* Granite consisting of a typical granular matrix in which are set numerous spheroidal crystalline aggregates composed of the constituent minerals arranged in concentric shells.

granite, Oriental   *meta.* Popular term applied to a wide-banded metamorphic gneiss, esp. one from Minnesota, used for trimming buildings.

granite, Plauensche Grund   *ig. pluton.* Very pure hornblende granite, originally considered to be a syenite. [Plauensche Grund, near Dresden, Saxony, Germany] (Werner in 1788)

granite, pudding   *ig. pluton.* Orbicular granodiorite (originally described as a granite) containing orbs rich in biotite; when the rock is broken these spherical black zones stand in relief and resemble plums in a pudding; original specimens from Craftsbury, Vermont. (Hitchcock and Hall in 1861)

granite, pyroxene   *ig. pluton., meta.?* Granite in which the mafic is almost exclusively pyroxene; these rocks are relatively rare, some contain augite while others contain orthopyroxenes (enstatite granite, bronzite granite, charnockite); some have been classified as metamorphic.

granite, pyroxene-biotite *ig. pluton.* Granite containing biotite with diopside or augite rimmed by primary hornblende (where the original pyroxene was partially resorbed and recrystallized as hornblende).

granite, quartz *ig. pluton.* Granite exceptionally rich in quartz; in appearance like other granites. (Johannsen in 1919)

granite, recomposed *sed.* Arkose so little reworked and so little decomposed that upon cementation the rock looks very much like a granite. Also named reconstructed granite.

granite, recomposed *meta.* Sediment (arkose, conglomerate) recrystallized by strong metamorphism into a rock that simulates a granite. Also named reconstructed granite.

granite, schorl *ig. pluton.* = granite, tourmaline

granite, schrift *ig. pluton., dike.* = granite, graphic. [G. *Schrift,* writing]

granite, sodaclase *ig. pluton.* = granite, albite

granite, Syene *ig. pluton.* Granite from near the ancient town of Syene on the eastern bank of the Nile, near the modern town of Aswan, Egypt; this pink granite, composed of quartz, red microcline, white oligoclase, biotite, and hornblende, was quarried as early as 5,000 years ago and supplied rock for many monuments and sarcophagi.

granite, syenite *ig. pluton.* Obsolete name for hornblende granite; Plauensche Grund granite is related. (von Cotta in 1862)

granite, tourmaline *ig. pluton.* Granite composed of orthoclase, quartz, and a little plagioclase and tourmaline (schorl). (von Lasaulx in 1875)

granite, true *ig. pluton.* = granite, binary (two-mica granite). (Rosenbusch in 1907)

granite, two-mica *ig. pluton.* = granite, binary

granite, uralite *ig. pluton., altered.* Rock composed of quartz, feldspar (usually altered orthoclase and plagioclase), uralite, and epidote; before alteration the rock may have been either quartz monzonite or granodiorite.

granitelle *ig. pluton.* 1. Granite with comparatively little mica or mafic, so that it consists almost entirely of quartz and feldspar. Compare alaskite. 2. Augite granite. 3. = granite, binary. [granite]

granitello *ig. pluton.* Obsolete term applied to a fine-grained granite. [granite] (Brückmann in 1778)

granitine *ig. pluton.* Obsolete term applied to a crystalline rock composed of any three minerals other than those of a granite. [granite]

granitite *ig. pluton.* Granite in which biotite is the only dark constituent; term has had various other meanings and is now falling into disuse. [granite] (Rose in 1849)

granitoid *ig. pluton.* 1. Any igneous rock closely related to granite in appearance or composition. 2. IUGS, preliminary field term for plutonic rock with 20-60 Q.

granitoid, quartz-rich *ig. pluton. IUGS,* preliminary field term for plutonic rock with 60-90 Q.

granodiorite *ig. pluton.* 1. Phaneritic rock containing quartz and both potassic feldspar and sodic-plagioclase, the latter in excess of the former; intermediate in composition between quartz diorite and quartz monzonite, containing quartz, plagioclase (oligoclase or andesine), and potassic feldspar (orthoclase, microcline, various perthites) with biotite, hornblende, and rarely pyroxene. [granite + diorite; originally considered intermediate between granite and quartz diorite] (Becker, pub. by Lindgren in 1893). 2. *IUGS,* plutonic rock with 20-60 Q, and 65-90 P/(A + P).

granodiorite, orbicular *ig. pluton.* Granodiorite with an orbicular structure; many rocks originally described as orbicular granite have compositions belonging with these, e.g., pudding granite.

granodiorite, quartz *ig. pluton.* Granodiorite in which quartz is very abundant; forms more than 50% of the felsic components.

granodiorite, sodaclase *ig. pluton.* Granodiorite normal in every respect except that the plagioclase is albite (sodaclase).

granodolerite *ig. dike.* Obsolete term for diabase (dolerite) containing interstitial quartz and orthoclase. [granite + dolerite] (Shand in 1917)

granofels *meta.* Field term for a medium- to coarse-grained, granular metamorphic rock with little or no foliation or lineation. [granular + G. *Fels,* rock] (Goldsmith in 1959)

granogabbro *ig. pluton.* Orthoclase-bearing quartz gabbro; analogous to granodiorite and differing only in having calcic, rather than sodic-plagioclase. [granite + gabbro] (Johannsen in 1917)

granolite *ig. pluton.* General term for any phaneritic plutonic igneous rock having a granitic (granular) rather than porphyritic texture. [granite or granular] (Pirsson in 1899)

granophyre *ig. extru.* 1. Rhyolite and rhyolite porphyries with holocrystalline groundmasses in which the constituents are intergrown like graphic granite on a microscopic sale. 2. Porphyry of granitic composition with a holocrystalline, granular groundmass.   3. Irregular microscopic intergrowth of quartz and potassic feldspar. [granular + porphyry] (Vogelsand in 1872)

granophyre, granodiorite   *ig. aschist.* Granodiorite porphyry with phenocrysts of green augite and oligoclase in a groundmass that is micrographic (granophyric).

granosyenite   *ig. pluton.* Igneous rock whose composition is intermediate between that of granite and that of syenite; term used esp. in Russian literature. [granite + syenite]

granule   *sed.* Fragment of any rock or mineral 2–4 mm in diameter; intermediate in size between sand and pebble and sometimes referred to as a very fine pebble. [L. L. *granulum* from L. *granum*, grain]

granulite   *ig. pluton.* Muscovite-bearing granite; term used esp. in French literature. [L.L. *granulum* from L. *granum*, grain] (Michel-Lévy, in 1874)

granulite   *sed.* Sedimentary rock composed of sand-sized aggregates of constructional (nonclastic) origin, simulating in texture arenite of clastic origin, e.g., rock formed of ooliths or of lapilli. [L.L. *granulum* from L. *granum*, grain] (Grabau in 1911)

granulite   *meta.* Metamorphic rock composed essentially of a fine-grained mosaic of feldspar, with or without quartz; ferromagnesian minerals, if present, are predominantly anhydrous; granulites typically contain lenticular (or elongate) grains or aggregates of grains. Some of the major mineralogical varieties include: biotite g., cordierite-biotite g., garnet g. (leptynite), garnet-cordierite g. (laanilite), garnet-kyanite g., garnet-pyroxene (diopside, hypersthene) g., garnet-sillimanite g., graphite g., hornblende g., magnetite-biotite g., pyroxene g., sillimanite g. [L.L. *granulum* from L. *granum*, grain] (Weiss in 1803)

granulite, felsic   *meta.* Metamorphic granulite consisting mainly of feldspars (orthoclase, microcline, cryptoperthite, oligoclase) and quartz, with lesser garnet (chiefly pyrope-almandine), sillimanite, kyanite, cordierite, diopside, hypersthene, hornblende, and graphite.

granulite, pyroxene   *meta.* Granulite mineralogically similar to charnockite; contains hypersthene, plagioclase (andesine to oligoclase), orthoclase, garnet, and others; some petrologists consider charnockite to be a pyroxene granulite, at least those felsic varieties containing high silica, considerable orthoclase, and sodic plagioclase.

grapestone   *sed.* Name given to clusters of small calcareous pellets or other grains, commonly of sand size, stuck together by microcrystalline cement or bound by organic matter, shortly after deposition; occurs in modern carbonate environments, especially on the Bahama Banks. (Illing in 1954).

graphiphyre   *ig. pluton.* Igneous rock having a granophyric (graphic intergrowths) groundmass in which the components are of microscopic size. Contrast with graphophyre. [G. *graphein*, to write; alluding to the microscopic graphic intergrowths] (Cross and others in 1906)

graphite rock   *meta.* Rock in which the only essential constituent is graphite; often represents the metamorphic residue of coals, or similar organic-rich sediments. Varieties include graphite schist, shungite, and tremenheerite.

graphitite   *meta.* = graphite rock. (Luzi in 1893)

graphitoid   *meta.* Impure, massive graphite-rich metamorphic rock; will burn in the Bunsen flame. (Sauer in 1885)

graphocite   *meta.* End product of coal metamorphism, composed mainly of graphitic carbon; metamorphosed anthracite. [graphite + anthracite]

graphophyre   *ig. pluton.* Igneous rock having a granophyric (graphic intergrowths) groundmass in which the components are of megascopic size. Contrast with graphiphyre. [Gk. *graphein*, to write; alluding to the megascopic graphic intergrowths] (Cross and others in 1906)

gravel   *sed.* Unconsolidated accumulations of rounded to subangular rock (and mineral) fragments coarser than sand; these include: granule g., pebble g., cobble g., boulder g. [O.F. *gravelle*, dim. of *grave, greve*, sand, gravel; originally from Celtic]

gravel, alluvial fan   *sed.* Alluvial fan accumulations; of vast extent, esp. in arid regions of bold relief; when cemented they become fanglomerate.

gravel, bound   *sed.* Hard, often lenticular, cemented mass of gravel and sand occurring in the region of the water table in otherwise unconsolidated sediments; often mistaken for bedrock.

gravel, cement   *sed.* Gravel consolidated by some binding material such as clay, calcite, silica, or iron oxide.

gravel, coal   *sed.* Coal deposit consisting of transported and redeposited coal gravel.

gravel, sandy   *sed.* 1. Unconsolidated sediment containing more particles of gravel size than of sand size; more than 10% sand and less than 10% of all other finer sizes. (Wentworth, 1922)   2. Unconsolidated sediment containing 50-75% sand

Agate (fortification). Brazil. 10 cm. *(Photograph by Carolina Biological Supply Company)*

Coprolite (mammalian). Locality unknown. 3 cm. *(Photograph by Carolina Biological Supply Company)*

Alnoite. Alnö, Sweden. 11.5 cm.

Ditroite (nepheline syenite). Ditro (Ditrau), Transylvania, Rumania. 11.5 cm.

Amber (retinite) with insect inclusions. Dominican Republic. Largest 1.3 cm. *(Photograph by Carolina Biological Supply Company)*

Eclogite. Silberberg, Bodenmais, Bavaria, Germany. 11.5 cm.

Concretions (ferruginous). Locality unknown. 2.5 cm. *(Photograph by Carolina Biological Supply Company)*

Granite (granitite). Kinzigtal, Black Forest, Germany. 11.5 cm.

Granite porphyry. Locality unknown. 11.5 cm.

Lumachelle (fire marble). Bleiburg, Carinthia, Austria. 11 cm.

Hälleflinta. Dannemora, Sweden. 11.5 cm.

Meteorite (octahedrite). Odessa, Texas. Largest, 50 grams. *(Photograph by Carolina Biological Supply Company)*

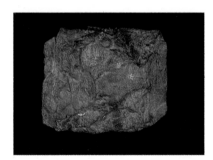

Jaspilite. Ishpeming, Marquette County, Michigan. 11.5 cm.

Monchiquite. Bohemia, Czechoslovakia. 11.5 cm.

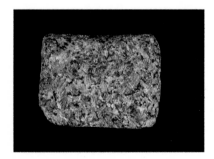

Larvikite. Larvik, Norway. 11.5 cm

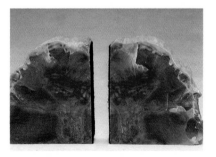

Petrified wood (silicified, jasperized). Arizona. 10 cm. *(Photograph by Carolina Biological Supply Company)*

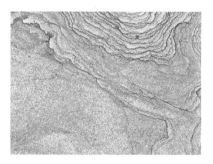

Picture rock (sandstone). Utah. 18 cm. *(Photograph by Carolina Biological Supply Company)*

Schist (actinolite-sericite). Greiner, Tyrol, Austria. 11.5 cm.

Rhyolite (spherulitic or bird's-eye). American West. 7.5 cm. *(Photograph by Carolina Biological Supply Company)*

Schist (kyanite). Baker Mountain, Prince Edward County, Virginia. 15 cm.

Sand crystals (calcite). Badlands of South Dakota. 7.5 cm. *(Photograph by Carolina Biological Supply Company)*

Schist (mica, with garnet metacryst). Stikine River, Wrangell, Alaska. 11 cm.

Sand spike. Locality unknown. 4 cm. *(Photograph by Carolina Biological Supply Company)*

Schist (ottrelite). Ottré, Ardennes, Belgium. 11.5 cm.

Serpentinite (ricolite). Redrock, Grant County, New Mexico. 8.5 cm.

Thunder egg. Priday agate beds, Jefferson County, Oregon. 6.5 cm. *(Photograph by Carolina Biological Supply Company)*

Shale (with fossil beetle). Green River formation, Colorado. *(Photograph by Carolina Biological Supply Company)*

Tinguaite porphyry. Skritin, Bohemia, Czechoslovakia. 11.5 cm.

Spherulite (pyrite). Illinois. 7.5 cm. *(Photograph by Carolina Biological Supply Company)*

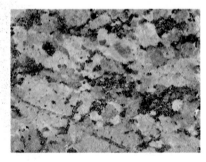

Unakite. Blue Ridge Mountains, Virginia. *(Photograph by Carolina Biological Supply Company)*

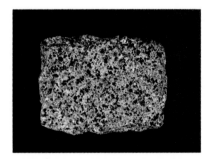

Syenite (biotite-bearing). Unterhof, Black Forest, Germany. 11.5 cm.

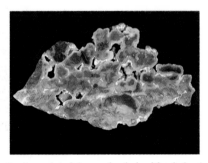

Youngite (brecciated jasper healed with chalcedony). Guernsey, Platte County, Wyoming. 23 cm.

and 25-50% pebbles. (Willman and others in 1942) 3. Unconsolidated sediment containing 30-80% gravel and having a ratio of sand to mud (silt + clay) greater than 9:1. (Folk in 1954)

gravel, volcanic *ig. extru.* Unconsolidated pyroclastic deposit in which the individual clasts are 0.5 cm and smaller but still distinguishable by the unaided eye.

gravelstone *sed.* 1. Conglomerate; cemented gravel. 2. One of the individual rounded rock fragments in a gravel or conglomerate.

gravitite *sed.* Bed of unsorted clastic fragments deposited by a sedimentary flow driven only by gravitational forces; there is no bedding and the paricle arrangement is random. [gravity] (Natland in 1976)

grayband *sed.* Variety of sandstone (flagstone) used for sidewalks.

graystone *ig. extru.* Grayish to greenish compact rock related to basalt, composed of augite and feldspar; anamesite.

graywacke *sed.* Dark gray, firmly indurated arenite composed of quartz, feldspar, and rock particles, cemented by a matrix consisting of fine-grained illite, sericite, chlorite, and silt-sized quartz and feldspars; generally the quartz content is 30-45%; feldspar, 10-50%; mica-clay matrix, greater than 20%; and rock fragments, 5-10%. [G. miners' term *Grau,* gray + *Wacke,* a variously defined rock] (Lasius in 1789)

graywacke, arkosic *sed.* Sandstone containing more feldspar grains than fine-grained rock fragments; feldspar content exceeds 25%; more feldspathic than feldspathic graywacke.

graywacke, calcareous *sed.* Graywacke containing carbonate sand grains, often showing relict organic fabrics; some have granular patches of calcite occupying gaps between silicate grains in the sand framework that may represent recrystallized carbonate sand grains.

graywacke, feldspathic *sed.* Graywacke showing a predominance of quartz and feldspar (with associated augite and hornblende) over lithic fragments; rock derived from felsic igneous rocks and high-rank metamorphics, with a secondary contribution from lower rank metamorphics and sediments; less feldspathic than an arkosic graywacke.

graywacke, high-rank *sed.* Graywacke containing abundant feldspar (at least 20%), usually sodic-plagioclase; a type of feldspathic graywacke. (Krynine in 1945)

graywacke, lithic *sed.* Graywacke containing less than 75% quartz (including chert) and 15-75% detrital clay-rich matrix and having rock fragments (primarily sedimentary or low-rank metamorphic types) in greater abundance than feldspar grains (chiefly sodic-plagioclase); low-rank graywacke. (Pettijohn in 1954)

graywacke, low-rank *sed.* Graywacke containing very little feldspar; essentially the same as lithic graywacke. (Krynine in 1945)

Graywacke (ferruginous). Portland, Connecticut. 12 cm.

graywacke, quartz *sed.* Graywacke containing abundant quartz and chert grains but less than 10% each of feldspar and rock fragments. (Williams, Turner, and Gilbert in 1954)

graywacke, quartzose *sed.* 1. Graywacke that has lost its micaceous and clay constituents through abrasion and thus approaches an orthoquartzite; subgraywacke. (Krynine in 1951) 2. Sandstone containing 50-85% quartz (including chert and quartzite), 15-25% micas and micaceous metamorphic rock fragments, and 0-25% feldspars and feldspathic crystalline rock fragments. (Hubert in 1960)

graywacke, volcanic *ig. extru.* = wacke, volcanic

greenalite rock *sed., meta.* Dull, dark-green, fine-grained rock containing ovoid granules of greenalite in a fine-grained matrix of chert, carbonate minerals (esp. siderite), and ferruginous amphiboles; originally sedimentary, some of these rocks have been subjected to low-grade regional metamorphism; associated with taconite in the Lake Superior iron district.

greenhalghite *ig. extru.* Extrusive igneous rock, related to quartz latite, consisting of sanidine phenocrysts, andesine, quartz, and biotite accompanied by minor apatite, zircon, and glass. [Greenhalgh Mountain, near Silverton, Colorado] (Niggli in 1923)

greensand *sed.* 1. Unconsolidated greenish marine sediment consisting largely of dark greenish grains of glauconite, often mixed with clay or sand (quartz often dominant). See glauconite rock. 2. Poorly cemented sandstone composed of greensand.

greensand of Peru *sed.* Atacamite sand from Peru and from the Atacama desert of northern Chile.

greenschist *meta.* Greenish schistose metamorphic rock containing chlorite, actinolite, and/or epidote.

greenschist, actinolitic *meta.* Greenschist in which actinolite is a principal constituent, often accompanied by chlorite, epidote, albite, magnetite, and others.

greenstone *ig. hypab.* General name for any intrusion of igneous rock in the Coal Measures in Scotland.

greenstone *sed.* Nonoolitic, compact, relatively pure chamosite mudstone interbedded with oolitic ironstone in the Lower Jurassic of Great Britain.

greenstone *meta.* 1. Field name for any compact, dark-green, altered or metamorphosed mafic igneous rock (basalt, gabbro, diabase) that owes its color to chlorite or to combinations of chlorite with actinolite or epidote; albite, magnetite,

and other minerals are often present; when foliated it may be named greenschist. 2. Informal name for the various green jade rocks and for other green rocks: californite, prehnite rock, verdite, and others.

greenstone, New Zealand *meta.* Originally nephrite from New Zealand; now usually serpentinite.

greisen *ig. pluton., autometa.* Feldspar-free rock of granitic or aplitic texture composed of quartz, lithian mica (muscovite, zinnwaldite, lepidolite), and usually, but not necessarily, topaz, tourmaline, fluorite, cassiterite, wolframite, and others; formed by autometamorphism, where an igneous rock (here, granite) is transformed by the action of its own volatile emanations. [G. (Saxony) miners' term, possibly related to G. *Greisstein,* old gray stone; another old spelling is *graisen*]

greisen, apo-carbonate *meta.* = wrigglite. (Govorov in 1958)

greisen, feldspar *ig. pluton.* Obsolete term for rock related to tarantulite and composed of much quartz with lesser feldspars. (Jokély in 1858)

greisen, granite *ig. pluton.* Rock composed essentially of quartz and feldspar (orthoclase, microcline, oligoclase) and a little muscovite. (Jokély in 1858)

greisen, tonalite *ig. pluton.* Obsolete name for a quartz-rich quartz diorite. (Johannsen in 1922)

grennaite *ig. pluton.* Fine-grained, porphyritic nepheline syenite containing catapleiite and eudialyte. [Grenna farm, Norra Kärr complex, Sweden] (Adamson in 1944)

griotte *sed.* Ornamental, fine-grained limestone of red color, often variegated with dashes of purple and spots or streaks of white or brown and containing goniatite (an ammonoid cephalopod) shells. [F. *griotte,* sour cherry]

griquaite *ig., meta.?* Coarse-grained garnet-diopside rock (with or without olivine or phlogopite), occurring as nodular xenoliths in kimberlite pipes and dikes; garnet pyroxenite. [Griqualand, South West Africa] (Beck in 1907)

grit *sed.* Coarse, compact sandstone that can be adapted for grindstones; particles are usually angular where the sharpness may be the result of secondary enlargement by quartz cement; many grits are actually graywackes in composition. [O.Eng. *greot,* sand, dust]

grit, millstone *sed.* Coarse quartzose sandstone used for millstones (for grinding); in England, usually refers to the sandstone at the base of the Carboniferous group.

grit, pea *sed.* Limestone composed of calcareous pisoliths; calcitic pisolite.

grit, schistose   *meta.* Quartz-rich rock derived from an arenaceous sediment (often graywacke) by varying degrees of metamorphism; some are merely cataclastically deformed, while in others, quartz and potassic feldspar have been recrystallized, some albite may have been formed, and the clayey matrix has been recrystallized to form a sericite-chlorite-magnetite aggregate; term frequently applied to Scottish rocks.

gritstone   *sed.* Hard, coarse-grained siliceous sandstone, esp. one suitable for grindstones and millstones; also named gritrock and essentially equivalent to grit.

gröbaite   *ig. pluton.* Light-colored mangerite containing zoned plagioclase (oligoclase-andesine), orthoclase, augite, and biotite, with lesser hornblende, quartz, apatite, and others. [Gröba, north of Meissen, Saxony, Germany] (Reinisch in 1927)

grønlandite   *ig. pluton.* Hypersthene-containing hornblendite in which hornblende exceeds hypersthene in amount. [Dan. *Grønland,* Greenland] (Machatschki in 1927)

grorudite   *ig. diaschist.* Acmite-rich sodic granite, composed of phenocrysts of microcline or microcline-perthite, acmite, and less kataphorite in a tinguaitic groundmass of microcline or microperthite, acmite, and abundant quartz. [Grorud, Norway] (Brögger in 1890)

grospydite   *ig.* Ultramafic rock containing garnet (esp. grossular), plagioclase, pyroxene, and at high pressures, kyanite, rather than spinel or olivine; occurs in nodules in kimberlite pipes of Yakutia, Soviet Union. [grossular + pyroxene + disthene (kyanite)] (Sobolev and others in 1966)

grossouvreite   *sed.* Pulverulent variety of opal from Vierzon, France, formed by the weathering of Cretaceous sediments. Formerly named vierzonite. [Albert Durand de Grossouvre, 1849-1932, French geologist] (Meunier in 1902)

growan   *sed.* Old term used in Cornwall for gravel or rough sand; hard growan is moorstone or granite; soft growan is the same material in a lax and sandy state; growan lode is a tin deposit abounding with this gravel. [Corn. *grow,* gravel, sand]

grunerite-magnetite rock   *meta.* Banded metamorphic rock composed of magnetite and grunerite with subordinate quartz and variable garnet, hypersthene, and hedenbergite.

grünstein   *ig. extru.* Obsolete term supplanted by diabase in 1807; at present, equivalent to greenstone but not used in English. [G. *grün,* green + *Stein,* stone]

grus   *sed.* Fragmental rubble formed from the surface weathering (disintegration with little

chemical decomposition) of granitic rocks; accumulates essentially in situ and corresponds to an unconsolidated basal arkose. Also spelled gruss and grush. [G. *Grus,* grit, fine gravel]

guano   *sed.* Phosphate-rich sediment resulting from the accumulation of excrements and bodies of organisms in regions sufficiently dry to delay bacterial decomposition and inhabited by a great number of animals, mostly migrating birds (e.g., on oceanic islands) or bats (cave accumulations); the material is a complex mixture of calcium phosphates, ammonium phosphate, and ammonium acid phosphate, along with uric acid, sodium acid urate, and various sulfates, oxalates, and other complex organic substances. [Peru. *huanu,* dung]

guardiaite   *ig. extru.* Nepheline-containing latite, consisting of phenocrysts of sanidine (some as mantles around plagioclase) and plagioclase (zoned bytownite to andesine) in a groundmass of nepheline, andesine, augite, biotite, apatite, and others. [Guardia point, Ponza, Ponziane Island, Italy] (Bieber in 1924)

guayaquilite   *sed.* Soft, pale yellow, amorphous fossil resin found in large masses and layers in Ecuador. [Guayaquil, Ecuador] (Johnston in 1838)

guhr   *sed.* Loose, earthy, water-laid deposit found in the cavities or clefts of rocks; mostly white but sometimes red or yellow from a mixture of clay or ochers. [G. *Gur,* from *gähren* or *gären,* to ferment, from its resemblance to buttermilk]

gum animé   *sed.* = animé

gumbo   *sed.* Clay soil yielding a sticky mud when wet; used esp. in the Southern and Western (prairie mud) areas of the United States. [Louisiana Fr. *gombo,* from Bantu]

gumbotil   *sed.* = till, gumbo

gurhofite   *sed.* Dolomite rock, snow-white and subtranslucent, with a conchoidal fracture; resembles opal. Also named Gurhofian dolomite. [Gurhof, in lower Austria] (Karsten in 1807)

gyparenite   *sed.* = gypsarenite. [gypsum + arenite]

gypcrete   *sed.* Gypsum-cemented rock or crust, found in arid climates in some playa-lake beach environments. [gypsum + L. *crescere,* to grow]

gyprock   *sed.* = gypsum rock

gypsarenite   *sed.* Arenite composed almost entirely of detrital grains of gypsum; clastic texture may sometimes be seen macroscopically by differences in color between the grains. Also spelled gyparenite. [gypsum + arenite] (Ogniben in 1957)

gypsite   *sed.* Impure, earthy or sandy gypsum,

found in arid regions as an efflorescent deposit, esp. over outcrops of gypsiferous strata. [gypsum]

gypsum rock   *sed.* Fine- to coarse-grained, bedded or massive evaporite rock, consisting either nearly entirely of gypsum or with gypsum as the major essential component; other minerals that may be present include anhydrite, calcite, dolomite, halite, sulfur, quartz, and clays. Also named gyprock.

gypsum-anhydrite rock   *sed.* Evaporite rock containing both gypsum and anhydrite as essential components; transitional between gypsum rock and anhydrite rock.

gyttja   *sed.* Dark, pulpy, fresh-water anaerobic mud, characterized by abundant organic matter that is more or less determinable; deposited or precipitated in a marsh or in a lake whose waters are rich in nutrients and oxygen. [Sw. *gyttja,* mud, sludge] (Post in 1861)

H

hakutoite   *ig. extru.* Quartz-bearing alkali trachyte from Korea. [Hakuto San, on boundary between North Korea and Manchuria] (Yamanari in 1925)

halilith   *sed.* = halite rock. [halite] (Grabau in 1924)

halite rock   *sed.* Evaporite rock in which the principal constituent is halite; anhydrite, sylvite, and other minerals may be present, and variations in grain size may be extreme. Also named rock salt, halilith, and haloidite.

halite-anhydrite rock   *sed.* Evaporite rock consisting of halite and anhydrite; often with a banded structure consisting of thin dark layers of anhydrite alternating with thicker halite bands; polyhalite, kieserite, and other minerals may be present.

halite-anhydrite-dolomite rock   *sed.* Layered, to almost schistose, rock consisting of halite with anhydrite-dolomite layers folded and crenulated; the structure implies some recrystallization under pressure or movement under overburden.

halite-anhydrite-magnesite rock   *sed.* Evaporite rock consisting of wavy and contorted halite layers that alternate with anhydrite-magnesite laminae; anhydrite and halite pseudomorphs after gypsum are common.

halite-gypsum rock   *sed.* Fine- to coarse-grained evaporite rock containing essential quantities of halite and gypsum; anhydrite and polyhalite also may be present.

halite-sylvite rock   *sed.* Evaporite containing halite and sylvite in which the former is the most abundant; closely related to sylvite-halite rock. Also named sylvinhalite.

halite-sylvite-carnallite rock   *sed.* Evaporite rock consisting of granular halite (often colorless) and granular sylvite (often red from included hematite) with small grains of carnallite.

halitosylvine   *sed.* = sylvite-halite rock. [halite + sylvite]

hälleflinta   *meta.* Dense, porcelainoid hornfels, formed by contact metamorphism of a felsic igneous rock, such as rhyolite, resulting in a banded or blastoporphyritic rock; usually composed of microscopic quartz and feldspar. [Sw. *häll,* flat rock, flat stone + *flinta,* flint, hornstone]

hälleflintgneiss   *meta.* = leptite. [hälleflinta + gneiss]

haloidite   *sed.* = halite rock. [halite] (Wadsworth in 1891)

hamrongite   *ig. dike.* Dark, violet gray, fine-grained lamprophyre; phenocrysts of biotite are in a groundmass of biotite, andesine laths, and minor quartz; quartz kersantite. [Hamrånge Parish, Sweden] (von Eckermann in 1928)

haplite   *ig. diaschist.* = aplite. [Gk. *haploos,* simple] (Fletcher in 1895)

haplophyre   *ig. pluton.* Granite characterized by large quartz and feldspar grains in a mortar structure; occurs in the Alps. [Gk. *haploos,* simple]

harbolite   *sed.* Hard, lustrous variety of asphalt used as a fuel. [Harbol, southeastern Turkey] (Tasman in 1946)

hardebank   *ig. pipe, dike.* Miners' term for the unaltered kimberlite occurring below the zone of blue ground in diamond mines, esp. in South Africa.

hardground   *sed.* At the sea bottom, a horizon of synsedimentary cementation occurring at or just below the sediment surface; this hardened surface is often encrusted, discolored, bored, or solution-ridden.

hardpan   *sed.* 1. Dense subsurface layer of soil whose hardness and relatively slow permeability to water are chiefly due to cementation by silica, iron oxides, calcium carbonate, and/or organic matter; specific varieties include ironpan and limepan.   2. Layer in gravel, usually present a few feet below the surface, cemented by limonite or some similar bond.   3. Popular term used to designate any relatively hard layer of rock difficult to drill or excavate.

harrisite   *ig. hypab.* Granular rock composed of shining black crystals of olivine in a white groundmass composed of anorthite; olivine exceeds anorthite and is oriented at approximately right angles to the cumulate layering of the rock (harrisitic texture). [Harris, Isle of Rum, Scotland] (Harker in 1908)

hartleyite   *sed.* Oil shale from Australia. [Hartley, New South Wales, Australia] (Potonié in 1910) hartsalz *sed.* Sylvite-halite rock containing considerable kieserite (3-50%); kieserite may appear as ellipsoids in a sylvite matrix, and halite may be minor. [G. *hart,* hard + *Salz,* salt]

hartschiefer *meta.* = flinty crush-rock. [G. *hart,* hard + *Schiefer,* slate, schist]

harzburgite *ig. pluton., hypab.* 1. Peridotite composed of olivine and orthopyroxene (enstatite). [Harzburg, Germany] (Rosenbusch in 1887) 2. *IUGS,* plutonic rock with M equal to or greater than 90, 40-90 ol/(ol + opx + cpx), and cpx/(ol + opx + cpx) less than 5.

hatherlite *ig. pluton.* Biotite-hornblende syenite whose feldspar is anorthoclase instead of orthoclase; also named leeuwfonteinite, a name that more accurately relates to its original occurrence. [Hatherley, old powder factory, Transvaal] (Henderson in 1898)

hauynfels *ig. pluton.* = ditroite. [hauyne (a mineral not important in the rock; a misnomer)] (Haidinger in 1861)

hauynitite *ig. extru.* Feldspathoidal extrusive rock composed of hauyne and a mafic other than olivine. [*hauyne,* formerly, hauynite] (Johannsen in 1939)

hauynolith *ig. extru.* Feldspathoidal extrusive rock composed entirely of hauyne. [hauyne] (Johannsen in 1939)

hauynophyre *ig. extru.* Extrusive rock similar to leucitrophyre but in which some of the leucite is replaced by hauyne; composed of nepheline, leucite, hauyne, augite, and in some instances, magnetite, apatite, melilite, and mica. Also spelled hauyn-porphyr. [hauyne] (Abich in 1839)

hawaiite *ig. extru.* Olivine-bearing andesine andesite resembling basalt; generally lacks normative quartz, has a soda potash ratio greater than 2:1 and commonly contains normative and modal olivine; rock intermediate between alkali-olivine basalt and mugearite. [Hawaii] (Iddings in 1913)

haytorite *altered.* Variety of chalcedony formed from the alteration of datolite; from the Haytor iron mines, Devonshire, England. [Haytor mines] (Tripe in 1827)

hedrumite *ig. hypab.* Coarse-grained, light-colored pulaskite porphyry with a trachytic texture; contains tabular feldspar (orthoclase), augite, biotite, less than 5% nepheline, and no quartz. [Hedrum, Norway] (Brögger in 1890)

helenite *sed.* Light to dark yellow variety of ozokerite from Galicia. [Helena shaft, petroleum region of Ropa, Galicia] (Nawratil in 1883)

helictite *sed.* Twisted, wormlike or rootlike, speleothem with a small capillary central canal; growth appears to be random with little or no control by gravity. [Gk. *helix,* spiral]

helictite, antler *sed.* Popular name for a thick, branching helictite. [For resemblance to an antler]

helictite, dendritic *sed.* Branching helictite; nearly equivalent to an antler helictite. [Gk. *dendron,* tree]

heligmite *sed.* Helictite extending upward from a cave floor or from another speleothem. [helictite + stalagmite]

heliotrope *sed.* Dark green chalcedony containing spots or patches of red jasper; bloodstone is similar but consists of green jasper with red jasper spots. [Gk. *helios,* sun + *tropos,* turn; alluding to the red appearance obtained when immersed in water in sunlight]

helsinkite *ig. dike.* Equigranular, medium-grained rock whose chief constituents are sodic-plagioclase and red epidote (red from hematite?); igneous origin of the rock has been questioned. [Helsinki, Finland] (Laitakari in 1918)

hemachate *sed.* Light-colored agate spotted with red jasper. Also spelled haemachates. [Gk. *haimatitis,* blood-red (or the mineral hematite) + G. *Achat,* agate]

hematite rock *sed.* Hematite-rich sedimentary rock containing 10% or more of iron (or 15% or more Fe_2O_3); often fossiliferous or oolitic and may contain quartz, calcite, siderite, and iron silicates.

hematite rock, fossiliferous *sed.* Sedimentary rock consisting of flattened and elongated shell fragments in subparallel arrangement, which are coated, or replaced, by hematite and are cemented by calcite, hematite, or rarely by quartz (including chalcedony); illustrated by the Clinton (Silurian) ores of the Appalachian region, United States.

hematite rock, oolitic *sed.* Sedimentary rock consisting of hematite ooliths in a cement of hematite and calcite or, more rarely, of quartz and siderite; ooliths are rounded or flattened and normally are concentric.

hematoconite *sed.* Limestone colored by ferric oxide (hematite). Also named haematoconite. [Gk. *haimatitis,* blood-red (or the mineral hematite) + *konis,* powder] (Hausmann in 1847)

hemiopal *sed.* = semiopal. [Gk. *hemi,* half + opal] (Francke in 1890)

hemipelagite *sed.* Sediments formed on the sea floor by the slow accumulation of biogenic materials and fine-grained terrigenous particles deposited on top of the pelitic interval of a turbidite. [Gk. *hemi,* half + *pelagos,* sea] hemithrene   *ig. pluton.* Diorite containing a large amount of calcite, presumably as an alteration product; term esp. used in France. [Possibly L. (from Gk.) *hemi,* half + *threnus,* lamentation] (Brongniart in 1813)

heptorite   *ig. dike.* Dark lamprophyre composed of hornblende (barkevikitic), olivine, titanaugite, and hauyne phenocrysts in a pitchy groundmass. [Gk. *hepta,* seven + *oros,* mountain; for locality at Siebengebirge, Rheinland, Germany] (Busz in 1904)

hermatolith   *sed.* = reef rock. [Gk. *hermato,* sunken reef]

heronite   *ig. dike.* Dark red, felsitic rock composed of analcime, orthoclase, and minor plagioclase and aegirine; apparently an altered tinguaite. [Heron Bay, Lake Superior, Canada] (Coleman in 1899)

heumite   *ig. dike.* Dark brownish black, fine-grained rock characterized by a granular texture and composed of feldspar (sodium orthoclase or sodium microcline), barkevikite, abundant dark-brown biotite, and small amounts of nepheline, sodalite, and others. [Heum, southern Norway] (Brögger in 1898)

hexahedrite   *meteor.* Iron meteorite (generally with 4-6% Ni) normally consisting of large crystals of kamacite, which are cubic; these crystals have cleavage parallel to the faces of the cube and, after a polished surface is etched, they show fine lines (Neumann lines) as a result of twinning on a trapezohedron face. [hexahedron, synonym for cube] (Rose, date uncertain)

highwoodite   *ig. pluton.* Dark-colored monzonitic rock consisting of sodium orthoclase, labradorite, pyroxene, biotite, iron oxide minerals, apatite, and possibly small amounts of nepheline. [Highwood Mountains, Montana] (Johannsen in 1938)

hilairite   *ig. pluton.* Porphyritic rock showing large phenocrysts of sodic-plagioclase and nepheline, and lesser amounts of sodalite, aegirine, and eudialyte in a trachytic groundmass of aegirine, nepheline, sodic-plagioclase, orthoclase, and aegirine; generally the rock contains more sodic-plagioclase than orthoclase, more feldspathoid than feldspar, and the feldspathoid is nepheline. Not to be confused with the mineral of the same name. [Mt. St. Hilaire, Quebec, Canada] (Johannsen in 1938)

hilairite, sodalite   *ig. pluton.* Hilairite containing, in addition to the major feldspathoid nepheline, considerable sodalite. (Johannsen in 1938)

hircite   *sed.* Yellowish brown, amorphous hydrocarbon found in Burma; emits a bad smell on being burned. [Gk. *hircus,* goat] (Piddington in 1855)

hirnantite   *ig. extru., altered.* Grayish to greenish gray keratophyre, composed of sodic plagioclase (albitized andesine) with interstitial chlorite, clinopyroxene, and small amounts of quartz and others. Also spelled hirnandite. [Hirnant, northern Wales] (Travis in 1915)

högbomitite   *ig. pluton.* Magnetitite containing considerable högbomite. The name högbomite magnetitite is more appropriate. [högbomite]

högbomitite, magnetite   *ig. pluton.* = högbomite magnetitite (see magnetitite). (Gavelin in 1916)

hollaite   *ig. pluton.* Hybrid carbonatite produced by the interaction between sövite and rocks of the ijolite-melteigite series; major minerals are green pyroxene (55%), calcite (16%), and nepheline (10%). [Holla Church, Fen complex, Norway] (Brögger in 1921)

holmite   *ig. dike.* Grayish green, fine-grained, nonporphyritic rock similar to monchiquite except that the groundmass consists almost wholly of melilite instead of analcime; pyroxene-rich alnoite. [Holm, Orkney Islands, Scotland] (Johannsen in 1938)

holosiderite   *meteor.* Meteorite consisting of metallic iron without stony matter. [Gk. *holos,* whole + *sideros,* iron] (Daubrée in 1867)

holyokeite   *ig. dike.* Fine-grained, amygdaloidal albitite with an ophitic texture. [Holyoke trap, Mt. Tom, Massachusetts] (Emerson in 1902)

hooibergite   *ig. pluton.* Dark-colored, hornblende-orthoclase gabbro with green hornblende (75%), labradorite (12%), orthoclase (7%), and others. [Hooiberg, Aruba Island, Lesser Antilles] (Westermann in 1932)

hörmannsite   *ig. pluton.* Igneous rock apparently formed by the resorption of marble, occurring in crystalline granite; consists of andesine (37%), microcline (20%), chloritized biotite (24%), and calcite (17%). [Hörmanns, northwestern Waldviertel, Austria] (Ostadal in 1935)

hornberg   *ig. pluton.* Old name for rock consisting only of a mixture of quartz and mica; essentially the same as esmeraldite. (Brückmann in 1773)

hornblendegarbenschiefer   *meta.* Kind of hornblende schist containing large poikiloblastic hornblende blades, including quartz, feldspar, and magnetite and flattened in decussate or stellate patterns on foliation planes. [hornblende + G. *garben,* to bundle + *Schiefer,* schist]

hornblendite   *ig. pluton.* 1. Igneous rock composed almost entirely of hornblende. [hornblende] (Dana in 1875) 2. *IUGS,* plutonic rock with M equal to or greater than 90 and hbl/(hbl + px + ol) greater than 90.

hornblendite, olivine   *ig. pluton.* 1. Igneous rock composed of hornblende with olivine as a characterizing accessory.   2. *IUGS,* plutonic rock with M equal to or greater than 90 and ol/(ol + hbl + px) less than 90.

hornblendite, olivine-pyroxene   *ig. pluton.* 1. Igneous rock composed of hornblende with olivine and pyroxene present as characterizing accessory minerals.   2. *IUGS,* plutonic rock with M equal to or greater than 90, 5-40 ol/(ol + hbl + px), and amphibole more abundant than pyroxene.

hornblendite, osannite   *ig. pluton.* Perknite composed of 97% osannite (a variety of riebeckite amphibole) with magnetite; the same as pedrosite; occurs in Portugal. (Osann in 1923)

hornblendite, pyroxene   *ig. pluton.* 1. Igneous rock composed of hornblende with pyroxene as a characterizing accessory.   2. *IUGS,* plutonic rock with M equal to or greater than 90, ol/(ol + hbl + px) less than 5, and 50-90 hbl/(px + hbl).

hornfels   *meta.* Dark, fine-grained, contact-metamorphic rock composed of a mosaic of equidimensional grains without preferred orientation. Traditionally, term refers to a pelitic hornfels. [G. *Horn,* horn (alluding to luster) + *Fels,* rock]

hornfels, andalusite   *meta.* Pelitic hornfels containing andalusite with quartz, feldspar (potassic and plagioclase), mica (biotite), and graphite; andalusite occurs as network porphyroblasts stuffed with quartz, biotite, and graphite.

hornfels, arenaceous   *meta.* Contact metamorphic rock derived from feldspathic sandstone, arkose, and felsic volcanic rocks and containing granoblastic quartz, feldspars, and biotite; usually fine-grained and nonfoliated.

hornfels, basic   *meta.* Hornfels derived from high-temperature contact metamorphism of rocks of the basalt and andesite families; dense, dark, and composed of diopside, labradorite, hypersthene, and accessory magnetite, apatite, and titanite. Also named mafic hornfels.

hornfels, calc-silicate   *meta.* Hornfels derived from the contact metamorphism of argillaceous limestone or dolomite and consisting entirely of calcium-bearing silicates with little or no calcite; commonly fine-grained, granoblastic and containing diopside, hedenbergite, andradite-grossular, calcic-plagioclase, vesuvianite, wollastonite, and others. Essentially the same as calc-flinta.

hornfels, calcic   *meta.* Term sometimes applied to marble formed by high-grade contact metamorphism of initially pure calcite rocks.

hornfels, cordierite   *meta.* Pelitic hornfels containing cordierite with quartz, feldspar (potassic and plagioclase), mica (biotite), graphite, and andalusite; cordierite forms highly poikilitic porphyroblasts that may show sector twinning.

hornfels, dolomitic   *meta.* Term sometimes applied to marble formed by the high-grade contact metamorphism of initially pure dolomite rocks.

hornfels, hypersthene   *meta.* Basic hornfels rich in hypersthene.

Hornfels. Bremerteich, Harz Mountains, Germany. 12 cm.

hornfels, mafic   *meta.* = hornfels, basic

hornfels, magnesian   *meta.* General term sometimes applied to magnesian-rich contact metamorphic rocks; they may be products of local magnesium metasomatism and contain magnesian amphiboles (anthophyllite or cummingtonite), cordierite, biotite, almandine, and others, or they may be derived from the metamorphism of serpentinite and contain olivine, enstatite, spinel, clinochlore, and others.

hornfels, micaceous   *meta.* Pelitic hornfels that contains characterizing mica; biotite is very common, and muscovite may occur, but it usually indicates a lower-grade of metamorphism.

hornfels, pelitic   *meta.* Dark, fine-grained, usually massive hornfels, consisting mainly of quartz, mica (biotite, muscovite), feldspars (orthoclase, microcline, plagioclase), graphite, cordierite, and andalusite; from the contact metamorphism of clays, shales, and graywackes in the innermost zone of batholithic contact aureoles. Often simply named hornfels.

hornfels, pyroxene   *meta.* Very high temperature hornfels, formed in the inner part of a contact aureole (temperature in excess of about 550° C, low pressure), containing pyroxenes such as diopside and hypersthene, with plagioclase.

hornfels, quartzo-feldspathic   *meta.* Term sometimes applied to a contact metamorphic rock derived from parent materials composed of quartz, plagioclase, potassic feldspar (sandstones, rhyolite, dacite) and consisting of a granoblastic mosaic of quartz and feldspars with minor biotite, cordierite, andalusite, and others.

hornfels, sillimanite   *meta.* Pelitic hornfels containing sillimanite, whose presence indicates a very high-temperature environment close to the batholith that caused the contact metamorphism.

hornstein   *sed., meta.* = hornstone. [G. *Horn,* horn (alluding to luster) + *Stein,* stone]

hornstone   *sed.* General term for compact, tough, siliceous rock having a splintery, subconchoidal, or conchoidal fracture, e.g., chert. [horn; alluding to its luster]

hornstone   *meta.* 1. = hornfels.   2. = sandstone, fritted

hornstone, andalusite   *meta.* = hornfels, andalusite

hortite   *ig. pluton.* Dark-colored, hybrid rock close to syenite; probably derived from gabbro by the assimilation of limestone. [Hortavaer, Norway] (Vogt in 1916)

hovlandite   *ig. extru., altered.* Poorly defined basic volcanic rock containing a zeolitelike mineral as white needles distributed through a green matrix with black crystals, from near Grand Marais, Cook County, Minnesota. [Hovland, Minnesota]

howardite   *meteor.* = achondrite, pyroxene-plagioclase. [Luke Howard, 1772-1864, English chemist and meteorologist] (Rose in 1863)

howlite rock   *sed.* Nodules of white, fine-grained, howlite, weighing up to 100 pounds, occurring in sedimentary beds, either with gypsum and anhydrite (as in Nova Scotia), or with borate rocks (as in California). A variety is named winkworthite.

hrafntinna   *ig. extru.* = obsidian. [Icel. *hrafn,* raven (for black color) + *tinna,* flint]

hsiu yen   *sed.* Chinese name for a coarse, green and white, jasper, often sold as jade.

hudsonite   *ig. pluton.* = cortlandtite. [Hudson River, New York] (Cohen in 1885)

humanthracite   *sed., meta.* Humic coal of anthracitic rank. [L. *humus,* ground + anthracite]

humanthracon   *sed.* Humic coal of bituminous rank. [L. *humus,* ground + Gk. *anthrax,* coal + *konis,* powder]

huminite   *ig.? dike.* Oxidized bitumen, resembling brown coal but occurring in a granite pegmatite vein in Sweden. [Possibly L. *humus,* ground]

humite   *sed.* = coal, humic. [L. *humus,* ground]

humocoll   *sed.* Humic material of the rank of peat; this rank lies between humopel and humodil. [Possibly humic + Gk. *kolla,* glue] (Heim and Potonié in 1932)

humodil   *sed.* Humic coal of lignite rank. [Possibly humic + Gk. *odme,* stink] (Heim and Potonié in 1932)

humodite   *sed.* Humic coal of subbituminous rank. [Possibly humic] (Heim and Potonié in 1932)

humolite   *sed.* = coal, humic. [L. *humus,* ground]

humolith   *sed.* = coal, humic. [L. *humus,* ground]

humonigritite   *sed.* Variety of nigritite occurring in sediments. [L. *humus,* ground + nigritite] (Potonié in 1950)

humopel   *sed.* Organic matter from which humic coals can be derived; the lowest rank in coal metamorphism. [Possibly humic + Gk. *pelos,* mud] (Heim and Potonié in 1932)

humulite   *sed.* = coal, humic. [L. *humus,* ground]

humulith   *sed.* = coal, humic. [L. *humus,* ground]

humus   *sed.* Dark-colored, fresh, decaying, or decayed organic matter occurring mainly in the upper part of soil profiles; decay products are mainly humic acids that can leach rock fragments and clays. [L. *humus,* ground]

hungarite *ig. extru.* Obsolete name for hornblende andesite. [Hungary] (Lang in 1877)

hurumite *ig. hypab.* Dark-colored, potassium-rich windsorite, containing andesine (37%), orthoclase (29%), biotite (12%), and quartz (11%). [Hurum, Oslo region, Norway] (Brögger in 1931)

husebyite *ig. pluton.* Plagioclase-bearing nepheline syenite whose dark mineral is pyroxene. [Huseby, Aker, Oslo region, Norway] (Brögger in 1933)

hyalite *sed., hydrotherm.* Variety of opal that is colorless (rarely light green or bluish) and transparent and usually occurs as small globules or botryoidal incrustations on rocks or other minerals; looks like drops of melted glass and sometimes exhibits a bright-green fluorescence when exposed to ultraviolet radiation. [Gk. *hualos,* glass] (Werner, before 1800)

hyaloandesite *ig. extru.* Andesite that is entirely or partly glassy; often brownish or yellowish with phenocrysts of plagioclase and mafics. [Gk. *hualos,* glass + andesite]

hyalobasalt *ig. extru.* Basalt glass containing very little crystallized material other than crystallites; some rarely contain well-formed phenocrysts in a groundmass almost entirely glass. Also named hyalomelane and tachylyte. [Gk. *hualos,* glass + basalt]

hyaloclastite *ig. extru.* Vesicular lava fragments ranging from a few millimeters to centimeters in diameter and generally consisting of flakes and chips of glass; the material forms as lava is extruded directly into shallow water, where there is rapid chilling, lava fragmentation, and exsolution of gases. Also named aquagene tuff. [Gk. *hualos,* glass + *klastos,* broken]

hyalodacite *ig. extru.* Dacite glass; it may contain phenocrysts of andesine, quartz, hornblende, biotite, and others. [Gk. *hualos,* glass + dacite]

hyalodiabase *ig. hypab.* Glassy diabase; usually occurs as a border phase of a diabasic intrusion, e.g., along the margins of a dike. [Gk. *hualos,* glass + diabase]

hyalomelane *ig. extru.* Old term for hyalobasalt; originally said to differ from tachylyte (another hyalobasalt) in being insoluble in acids. [Gk. *hualos,* glass + *melas,* black] (Hausmann in 1847)

hyalomicte *ig. pluton., autometa.* = greisen. [Gk. *hualos,* glass + *miktos,* mixed; possibly alluding to glassy quartz] (Brongniart in 1813)

hyalomylonite *meta.* Glassy mylonite formed by the fusion of granite, arkose, or similar rock, by frictional heat in zones of intense differential movement. Compare with flinty crush-rock. [Gk. *hualos,* glassy + mylonite]

hyalopsite *ig. extru.* Obsolete name for obsidian. [Possibly Gk. *hualos,* glass + *psilos,* mere, bare] (Gümbel in 1886)

hyalotourmalite *ig. pluton., autometa.* = tourmalite. [Gk. *hualos,* glass (from quartz content) + tourmaline] (Daubrée in 1841)

hyblite *ig., weathered.* Variety of palagonite from Sicily. [Mt. Hybla, Sicily]

hybrid rock *ig.* Igneous rock whose composition is the result of contamination of the magma by assimilation of materials from the wall rock.

hydroclastic rock *sed.* Clastic sedimentary rock deposited by the agency of water or one in which the clasts were broken by wave or current action. [Gk. *hudor,* water + *klastos,* broken]

hydrolite *sed.* 1. = enhydros. 2. = sinter, siliceous. [Gk. *hudor,* water] (Mackenzie, pub. by Allan in 1819)

hydrolith *sed.* 1. Rock chemically precipitated from solution in water, such as halite rock or gypsum rock; evaporite. [Gk. *hudor,* water] (Grabau in 1904) 2. Hydroclastic rock consisting of carbonate fragments. (Bissell and Chilingar in 1967)

hydrophane *sed.* Dehydrated yellowish, brownish, or greenish variety of common opal that, when immersed in water, becomes more translucent or transparent and sometimes may exhibit a play of colors. [Gk. *hudor,* water + *phanin,* to appear]

hydrotachylyte *ig. extru.* Hyalobasalt (tachylyte) containing as much as 13% water. [Gk. *hudor,* water + tachylyte]

hyperite *ig. pluton.* Rock composed of orthopyroxene (usually hypersthene), clinopyroxene (augite or diallage), calcic-plagioclase, and at times olivine; intermediate in composition between gabbro and norite. [Possibly hypersthene] (Swedish term used before 1849)

hypersthenfels *ig. pluton.* = norite. [hypersthene] (Rose in 1835)

hypersthenite *ig. pluton.* Orthopyroxenite composed almost entirely of hypersthene. [hypersthene] (Williams in 1890)

hypersthenite, hornblende-bearing *ig. pluton.* Hypersthenite containing hornblende (as much as 40%), magnetite, and other minor minerals, from Madras, India. (Washington in 1916) hypoxenolith   *ig., sed., meta.* Xenolith derived from a source more remote than the adjacent wall-rock. [Gk. *hupo,* under + xenolith] (Goodspeed in 1947)

hysterobase   *ig. dike.* Rock related to diabase but consisting of plagioclase, quartz, brown biotite, and brown hornblende (sometimes replacing augite). [Gk. *husteron,* later + diabase; alluding to its late crystallization from magma and its similarity to diabase] (Lassen in 1888)

hysterogenite   *ig., sed.?* Last rock to be formed in a series of rock formations. [Gk. *husteron,* later + *genes,* born]

I

ice *sed., meta.* Crystallized water; solid condition assumed by water at or below 32° F or 0° C; may form from the freezing of liquid water, from condensation of water vapor into ice crystals (snow), or by recrystallization or compaction of fallen snow. [O. Eng. *is*]

ice, agglomerate *sed.* Aggregate of ice formed by the freezing of floating ice fragments. [L. *ad,* to + *glomerare,* to form into a ball; and similarity to conglomerate]

ice, blue *meta.* Unweathered, nonbubbly, coarse-grained glacial ice, often as bands, having a slightly bluish or greenish color.

icecrete *sed.* = concrete, ice. [ice + L. *crescere,* to grow]

icelandite *ig. extru.* Intermediate lava that, when compared to calc-alkaline andesite, is low in aluminium, high in iron, and has fewer mafic phenocrysts; occurs at Thingmuli volcano, Iceland. [Iceland] (Carmichael in 1964)

idiogenite *ig., sed., meta.* Rock or mineral deposit contemporaneous in origin with the wall rock. [Gk. *idios,* same + *genes,* born] (Posepny in 1893)

ignimbrite *ig. extru.* Volcanic rock formed by the widespread deposition and consolidation of ash flows and nuées ardentes; originally the term implied dense welding or fusing of the shards, but there is no longer such a restriction; term now includes welded tuff as well as sillar. [L. *ignis,* fire + *imber,* shower, rain] (Marshall in 1935)

ignispumite *ig. extru.* Volcanic rhyolite characterized by lenses and bands; believed to have been deposited as a foamy lava and to be transitional with true ignimbrite. [L. *ignis,* fire + pumice]

ijolite *ig. pluton.* 1. Group of feldspar-free phaneritic rocks containing nepheline and 30-60% mafic minerals (usually clinopyroxene). 2. *IUGS,* plutonic rock in which F is 60-100, M is 30-70, and sodium exceeds potassium. [Iiwaare, Iijarve, Iijoki, parish Ijo, localities of the origi-nal rock in Finland] (Ramsay and Berghell in 1891)

ijolite, carbonatitic *ig. hypab.* General term for a rock close to carbonatite; contains calcite and silicate minerals but silicates are more abundant.

ijussite *ig. pluton.* Teschenite composed of titanaugite and barkevikite with smaller amounts of bytownite, anorthoclase, and analcime. [Ijuss River, Siberia, Soviet Union] (Rakovski in 1911)

ilmenitite *ig. hypab.* Igneous rock composed almost exclusively of ilmenite. Varieties, characterized by accessory minerals, include the following: apatite i., augite i., bronzite i., hypersthene i., olivine i. [ilmenite] (Kolderup in 1897)

ilzite *ig. pluton.* Orthoclase-bearing biotite malchite. [Possibly Ilz stream or Ilzstadt, near Tittling bei Passau, Germany] (Frentzel in 1911)

imandrite *ig. pluton.* Hybrid rock composed of quartz and albite with minor chlorite, biotite, and rutile; formed by the interaction of a nepheline syenite magma with graywacke. [Imandra Sea, Umptek, Kola, Soviet Union] (Ramsay and Hackman in 1894)

Imatra stone *sed.* = marlekor. [Imatra, Finland] (Hoffmann in 1837)

impactite *impact.* General term given to glassy rocks associated with craters presumably produced by meteorite (or other) impact; most are highly vesicular, contain some crystalline inclusions, and have highly diversified compositions compared to tektites; well illustrated by materials found around the Wabar (in Arabia), Henbury (in Australia), and Canyon Diablo (in Arizona) craters. [meteorite impact; alluding to mode of origin]

impactite, Ries *impact.* = suevite. [Ries crater near Nördlingen, Schwaben, Germany]

impactite, Wabar *impact.* Silica-glass occurring in large quantity around the meteorite crater at Wabar in Arabia. Also named Wabar glass.

impsonite *sed.* Dull black, asphaltic pyrobitumen, similar to albertite but differing in being almost insoluble in turpentine; formed from the metamorphism of petroleum. [Impson Valley, Oklahoma] (Taff in 1899)

incretion *sed.* 1. Type of concretion whose growth has been directed inward from without. 2. Cylindrical concretion with a hollow core, e.g., rhizocretion. [L. *in,* in + *crescere,* to grow, and influence of concretion] (Todd in 1903)

indianaite *sed.* Impure, white, porcelainlike halloysite clay that occurs in beds 4-10 ft thick at Lawrence County, Indiana. [Indiana] (Cox in 1874)

indochinite *tekt.* General name applied to tektites from Indochina, including Laos, Cambodia, Thailand, and elsewhere.

indomalaysianite *tekt.* General name for tektites from the Far East, including indochinites, malaysianites, and others. (Beyer in 1933)

ingenite *ig., meta.* Obsolete general name for any rock originating below the earth's surface; applies to most igneous and metamorphic rocks. [L. *in,* in + Gk. *genes,* born] (Forbes in 1867)

inninmorite *ig. extru.* Porphyritic rock containing phenocrysts of plagioclase (labradorite to anorthite) and augite, in a groundmass of more sodic plagioclase, augite, and abundant glass. [Inninmore, Morven, Scotland] (Thomas and Bailey in 1915)

inolite *sed.* Obsolete name for tufa (calcareous tufa, calcareous sinter). [Gk. *inos,* muscle; for fibrous structure] (Gallitzin in 1801)

intercretion *sed.* Concretion growing by accretion (on the exterior) and by irregular and interstitial addition, causing a circumferential expansion and cracking of the interior of the concretion; e.g., septaria. [L. *inter,* between + *crescere,* to grow, and influence of concretion] (Todd in 1903)

intermediate rock *ig.* Collective name for igneous rocks between acidic (acidite) and basic (basite), generally having a silica content of 50-65% by weight; e.g., syenite and diorite.

intraclast *sed.* Component of limestone representing rock fragments of penecontemporaneous, generally weakly cemented, carbonate sediment that have been broken up and redeposited as clasts in a new framework. [L. *intra,* within + clast] (Folk in 1959)

intramicrite *sed.* Limestone composed of intraclasts (fragments of lithified or partly lithified sediment) in a matrix of micrite (carbonate mud). [L. *intra,* within + micrite] (Folk in 1959)

intramicrudite *sed.* Intramicrite limestone containing gravel-sized intraclasts. [intramicrite + rudite]

intrasparite *sed.* Limestone composed of intraclasts (fragments of lithified or partly lithified sediment) cemented by sparite, clear equant calcite. [L. *intra,* within + sparite] (Folk in 1959)

intrasparrudite *sed.* Intrasparite containing gravel-sized intraclasts. [intrasparite + rudite]

invernite *ig.* Granitic rock containing phenocrysts of orthoclase and minor plagioclase in a groundmass consisting of the same feldspars with sparse hornblende, mica, and interstitial quartz. Also spelled ivernite. [Iverness, County Limerick, Ireland] (Watt, 1895)

iolanthite *sed.?* Local name for a banded, reddish jasperlike material from Crooked River, Crook County, Oregon. [Possibly Iolanthe, a Gilbert and Sullivan operetta, 1882] (Maguire, pub. by Sterrett in 1914)

ionite *sed.* Resinous, brownish-yellow fossil hydrocarbon associated with lignite in Amador County, California. [Ione Valley, Amador County, California] (Purnell in 1878)

iron *meteor.* = iron meteorite

iron carbonate rock *sed.* = siderite rock

iron carbonate rock, cherty *sed.* = siderite rock, cherty

iron meteorite *meteor.* Second largest group of meteorites, composed of iron and nickel-iron alloys; major types include hexahedrite, octahedrite, and nickel-rich ataxite.

iron ore, Clinton *sed.* Name applied to Silurian sedimentary rocks rich in hematite. Major types include: oolitic ore, hematite ooliths (1-2 mm across) enclosed in hematite and calcite matrix; fossil ore, fossil fragments coated and partly replaced by hematite and enclosed in hematite and calcite; and flaxseed ore, flattened hematite concretions surrounded by hematite mud and replaced fossil fragments. Extensive deposits outcrop across Wisconsin and New York and south to Alabama. Also named hematite rock, fossiliferous hematite rock, oolitic hematite rock, and flaxseed ore. [Clinton, New York]

iron ore, micaceous *meta.* Soft, soapy or oily, variety of hematite schist; resembles graphite or mica schist in appearance.

iron oxide rock *sed.* General term for sedimentary rocks whose principal minerals are hematite or limonite (goethite) or both; some types include hematite rock, limonite rock, bog iron ore, gossan.

iron silicate rock *sed.* General term for sedimentary rocks containing iron silicates like chamosite and glauconite; included are chamosite mudstone, glauconite rock, greenalite rock, and others.

iron silicate rock, hydrous *meta.* General term for low-grade, regional metamorphic rocks, containing various combinations and amounts of greenalite, minnesotaite, and stilpnomelane (iron silicate minerals) with fine-grained quartz, magnetite, hematite, and siderite; normally fine-grained, banded or finely laminated. Taconite is a variety.

iron sulfide rock *sed.* General term applied to sedimentary rocks (limestones, shales, sandstones) containing authigenic pyrite or marcasite; these may appear as the major constituent of the rock, but this is rare. Also see marcasite rock and pyrite rock.

ironpan *sed.* Hardpan in soil in which iron oxides are the principal cementing agents. Also spelled iron pan.

ironstone *sed.* General term for iron-bearing sedimentary rocks displaying a great variety of mineral compositions and textures; some types are iron oxide rock, iron silicate rock, iron sulfide rock, and siderite rock.

ironstone, ball *sed.* Any sedimentary rock containing large argillaceous concretions of ironstone.

ironstone, banded *sed.* South African term for an iron formation consisting of iron oxides and chert occurring in prominent layers or bands of brown or red and black.

ironstone, blackband *sed.* Dark variety of clay ironstone (siderite) containing sufficient carbonaceous matter (10-20%) to make it self-calcining, without the addition of fuel.

ironstone, chamosite *sed.* = mudstone, chamosite

ironstone, clay *sed.* Dark gray to brown, fine-grained, concretion or relatively continuous irregular thin bed, composed of argillaceous material (up to 30%) and siderite; rarely, the term is applied to an argillaceous rock containing hematite or limonite.

ironstone, glauconitic *sed.* = greensand

isenite *ig. extru.* Trachyandesite; apatite contained in the original specimens was misidentified as hauyne. [Eis (L. *Isena*), stream in Westerwald, Germany] (Bertels in 1874)

isopyre *sed.* Impure opal, grayish black or velvet black, occasionally spotted red, like heliotrope; original specimens from St. Just, Cornwall, England. [Gk. *isos,* like + *pur,* fire; the fused material closely resembled the original material] (Haidinger in 1827)

issite *ig. dike.* Dark-colored, pure hornblende dike rock (with subordinate green pyroxene and labradorite) occurring in platinum-bearing dunite, southern Urals. [Isse River, Penza oblast, Soviet Union] (Duparc in 1910)

itabirite *meta.* Schistose rock composed essentially of quartz grains and scales of specular hematite (with lesser magnetite or martite). [Itabira, Brazil] (Eschwege in 1822)

itacolumite *meta., sed.?* Light-colored, micaceous sandstone or schistose quartzite in which, under the microscope, the majority of quartz grains show very irregular shapes that are strongly interlocked but separated from one another by voids, so that the rock exhibits flexibility when split or cut into thin slabs; mica, chlorite, and talc may be present but do not play a direct role in the flexibility. Also named articulite, flexible micaceous quartzite, and flexible sandstone. [Mt. Itacolumi, Minas Gerais, Brazil] (Humboldt, pub. by Eschwege in 1822)

italite *ig. pluton.* 1. White or pale yellowish foidite in which leucite (crystals 3-5 mm diameter) is the major constituent associated with very minor augite, melanite, biotite, and magnetite. [Italy] (Washington in 1920) 2. *IUGS,* plutonic rock in which F is 60-100, M is 10 or less, and potassium exceeds sodium.

itsindrite *ig. dike.* Potassium-rich nepheline syenite dike rock, composed of nonperthitic microcline

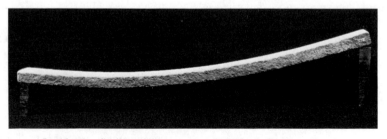

Itacolumite (flexible micaceous quartzite). North Carolina. 40 cm.

and nepheline graphically intergrown, biotite, aegirine, and zoned melanite. [Itsindra Valley, Madagascar] (Lacroix in 1922/23)

ivernite *ig.* = invernite

ivoirite *ig., meta.?* Obsolete varietal name for charnockite.

ivorite *tekt.* Black tektite from the Ivory Coast, western Africa. [Ivory Coast]

ixolyte *sed.* Amorphous, hyacinth red, greasy hydrocarbon that softens at 76° C; found at Oberhart, near Gloggnitz, Austria. Also spelled ixolite. [Gk. *ixos,* gluey (like birdlime) + *lutos,* loosed, dissolved] (Haidinger in 1842)

J

jacupirangite   *ig. pluton.* Group of rocks, some of which are part of the ijolite series, consisting at times of pure magnetite (magnetitite), or of magnetite with accessory pyroxene, or of pyroxene with accessory magnetite, or of pyroxene and nepheline with biotite and olivine in greater or lesser quantity. [Jacupiranga district, São Paulo, Brazil] (Derby in 1891)

jacutinga   *sed., meta.* Term used in Brazil for disaggregated, powdery itabirite; or variegated thin-bedded, high-grade, hematite iron ores associated with, and often forming the matrix of, gold ore. [*Pipile jacutinga*, a Brazilian bird] (Heusser and Claraz in 1859)

jade   *meta.* 1. Hard, fine-grained, tough rock composed either of the pyroxene mineral jadeite or of the amphibole mineral nephrite (a variety of tremolite-actinolite) and usually having a color ranging from dark green to greenish-white, although other colors are known. [O. Sp. *piedra de yjada,* stone of the side; the stone was supposed to cure side pains]   2. Term incorrectly applied to various other green, fine-grained materials suitable for carving and resembling true jade in appearance: agalmatolite, garnet (especially grossular and hydrogrossular), marble (dyed green), pectolite rock (larimar), prehnite rock, saussurite, serpentinite, sillimanite rock, soapstone, verdite, californite (vesuvianite rock), and others. These materials are not true jade and are often named pseudojade.

jade, California   *meta.* = californite

jade, garnet   *meta.* Fine-grained, jadelike varieties of massive grossular or hydrogrossular, e.g., Transvaal jade.

jade, Honan   *meta.* = agalmatolite. [Honan province, China]

jade, Manchurian   *meta.* Soapstone resembling jade in appearance.

jade, Oregon   *sed., meta.* 1. European misnomer for a variety of green jasper.   2. Massive green grossular rock found in Oregon; term also applied to other green, jadelike rocks, e.g., plasma, found in Oregon and California.

jade, puddingstone   *meta.* Jade rock composed of nodules of nephrite cemented together by a darker, olive green variety of nephrite.

jade, Shanghai   *meta.* 1. Any jadeite or nephrite jade from Shanghai, once China's largest jade market.   2. Soapstone (steatite) rock resembling jade.

jade, Soochow   *meta.* = agalmatolite

jade, South African   *meta.* = jade, Transvaal

jade, Tawmaw   *meta.* Jadeite from Tawmaw, in Upper Burma, one of the most important jadeite occurrences.

jade, Transvaal   *meta.* Compact, fine-grained, light green hydrogrossular garnet rock, used for ornamental objects; from near Pretoria, Transvaal, South Africa. Also named South African jade or garnet jade.

jade, vesuvianite   *meta.* = californite

jade, Yarkand   *meta.* Nephrite jade from the jade market and cutting center of Yarkand, Russian Turkestan.

jade-albite   *meta.* = maw-sit-sit. (Gübelin in 1965)

jadeite-albite rock   *meta.* Rare sodium-rich metamorphic rock consisting entirely of nearly chemically pure jadeite and albite; occurs in blocks enclosed in serpentinite in San Benito County, California. A closely related schistose rock composed of the same minerals also contains light-colored glaucophane.

jadeitite   *meta.* Fine-grained, dark green to light green, rock consisting principally of jadeite, with subordinate to trace amounts of albite, muscovite, actinolite, wollastonite, nepheline, magnetite, titanite, analcime, picotite, quartz, prehnite, natrolite, and others; occurs as lenses or pods in serpentinites or schists. [jadeite]

jadeolite   *ig. pluton.* Deep green, chromiferous syenite resembling jade that can be cut as a gem; occurs at jadeite mine at Bhamo, Burma. [jade] (Kunz in 1908)

jarrowite   *sed.?* Local name for calcite pseudomorphs, apparently after celestite; from the Jarrow Docks, Durham, England. [Jarrow Docks] (pub. by Lebour in 1888)

jasp-agate   *sed.* Rock in which bands of translucent chalcedony occur between opaque bands of jasper. Also named agate jasper.

jasp-onyx *sed.* Narrow, straight-banded jasper, with alternating layers of light and dark color.

jasp-opal *sed.* Opaque opal, having the color of jasper (usually brown, yellow, red) but the luster, hardness, and other properties of opal. (Karsten in 1808)

jasper *sed.* Variety of impure, opaque chert, consisting of cryptocrystalline quartz and usually colored brown, yellow, or red by iron oxides; green, black, grayish blue, and other colors of impure chert are also included under the name. Varieties are numerous, and for the most part, descriptive, e.g., brecciated j. (fragments usually cemented by chalcedony), green j., moss j., orbicular j., ribbon j., spherulitic j., striped j., variegated j., zebra j. Other varieties have been given locality designations: Bruneau j. (Idaho), Cave Creek j. (Arizona), Egyptian j., Nunkirchen j. (Germany), Russian j., Stone Canyon j. (California), as well as catalinaite, oregonite, and zonite (Arizona). Some varieties have been named for persons: boakite, chapinite, kinradite, morrisonite, youngite, and others. [L. *iaspis,* jasper]

jasper, agate *sed.* = jasp-agate

jasper, basalt *meta.* Hard, opaque, blue, yellow or black jasperlike rock with a conchoidal fracture, formed by the contact metamorphism of a shale or marly sandstone by a basalt intrusion. Also named basalt jaspis or systyl.

jasper, onyx *sed.* = jasp-onyx

jasper, opal *sed.* = jasp-opal. Also named jasper opal.

jasper, orbicular *sed.* Jasper of one color containing scattered spherical masses of jasper of a contrasting color.

jasper, poppy *sed.* Solid, fine-grained jasper usually displaying bright red, yellow, or orange "flowers" (orbicular areas) against a lime-yellow background.

jasper, porcelain *meta.* Hard, naturally baked, impure clay or porcellanite, which, because of its red color, resembles jasper.

jasper, Sioux Falls *meta.* Decorative, jasperlike, fine-grained, brown quartzite from Sioux Falls, South Dakota; used for table tops and other ornamental purposes.

jasper, xyloid *sed.* Wood that has been replaced by jasper. [Gk. *xylon,* wood]

jasperine *sed.* Uncommon term for a banded jasper of various colors and shades.

jasperoid *meta., hydrothermal.* Metasomatic, usually gray, quartzite (chertlike) rock in which cryp-

tocrystalline quartz has replaced limestone or dolomite by hydrothermal solutions genetically associated with various types of ore deposits; esp. applied to mining districts of Oklahoma, Missouri, and Kansas. [jasper]

jaspilite *meta.* Banded, or mottled, compact siliceous rock consisting of hematite (commonly specularite) and quartz resembling jasper; occurs with iron ores, esp. in the Lake Superior region. Also spelled jaspillite and jaspylite. [L. *iaspis,* jasper] (Wadsworth in 1881)

jaspis, basalt *meta.* = jasper, basalt

jaspoid *ig. extru.* Old name for tachylyte.

jaspoid *sed.* General term for any rock resembling jasper.

jaulingite *sed.* Hyacinth-red, amberlike resin, occurring near St. Viet, Austria. [Jauling, near St. Viet] (Zepharovich in 1855)

javanite *tekt.* Term applied to the tektites from Java. [Java]

jelinite *sed.* = kansasite. (Buddhue in 1938)

jelly rock *sed.* = wilkinite

jenzschite *sed.* Opaline silica with the specific gravity of quartz but soluble in a hot solution of caustic potash; it is associated with chalcedony and forms a rock closely related to cacholong. [Gustav Julius Siegmund Jenzsch, 1830-1877, German mineralogist] (Dana in 1868)

jerseyite *ig. dike.* Quartz-rich minette that also contains biotite and orthoclase, from Jersey, Channel Islands. [Jersey Island] (Lacroix in 1933)

jet *sed.* Dense, compact, homogeneous, pitch black variety of lignite or brown coal that takes a good polish and is often used for beads, ornaments, jewelry, and the like. [Traceable to Gk. *gagates* (of Dioscorides and Pliny) and to Gagas in Lycia, Asia Minor]

jet, Montana *ig. extru.* Local name for obsidian, esp. that from Yellowstone Park, Montana-Wyoming.

jet, Whitby *sed.* Jet from the coal mines near Whitby, Yorkshire, England; considered to be the most desirable quality of jet when the material was in vogue as a gem and ornamental stone.

jet rock *sed.* = shale, jet

josefite *ig. dike.* Altered rock from the island of Philae, Egypt, composed of augite, olivine, magnetite, ilmenite, and apatite, with secondary serpentine, carbonates, biotite, riebeckite, chlorite, and magnetite; term no longer used. [Possibly Franz Josef, 1830-1916, emperor of Austria] (Szadeczky in 1899) josephinite   *ig.?* Highly magnetic, nickel-iron alloy (up to 75% nickel) occurring as grayish white pebbles, having a dark, brownish gray coating, in Josephine Creek, Josephine County, Oregon; once thought to be meteorites but now considered to be terrestrial. [Josephine County, Oregon] (Melville in 1892)

jotunite   *ig. pluton., meta.?* Rock similar to charnockite, intermediate between monzonite and norite and containing orthopyroxene, plagioclase, and microperthite. [Jotunheim, Norway] (Goldschmidt in 1916)

jumillite   *ig. extru.* Fine-grained variety of leucitite composed of phenocrysts of barium-bearing sanidine, olivine, and phlogopite in a fine-grained groundmass composed of diopside, aegirine-augite, leucite, sanidine, and kataphorite. [Jumilla, province of Murcia, Spain] (Osann in 1906)

juvite   *ig. pluton.* Nepheline syenite whose feldspar is exclusively or predominantly orthoclase and whose K_2O is greater in amount than Na_2O; the type rock also contains cancrinite, muscovite, aegirine-diopside, and others. [Juvet, Telemark, Norway] (Brögger in 1921)

K

kabaite   *meteor.* Waxy hydrocarbon occurring in a meteorite that fell in Hungary. [Kaba, Hungary] (Shepard in 1867)

kahusite   *meta.* Silica enriched magnetite-bearing rhyolitelike rock containing quartz (60%), magnetite and hematite (30%), and tourmaline, biotite, and graphite; probably derived from a rhyolite tuff or from the metamorphism of an iron-rich quartzite. [Kahusi volcano, southern Kivu district, eastern Africa] (Sorotschinsky in 1934)

kainite rock   *sed.* Saline evaporite rock in which halite and kainite are major constituents, accompanied by epsomite, bloedite, leonite, and picromerite.

kaiwekite   *ig. extru.* Dull green trachytic rock consisting of phenocrysts of anorthoclase and titanaugite (often with acmite-augite rims) in a groundmass of anorthoclase, oligoclase, augite, and minor olivine (often altered to serpentine), magnetite, apatite, and amphibole; in composition it is close to an anorthoclase syenite (e.g., larvikite). [Kaiweke, native name for Long Beach, Kawekorai Valley, New Zealand] (Marshall in 1906)

kajanite   *ig. extru.* Dark-colored leucitite consisting of phenocrysts of bronze-colored mica and olivine in a groundmass of leucite, diopside, and titaniferous iron oxides but no feldspar. [Oele Kajan, East Borneo] (Lacroix in 1926)

kakirite   *meta.* Sheared and brecciated cataclastic rock in which fragments of the original material are surrounded by numerous gliding surfaces in which intense granulation and some recrystallization have taken place. [Lake Kakir, Swedish Lapland] (Svenonius in 1894)

kakortokite   *ig. pluton.* Nepheline syenite composed of alkali feldspar, nepheline, eudialyte, arfvedsonite, and aegirine with accessories and secondary zeolites; eudialyte may form one-third of the total volume in some specimens. [Kakortok, Greenlandic name for the colony of Julianehaab, Greenland] (Ussing in 1911)

kali-   Common prefix used with rock names to indicate a high potassium content; esp. rocks with the potassic feldspars orthoclase, microcline, microperthite, and others. [L. *kalium*, potassium, from Arab. *qili*, saltwort and its ashes]

kaliakerite   *ig. hypab.* = humurite. (Brögger in 1931)

kalialaskite   *ig. pluton.* Alaskite without sodic-plagioclase; essential minerals are quartz and potassic feldspar (orthoclase, microcline, microperthite), accompanied by accessory zircon, amphiboles, biotite, apatite, and others. [L. *kalium*, potassium + alaskite] (Johannsen in 1932)

kaligranite   *ig. pluton.* Granite containing practically no plagioclase; differs from kalialaskite in having a greater abundance of mafics, e.g., biotite, amphiboles, and pyroxenes. [L. *kalium*, potassium + granite] (Johannsen in 1932)

kaligranite, quartz-leuco   *ig. pluton.* Light-colored, quartz-rich kaligranite; composed essentially of quartz, potassic feldspar, and minor mafics. [Gk. *leukos*, white] (Johannsen in 1932)

kaligranite, quartz-mela   *ig. pluton.* Dark-colored, quartz-rich kaligranite; composed essentially of quartz, microcline, biotite, and amphibole; mafics make up more than 50% of the dark-colored minerals. [Gk. *melas*, black]

kalikeratophyre   *ig. extru.* Keratophyre containing no sodic-plagioclase; all the feldspar is potassic. [L. *kalium*, potassium + keratophyre]

kaliliparite   *ig. extru.* Potassium-rich rhyolite, esp. glassy varieties. [L. *kalium*, potassium + liparite, synonym for rhyolite]

kaliphite   *sed.* Mixture of limonite, manganese oxides, and zinc and calcium silicates; original material was from Hungary. [Origin unknown] (Ivanov in 1844)

kalirhyolite   *ig. extru.* Extrusive rock whose composition corresponds to a kaligranite, in which the feldspar is potassic (sanidine, orthoclase, microcline, microperthite, cryptoperthite, or anorthoclase) and quartz is essential. [L. *kalium*, potassium + rhyolite] (Johannsen in 1932)

kalisyenite   *ig. pluton.* Syenite with orthoclase, microperthite, or anorthoclase but in which less than 5% of the total feldspar is plagioclase. [L. *kalium*, potassium + syenite]

kalisyenite, mela   *ig. pluton.* Rock differing from a normal kalisyenite in containing a greater amount of mafic constituents; total amount of mafics exceeds total amount of light-colored components. [Gk. *melas,* black]

kalitordrillite *ig. extru.* Tordrillite containing no sodic-plagioclase, potassic feldspars being major. [L. *kalium,* potassium + tordrillite]

kalitrachyte *ig. extru.* Trachyte containing no sodic-plagioclase, potassic feldspars being major. [L. *kalium,* potassium + trachyte]

kalitrachyte, leuco *ig. extru.* Light-colored kalitrachyte composed of 95% or more of potassic feldspar; there is no important sodic-plagioclase or quartz. [Gk. *leukos,* white]

kalmafite *ig. extru.* Igneous rock composed of kalsilite (or any other polymorph of $KAlSiO_4$) and mafic minerals. [kalsilite + mafic] (Hatch, Wells, and Wells in 1961)

kamafugite *ig. extru.* General term of potassium-rich, silica-undersaturated extrusive igneous rocks. [katungite + mafurite + ugandite] (Sahama in 1974)

kammgranite *ig. pluton.* Dark-colored kaligranite containing orthoclase perthite, quartz, albite, biotite, and minor hornblende, apatite, and zircon; occurs near Cornimont, La Bresse, the Voges ridge, France. [G. *Kamm,* ridge + granite] (Groth, redefined by Niggli in 1931)

kamperite *ig. dike.* Black, fine- to medium-grained dike rock related to syenite and composed of orthoclase, oligoclase, and biotite. [Kamperhough Valley, Fen district, Norway] (Brögger in 1921)

kankan-ishi *ig. extru.* Homogeneous, black, resinous, and flinty variety of sanukite composed of microphenocrysts of hypersthene, oligoclase, and resorbed hornblende in glass containing very tiny bronzite needles. [Jap. *kankan,* clang, ring + *ishi,* stone; possibly for ringing property] (Koto in 1916)

kankar *sed.* Concretionary masses of calcium carbonate occurring in alluvium in India. Also spelled kunkar. [Ind. (Hindi) *kankar,* stone]

kansasite *sed.* Variety of fossil resin occurring in shale associated with coal. [Kansas] (pub. by Buddhue in 1938)

kanzibite *ig. extru.* Normal rhyolite porphyry containing orthoclase phenocrysts, from eastern Africa. [Lake Kanzibi, southern Kivu district, eastern Africa] (Sorotschinsky in 1934)

kaolin *sed.* Unconsolidated or poorly indurated natural earth composed of kaolinite, with variable minor amounts of other clay minerals, quartz, feldspars, rock fragments, and others. [Chin. *kao,* high + *ling,* hill; from hill east of Ching-te-chen, China, where substance was originally obtained and sent to Europe by the French Jesuit missionary Père d'Entrecolles, early in the 18th century]

karite *ig. diaschist.* Variety of grorudite containing approximately 50% quartz. [Kara River, Soviet Union]

Karlsbad spring stone *sed.* Banded, red, white, and brown, aragonite satin spar, used for carving objects and cheap jewelry, from the hot springs at Karlovy Vary (Karlsbad) Bohemia, Czechoslovakia.

karlsteinite *ig. pluton.* Microcline-rich alkali granite. [Karlstein, on the Thaya bei Raabs, lower Austria] (Waldmann in 1935)

kärnäite *ig. extru., impact.?* Rock close to dacite in composition with a glassy groundmass and numerous inclusions consisting of agglomeratelike tuff; considered by some to be an impactite. [Kärnä, island in Lake Lappajärvi, central Finland]

kasanskite *ig. pluton.* = kazanskite

kåsenite *ig. hypab.* Pyroxene-rich, nepheline-bearing carbonatite. [Kåsene, Fen region, Telemark, Norway] (Brögger in 1921)

kassaite *ig. hypab.* Fine-grained, greenish-black, dike rock containing phenocrysts of hauyne, labradorite, barkevikite, and augite in a holo-crystalline, tinguaitic groundmass of hastingsite needles and crystals of andesine. [Kassa Island, Los Archipelago, Guinea] (Lacroix in 1917)

kassianite *sed.* Variety of sapropelic coal composed mainly of structureless sapropel with a few indefinite remains of algae. Also spelled cassianite. [Kassianovka, Irkutsk, Siberia, Soviet Union] (Zalessky in 1928)

katabugite *ig., meta.?* Charnockite or hypersthene diorite, composed of andesine (62%), hypersthene (27%), and minor quartz, biotite, and accessories. [Gk. *kata,* down, under + bugite] (Bezborodko in 1931)

katungite *ig. extru.* Gray, compact, fine-grained porphyritic leucite-melilite basalt that is pyroxene-free. [Katunga volcano, southwestern Uganda] (Holmes in 1937)

katzenbuckelite *ig. dike.* Nepheline-nosean rock composed of phenocrysts of nepheline, nosean, biotite, olivine, leucite, and apatite in a very fine-grained groundmass of nepheline, leucite, and aegirine. [Katzenbuckel, Odenwald, Germany] (Osann in 1902)

kauaiite *ig. pluton.* Dark-colored oligoclase-augite diorite; a dark variety of syenodiorite. [Kauai, Hawaiian Islands] (Iddings in 1913) kaukasite *ig. extru.* Relatively young volcanic granite in which the potassic feldspar is sanidine. Also spelled caucasite. [Possibly the Caucasus (Kaukasus)] (Beljankin in 1924)

kaulaite *ig. extru.* = pacificite, olivine. [Kaula gorge, Hawaii] (Niggli in 1936)

kauri gum *sed.* Light-colored, yellow to brown copal, usually found as a fossil resin; from the kauri pine, esp. from *Agathis australis,* a timber tree of New Zealand. [New Zealand Maori *kauri,* kauri tree, its resin]

kawakawa *meta.* = tangawaite. [New Zealand Maori *kawakawa,* channel, depression]

kaxtorpite *ig. pluton.* Nepheline syenite containing pectolite, eckermannite, and sodic clinopyroxene. [Kaxtorp, Norra Kärr complex, Sweden] (Adamson in 1944)

kazanskite *ig. pluton.* Black, fine-grained dunite containing some magnetite and plagioclase; related to cumberlandite. [Kazansky, Nicolai-Pawda, Urals, Soviet Union] (Duparc and Grosset in 1916)

kedabekite *ig. pluton.* Eukritelike rock, consisting of bytownite, Ca-Fe garnet, and pyroxene (probably hedenbergite). [Kedabek, Jelisabetpol, Soviet Union] (Federov in 1901)

keffekilite *sed.* Variety of fuller's earth. [Kaffa, old name for Theodosia (Feodosiya), Crimea, Soviet Union + Turk. *kil,* clay]

kemahlite *ig. hypab.* Igneous rock related to theralite, containing pseudoleucite. [Kemahl, Asia Minor] (Lacroix in 1933)

kentallenite *ig. pluton.* Olivine-bearing monzonite. [Kentallen quarry near Ballachulish, Argyllshire, Scotland] (Hill and Kynaston in 1900)

kentsmithite *sed.* Local name for a black, vanadium-bearing sandstone in Paradox Valley, Colorado. [J. Kent Smith, fl. 1920, of Paradox Valley, Colorado, on whose claim the rock occurred] (McMillan in 1910)

kenyte *ig. extru.* Olivine-bearing, phonolitic trachyte, occurring as massive lava flows. [Mt. Kenya, Teleki Valley, British East Africa] (Gregory in 1900)

keralite *meta.* Variety of hornfels having quartz and biotite as its essential minerals. [Gk. *keras,* horn] (Cordier in 1868)

keratophyre *ig. extru.* Dense, light-colored, trachyte porphyry that is much altered and contains albite (in part, secondary), chlorite, epidote, and calcite; originally had biotite, amphibole, and pyroxene. [Gk. *keras,* horn] (Gümbel in 1874)

keratophyre, quartz *ig. extru.* Highly silicic alkali rhyolite that, because of its great age (usually pre-Tertiary), shows devitrification, albitization, and other alterations. (Gümbel in 1874)

keratophyre, sodaclase *ig. extru.* Completely albitized keratophyre in which orthoclase is absent. [sodaclase, synonym for pure albite]

keratophyrspilite *meta.* Keratophyre that has been spilitized; a typical example contains orthoclase (67%), iron-rich chlorite (21%), and minor calcite, quartz, titanite, and apatite. [keratophyre + spilite] (Lehmann in 1933)

keratospilite *meta.* Spilitized albite diabase resembling a keratophyre; may be composed of albite (75%), chlorite (11%), iron oxides, calcite, and apatite. [keratophyre + spilite] (Loewinson-Lessing in 1933)

kerogen *sed.* Fossilized insoluble organic material found in sedimentary rocks, usually shales (oil shale), that can be converted by distillation to petroleum products. [Gk. *keros,* wax + *genes,* born]

kerogenite *sed.* = shale, oil. [kerogen]

keronigritite *sed.* Type of nigritite derived from kerogen. [kerogen + nigritite]

kerosene shale *sed.* = shale, oil

kersantite *ig. dike.* Lamprophyre composed of plagioclase (oligoclase, andesine), biotite, and augite with minor minerals such as alkali feldspar, hornblende, olivine, calcite, and others. [Kersanton, France] (Delesse in 1851)

kersantite, hornblende *ig. dike.* Kersantite containing primary hornblende (usually green) in addition to biotite. (Andreae in 1892)

kersanton *ig. dike.* Obsolete term for kersantite.

kerzinite *sed.* Lignite impregnated with hydrated nickel silicate and used as nickel ore, from Upper Ufalei region (central Urals), Soviet Union. [N. A. Kerzin, mining engineer, who discovered the deposit in 1913] (Shadlun in 1923)

keystoneite *altered ore.* Green or greenish-blue chalcedony colored by chrysocolla and possibly other copper silicate minerals. [Possibly Keystone mine, near Globe, Gila County, Arizona]

khagiarite *ig. extru.* Black, vitreous variety of pantellerite containing phenocrysts of soda-microcline, diopside, aegirine-augite, in a groundmass of brown glass. [Khagiar, on Island of Pantelleria, southwest of Sicily] (Washington in 1913)

khibinite *ig. pluton.* = chibinite

khondalite *meta.* Metamorphic granulite composed of garnet, quartz, and sillimanite. [Khond (Kandh) tribe, India, in whose area these rocks occur] (Walker in 1902)

kieselguhr *sed.* = diatomite. [G. *Kiesel*, flint, silica + *Gur*, sediment]

kieserite rock *sed.* Evaporite saline rock in which the mineral kieserite is a major component; often with halite, sylvite, and carnallite.

kiirunavaarite *ig. pluton.* = magnetitite. Also spelled kirunavaarite. [Kiirunavaara (Schneehuhnberg) district, Norrbotten province, Swedish Lapland] (Rinne in 1921)

kiirunavaarite, spinel *ig. pluton.* Magnetitite (kiirunavaarite) containing up to 20% ferroan spinel (pleonaste). Also named spinel magnetitite.

kilaueite *ig. extru.* Term originally applied to a supposed hornblende-rich basalt from Hawaii; more recent studies have failed to find hornblende in these lavas. [Kilauea volcano, Hawaii] (Silvestri in 1888)

killas *meta.* Miners' term for shale, clay-slate, and schist; esp. those forming the country rock associated with the tin veins of Cornwall, England. Also spelled kellas, kellus. [Corn. *killas*]

kimberlite *ig. dike, pipe.* Porphyritic peridotite or peridotite breccia containing phenocrysts of olivine, usually partly altered, and phlogopite, commonly chloritized, with pyrope and possibly geikielite, in a fine-grained groundmass of calcite and second-generation olivine and phlogopite with accessory ilmenite, serpentine, chlorite, magnetite, and perovskite; some are diamond-bearing. [Kimberley, South Africa] (Lewis in 1887)

kindchen *sed.* Concretion or nodule resembling the head of a small child, usually found in loess. [G. *Kindchen*, little child, baby]

kinradite *sed.* Orbicular jasper rock containing spherical inclusions of colorless or nearly colorless quartz; esp. used for specimens found at Point Bonita, Golden Gate Bridge, California, and less commonly for specimens from Oregon. [J. J. Kinrade, of San Francisco, who discovered the material ca. 1880] (pub. by Sterrett in 1912)

kinzigite *meta.* Quartz-oligoclase gneiss containing prominent garnet and usually biotite. [Kinzig Valley, Black Forest, Germany] (Fischer in 1860)

kirunavaarite *ig. pluton.* = magnetitite. Also spelled kiirunavaarite.

kirwanite *sed., meta.* Variety of anthracite with a metallic luster. [Richard Kirwan, 1733-1812, Irish chemist and mineralogist of Dublin] (Pinkerton in 1811)

kiscellite *sed.* Brown, amberlike fossil resin containing sulfur in place of oxygen. [Kis-Czell, near Budapest, Hungary] (Zechmeister, Toth, and Koch in 1934)

kivite *ig. extru.* Dark-colored leucite basanite containing phenocrysts of leucite, bytownite, olivine, biotite, and others. [Kisi volcano, Lake Kiva, East Africa] (Lacroix in 1923)

kjelsasite *ig. pluton.* Syenodiorite similar to larvikite but higher in CaO and lower in the alkalies. [Kjelsås, Sörkedal, Oslo region, Norway] (Brögger in 1933)

klapperstein *sed.* Hollow iron oxide concretionary bodies whose interiors are partially filled with sand that rattles when the rock is shaken. Also named rattlestone, rattle rock. [G. *Klapper*, rattle + *Stein*, rock]

klausenite *ig. hypab.* Hypersthene-bearing rock, intermediate between diorite and gabbro, containing andesine-labradorite (65%), hypersthene (20%), and minor orthoclase, quartz, and mafics. [Tinnebach, Klausen, southern Tyrol] (Cathrein in 1898)

klinghardtite *ig. hypab.* Nepheline phonolite porphyry containing large (12 cm) sanidine phenocrysts. [Klinghardt Mountains, South West Africa] (Kaiser in 1913)

klintite *sed.* Massive core of a biohermal or reef limestone. [Dan. and Sw. *klint*, Scot. *clint*, knob, hill]

knotenschiefer *meta.* Contact metamorphic spotted slates in which micas, esp. biotite, appear as aggregates of coarse flakes. [G. *Knoten*, knot, knob + *Schiefer*, slate, schist]

kochenite *sed.* Fossil resin resembling amber; from Tyrol. [Kochenthal, Tyrol] (Pichler in 1868)

kodurite *meta.* Granulite consisting of orthoclase, quartz, spessartine-andradite, and apatite. [Kodur, Madras, India] (Fermor in 1907)

koellite *ig. dike.* Alkali basalt composed of olivine, lepidomelane, barkevikite, apatite, magnetite, anorthoclase, and nepheline. [Possibly Kjølen (Koelen) Mountains, Norway] (Brögger, date uncertain)

köfelsite *impact.?* Frothy, pumiceous, silica-rich glass, occurring as small veins in fractured gneiss; apparently formed by meteorite impact. [Köfels structure, Austria] köflachite   *sed.* Dark brown variety of retinite found in brown coal at Köflach in Styria, Austria. (Doelter in 1878)

kohalaite   *ig. extru.* Andesite with normative oligoclase and with or without normative or modal olivine. The name is not generally used, since it is based on chemical analyses. [Kohala Mountain, Waimea, Hawaiian Islands] (Iddings in 1913)

kolm   *sed.* Variety of cannel coal occurring as lenticles in Swedish alum shales and containing about 30% ash that is high in uranium, radium, and other rare elements. [Sw. *kol,* coal]

komatiite   *ig. extru.* Ultramafic lava; resulting rocks include a wide range of compositions; noncumulate rocks range from peridotite to basalt to andesite, and cumulate rocks range from peridotite to mafic gabbro; a spinifex texture is often exhibited by these rocks. [Komati River, Transvaal, South Africa] (Viljoen and Viljoen in 1969)

konite   *sed.* Freestone composed of limestone, rather than the more common sandstone. [Gk. *konia,* lime] (Theophrastus, *On Stones*)

könlite   *sed.* Soft, reddish brown to yellow hydrocarbon, occurring as plates, grains, or stalactitic forms in brown coal at Uznach, Switzerland. Also named koenleinite and koenlite. [Mr. Könlein, superintendent of the coal works at Uznach] (Kraus in 1838)

koreite   *meta.* = agalmatolite. Also spelled koireiite. [Gk. *choireios,* of a swine; alluding to greasy feel] (Beaudant, date uncertain)

kornite   *sed., meta.* = hornstone. [L. *cornu,* horn]

koswite   *ig. pluton.* Magnetite-peridotite composed of olivine, diallage, magnetite, and subordinate hornblende as rims around magnetite. [Koswinsky Kamen, the Urals] (Duparc and Pearce in 1901)

krablite   *ig. extru.* Rhyolite tuff containing abundant sanidine, with plagioclase, augite, and quartz in smaller proportions. [Krafla (Krabla), volcano in Iceland] (Forchhammer in 1843)

krageröite   *ig. dike.* Aplitic rock composed almost entirely of albite and rutile. [Kragerö, Norway] (Brögger in 1904)

krantzite   *sed.* Light yellow to greenish-yellow substance related to amber, occurring as small grains disseminated in brown coal near Nienburg, Hanover, Germany. [August Krantz, 1809-1872, German mineral dealer in Berlin and in Bonn] (Bergemann in 1859)

KREEP   *lunar.* Basaltic lunar rock, first found in Apollo 12 fines and breccias, characterized by an unusually high content of potassium (K), rare-earth elements (REE), phosphorus (P), and other trace elements in comparison to other lunar rocks. [K (potassium) + REE (rare-earth elements) + P (phosphorus)]

kristianite   *ig. pluton.* Local name for red biotite granite occurring in the Oslo region of Norway. [Kristiania, former name of Oslo] (Brögger in 1921)

kryokonite   *uncertain.* = cryoconite

ktypeite   *sed.* Pisoliths of calcium carbonate (aragonite) from Karlsbad, Bohemia, Czechoslovakia. [Gk. *ktupos,* noise; alluding to decrepitation upon heating] (Lacroix in 1898)

kuckersite   *sed.* = kukersite

kugelsandstein   *sed.* = concretion, sandstone. [G. *Kugel,* sphere, ball + *Sandstein,* sandstone]

kukersite   *sed.* Organic-rich sediment containing the alga *Gloexapsamorpha prisca;* occurring in the Ordovician of Estonia. Also spelled kukkersite, kuckersite. [Kukers, Estonia] (Zalessky in 1916)

kulaite   *ig. extru.* Extrusive rock characterized by the presence of both orthoclase and calcicplagioclase, the latter slightly more abundant than the former, and also containing a moderate amount of nepheline and olivine. [Kula, Lydia, Asia Minor] (Washington in 1894)

kulaite, leucite   *ig. extru.* Leucite andesite; rock originally given this name is not now considered to be a kulaite: the term is obsolete. (Washington in 1894)

kullaite   *ig. dike.* Much altered, syenodiorite porphyry with phenocrysts of plagioclase (oligoclaseandesine) and soda-orthoclase in an ophitic groundmass of oligoclase-andesine, chlorite (from augite), soda-microcline, and magnetite. [Kullagården, Sweden] (Henning in 1899)

kundaite   *sed.* Variety of grahamite characterized by a brown-colored powder and by its good solubility in turpentine or chloroform. [Kunda, Estonia] (Doss in 1914)

kunkar   *sed.* = kankar

kupferschiefer   *sed.* Dark-colored Permian shale of Mansfeld, Germany, worked for copper (from copper sulfides) and other metals. [G. *Kupfer,* copper + *Schiefer,* shale]

kuselite   *ig. dike.* = cuselite

kuskite   *ig. dike.* Obsolete name for granite porphyry originally believed to consist of quartz and scapolite phenocrysts in a fine-grained ground-mass composed of quartz, orthoclase, and mus-covite; when it was discovered that the scapolite was actually quartz, the name was withdrawn. [Kuskokwin River, Alaska] (Spurr in 1900)

kvellite *ig. dike.* Dark-colored, ultrabasic, por-phyritic dike rock, composed of lepidomelane, olivine, barkevikite, apatite, ilmenite, and mag-netite in a groundmass of anorthoclase laths.

[Kvelle, north of Larvik, Norway] (Brögger in 1906)

kylite *ig. pluton.* Theralite rich in olivine. [Kyle district, Ayrshire, Scotland] (Tyrrell in 1912)

kyschtymite *ig. dike.* Dike rock composed of euhe-dral crystals of corundum and a little biotite in a groundmass of calcic-plagioclase (bytownite). [Kyschtym, the Urals] (Morozewicz in 1899)

L

laachite   *meta.* Rock formed through contact metamorphism of crystalline schist and composed
of anorthoclase (64%), biotite (26%) and some
hypersthene and magnetite. [Laacher See,
Rheinland, Germany] (Kalb in 1936)

laanilite   *meta.* Granulite composed of almandine,
cordierite, quartz, and biotite. [Laanila, Inari,
Finland] (Hackman in 1905)

labradite   *ig. pluton.* = labradoritite. [labradorite]
(Turner in 1900)

labradophyre   *ig. pluton.* Porphyritic anorthosite
(labradoritite) showing phenocrysts of labradorite in a groundmass of the same mineral. Also
named labradorite porphyry, labradorite porphyrite. [labradorite] (Coquand in 1857)

labradorfels   *ig. pluton.* = labradoritite. [labra-
dorite] (Kolderup in 1902)

labradorite   *ig. extru., pluton.* Term used by French
petrologists for a light-colored labradorite-rich
basalt and by Soviet petrologists for a light-
colored gabbro or norite. Not to be confused
with the mineral labradorite.

labradoritite   *ig. pluton.* Anorthosite composed
nearly entirely of labradorite. Also named
labradite, a term less preferred. [labradorite]
(Johannsen in 1919)

lahar deposit   *ig. extru., sed.* Volcanic mudflow
occurring on the slopes of some subaerial volcanoes; the flow, which lacks sorting, arises either
from heavy rain falling on the unconsolidated
ash, or from the eruption of ash into a crater
lake. [Indonesian *lahar,* lava]

lakarpite   *ig. pluton.* Dark, fairly coarse, microcline-
albite nepheline syenite also containing an
arfvedsonitelike amphibole and sometimes
aegirine, titanite, apatite, and secondary natrolite.
[Lakarp, Nord Småland, Sweden] (Törnebohm
in 1906)

laminite   *sed.* Finely laminated detrital rock in a
flysch deposit, occurring in geosynclinal succes-
sions in natural sequences complementary to
typical turbidite; finer-grained and thinner-bedded
than a turbidite and believed to form seaward
from turbidites as a bottomset bed of a large
delta. [laminated, alluding to its structure]

lampadite   *sed.* Variety of wad containing 4-18%

copper oxide as well as cobalt and other oxides.
[Wilhelm August Lampadius, 1772-1842, German chemist who first investigated it] (Huot in
1841)

lamproite   *ig. extru.* Collective name for potassium-
and magnesium-rich extrusive igneous rocks,
e.g., fortunite, gaussbergite, madupite, orendite,
verite, and wyomingite. Also named lamprophyric
extrusive rock. [lamprophyre] (Niggli in 1923)

lamprophyre   *ig. diaschist.* Group of dark porphyritic dike rocks in which mafics (esp. biotite,
hornblende, and pyroxenes) form abundant phenocrysts and in which a fine-grained groundmass is composed of the same mafics in addition
to feldspars and/or feldspathoids, e.g., camptonite,
kersantite, minette, spessartite, and vogesite. [Gk.
lampros, bright; alluding to prominent reflec-
tive biotite flakes in some examples] (Gümbel in
1879)

lamprophyre, diorite   *ig. diaschist.* General term
for lamprophyres whose major feldspar is
plagioclase, e.g., kersantite and spessartite.

lamprophyre, dioritic   *ig. diaschist.* Group of dark-
colored dioritic rocks in which mafics equal or
exceed the amount of plagioclase, including rocks
such as kersantite, camptonite, spessartite. Essentially the same as diorite lamprophyre.

lamprophyre, syenite   *ig. diaschist.* General term
for lamprophyres whose major feldspar is potassic (usually orthoclase), e.g., minette and vogesite.

lamproschist   *meta.* Metamorphosed lamprophyre
consisting of brown biotite and green hornblende
and having a schistose structure. [lamprophyre
+ schist]

lancasterite   *meta.* Fine mixture of brucite and
hydromagnesite, once thought to be a single
mineral. [Lancaster County, Pennsylvania]
(Silliman in 1850)

lapidite   *ig. extru.* Variety of ignimbrite with a
stony texture, as compared to one that is pulverulent (pulverulite). [L. *lapidosus,* stony] (Marshall
in 1935)

lapilli   *ig. extru.* 1. Fragments ejected during explosive volcanic eruptions, measuring 4-32 mm in
diameter (some petrologists use 2-64 mm range);
materials may be scoriaceous, crystalline,

threadlike, or droplike.   2. *IUGS,* pyroclasts of any shape that have diameters 2-64 mm. Lapilli is the plural form, and lapillus is the singular. [L. *lapillus,* small stone; dim. of *lapis,* stone]

lapilli, accretionary   *ig. extru.* Lapilli composed of pellets exhibiting a somewhat concentric texture formed by a gathering of clots of volcanic ash around wet nuclei, e.g., raindrops falling through ash; often in rhyolitic tuffs.

lapilli, filiform   *ig. extru.* = Pele's hair. [L. *filum,* thread]

lapillistone   *ig. extru.* Pyroclastic rock composed largely of lapilli fragments.

lapillite   *ig. extru.* = tuff, lapilli

lapis lazuli   *meta.* Semitranslucent to opaque blue rock, composed principally of lazurite and hauyne with variable amounts of pyrite, calcite, diopside, and other minerals; usually formed by the contact metamorphism of limestone. [L. *lapis,* stone + Pers. *lazhward,* blue]

lard stone   *meta.* = agalmatolite. Also named lardite. [lard, alluding to greasy luster]

lardalite   *ig. pluton.* Very coarse nepheline syenite, characterized by large crystals of nepheline (often the size of a fist) in a mass of finely twinned sodic-plagioclase (once thought to be anorthoclase or cryptoperthitic orthoclase) of rhombic form; like larvikite but containing essential nepheline (more than 10%). Also spelled lardallite, laurdalite. [Lardal, near Larvik, Norway] (Brögger in 1890)

lardalite, olivine   *ig. pluton.* Variety of lardalite rich in olivine and iron oxides.

larimar   *ig., altered.* Trade name for a pectolite rock used as a gem; name is esp. used in the Dominican Republic for specimens from the Hispanola Mountains.

larvikite   *ig. pluton.* Coarse, sometimes opalescent, pearl gray alkali syenite, grading to monzonite; composed of predominant cross-sections of rhombic, very finely twinned sodic-plagioclase (once believed to be anorthoclase or cryptoperthitic orthoclase) and alkali feldspar, with diopsidic augite and titanaugite as the chief mafics and accessory apatite, ilmenite, titaniferous magnetite, nepheline (less than 10%), olivine, biotite (lepidomelane), and others; used as a decorative stone because of beautiful opalescence when polished. Also spelled laurvikite. [Larvik, Norway] (Brögger in 1890)

lassenite   *ig. extru.* Fresh, unaltered glassy trachyte obsidian. [Lassen Peak, California] (Wadsworth in 1893)

lassolatite   *sed.* = siliceous sinter. [Puy-de-Lassolas, Puy-de-Dome, France] (Gonnard in 1876)

laterite   *sed.* Red, residual soils, or surface products, composed largely of aluminum and ferric hydroxides, with more or less free silica and clays, and originated in situ from the atmospheric weathering of rocks, esp. in the tropics or in forested warm to temperate climates; can be used for bricks. [L. *later,* brick] (Buchanan in 1807)

lateritite   *sed.* Sediment composed of reconstructed laterite detritus. [laterite] (Fermor in 1911)

lateritoid   *sed.* Lateritic rock formed by the metasomatic replacement of some older rock at its outcrop. [laterite] (Fermor in 1911)

latite   *ig. extru.* Aphanitic extrusive equivalent of monzonite, in which orthoclase and plagioclase are both present in about equal amounts; textures may be glassy, felsitic, porphyritic, or vitrophyric. [Latium or Latia, Italian province] (Ransome in 1898)

latite, quartz   *ig. extru.* Aphanitic extrusive equivalent of granodiorite (for some petrologists, quartz monzonite), containing sodic-plagioclase (calcicandesine to oligoclase), potassic feldspar, essential quartz (over 10%), and mafics; usually porphyritic with a holocrystalline to vitreous matrix.

laugenite   *ig. pluton.* Unnecessary term given to an oligoclase-containing diorite to distinguish it from the more common andesine-containing diorite. [Laugendal, Norway] (Iddings in 1913)

laurdalite   *ig. pluton.* = lardalite

laurvikite   *ig. pluton.* = larvikite

lava   *ig. extru.* General name for the molten outpourings of magma from volcanic craters or fissures; also the rock solidified from it. [It. *lava,* stream]

lava, block   *ig. extru.* Solidified lava having a surface composed of angular blocks; fragments are more regular in shape, somewhat smoother, and less vesicular than in aa.

lava, ellipsoidal   *ig. extru.* General term for any lava flow with an ellipsoidal pattern, e.g., pillow lava.

lava, pillow   *ig. extru.* General term for lava displaying a pillow structure (close-fitting pillow-shaped masses, where the concavities of one part match the convexities of another); usually poor in silica (often basalt) and considered to have formed in a subaqueous environment. (Bonney in 1893)

lava, tuff   *ig. extru.* = tufflava

lavialite   *meta.* Metamorphosed basalt porphyry or tuff in which relict labradorite phenocrysts contain quartz, microcline, biotite, and hornblende, and these are set in an amphibolitelike groundmass composed of the same minerals among which green hornblende is the most conspicuous. [Kirchspiel, Lavia, Findland] (Sederholm in 1899)

laxite   *sed., meta.* General name for fragmental or mechanically broken rocks, esp. when they are unconsolidated. [L. *laxus,* loose] (Wadsworth in 1892)

leachate   *sed.* Water that has percolated through soil or other matter containing soluble substances and that contains certain amounts of these substances in solution. [O. Eng. *leccan,* to moisten]

ledmorite   *ig. pluton.* Dark-colored malignite or nepheline syenite composed of orthoclase, aegirine-augite, nepheline, and melanite (10%), with some biotite, apatite, and secondary calcite and chlorite. [Ledmore River, Assynt, Scotland] (Shand in 1910)

leeuwfonteinite   *ig. pluton.* = hatherlite. [Leeuwfontein, Bushveld, South Africa] (Brouwer in 1917)

lehmanite   *ig. pluton., ig. altered.* 1. Granite composed only of quartz and potassic feldspar; the term never came into general use. (Pinkerton in 1811)   2. = saussurite. [Lake Lehman (Leman), Switzerland] (Delamétherie in 1797)

leidleite   *ig. extru.* Pitchstone of rhyodacite composition containing microlites (not phenocrysts) of andesine, augite, apatite, and iron oxides in a glassy matrix whose composition is close to a mixture of andesine, quartz, and orthoclase. [Glen Leidle, Mull, Scotland] (Thomas and Bailey in 1915)

lenticulite   *ig. extru.* Variety of ignimbrite with a lenticular texture. [L. *lenticula,* lentil] (Marshall in 1935)

lenzinite   *sed.* Somewhat decomposed semiopal that is earthy, compact, white, and translucent. [Johann Georg Lenz, 1748-1832, German mineralogist] (John in 1816)

leopard rock   *ig. hypab.* Light-colored variety of syenite, from Ottawa County, Ontario, Canada, containing cylindrical seggregations of mafics, esp. green pyroxene; when cut properly the rock has a leopard-spot appearance.

leopardite   *ig. dike.* Light-colored, aphanitic felsite, usually rhyolite, with pencil-like cylindrical stains of iron and/or manganese oxides; when cut properly the rock has a leopard-spot appearance; an important locality is within Charlotte, Mecklenburg County, North Carolina. (Hunter in 1853)

leptite   *meta.* Fine-grained, granular, metamorphic rock composed mainly of quartz and feldspar (oligoclase-andesine, microcline, or orthoclase) with subordinate mafic minerals; nearly equivalent to fine-grained felsic granulites but perhaps of lower metamorphic grade. [Gk. *leptos,* fine, small] (Eskola in 1914)

leptopel   *sed.* Fine particulate matter occurring suspended in natural waters, composed mainly

Leopardite (stained rhyolite). Mecklenburg County, North Carolina. 7.5 cm.

of colloidal organic and inorganic matter (hydrous oxides, silicates, insoluble carbonates). [Gk. *leptos,* fine, small + *pelos,* mud, clay] (Fox in 1957)

leptynite   *meta.* French equivalent of leptite. Also spelled leptinite. [Gk. *leptunein,* to make fine] (Haüy in 1822)

leptynolite   *meta.* Fissile or schistose variety of hornfels composed of mica, quartz, and feldspar, with or without cordierite and andalusite. [Gk. *leptunein,* to make fine] (Cordier in 1868)

lestiwarite   *ig. diaschist.* White, saccharoidal aplitic dike rock, composed of microperthite, aegirine, and arfvedsonite, with accessory albite, titanite, biotite, eudialyte, fluorite, and quartz (possibly secondary). [Lestiware, Umptek, Kola Peninsula] (Rosenbusch in 1898)

leucilite   *ig. extru.* Obsolete term roughly equivalent to either leucitite or leucite basalt. [leucite]

leucitite   *ig. extru.* Extrusive, gray to black, igneous rock composed of leucite and pyroxene (titaniferous augite, diopside, aegirine, aegirine-augite) but without olivine. [leucite] (Senft in 1857)

leucitite, olivine   *ig. extru.* = basalt, leucite. (Watts and others in 1921)

leucitite-basanite   *ig. extru.* Rock intermediate in composition between leucitite and basanite; differs from leucite basanite in containing more leucite than feldspar. (Johannsen in 1938)

leucitite-tephrite   *ig. extru.* Rock intermediate in composition between leucitite and tephrite; differs from leucite tephrite in containing more leucite than feldspar. (Johannsen in 1938)

leucitolith   *ig. extru.* Extrusive igneous rock composed entirely of leucite. [leucite] (Johannsen in 1938)

leucitophyre   *ig. extru.* Extrusive rock transitional between leucitite and nephelinite; composed essentially of both leucite and nepheline with clinopyroxene (augite). [leucite] (von Humboldt in 1837)

leuco-   *ig.* Common prefix meaning light-colored or white; term should never mean that the rock contains leucite; rocks with this prefix usually have less than 5% mafic components, but this is not a rigid requirement. Examples include the following: leucodiorite (oligoclasite), leucogranite, leucogranodiorite, leucogranogabbro, leuco-kalitrachyte, leucolitchfieldite, leucopulaskite, leucorhyolite, leucosyenite, leucosyenodiorite, leucosyenogabbro, leucotonalite, leucotrachyte. [Gk. *leukos,* white]

leucogabbro   *ig. pluton.* Gabbro whose color index is below 40; light-colored rock.

leucogabbroid   *ig. pluton.* IUGS, plutonic rock satisfying the definition of gabbroid; pl/(pl + px + ol) is 65-90.

leucophyre   *ig. dike.* 1. Altered diabase containing saussuritized feldspar (the most abundant constituent), pale green and purple pyroxenes, ilmenite, and abundant chlorite. (Gümbel in 1874) 2. Any light-colored, hypabyssal, porphyritic igneous rock; the antithesis of lamprophyre. Also named oxyphyre. [Gk. *leukos* white + porphyry]

leucophyride   *ig. hypab.* General field term for any light-colored porphyritic igneous rock with an aphanitic groundmass. [leucophyre]

leucotephrite   *ig. extru.* 1. light-colored tephrite. [Gk. *leukos,* white + tephrite] 2. tephrite containing leucite but without a sodium feldspathoid like nepheline; the term leucite tephrite is preferable. [leucite + tephrite] (Fouqué and Michel-Lévy in 1879)

leumafite   *ig. extru.* General name for igneous rock composed of leucite and mafic minerals. [leucite + mafic] (Hatch, Wells, and Wells in 1961)

lherzite   *ig. pluton.* Hornblendite composed essentially of brown hornblende (80%), with minor biotite, ilmenite and garnet. [Lherz, Pyrenees, France] (Lacroix in 1917)

lherzolite   *ig. pluton.* 1. Peridotite composed of olivine and of both orthopyroxene and clinopyroxene. [Lac de Lherz, Pyrenees, France] (de Lamétherie in 1795) 2. IUGS, plutonic rock with M equal to or greater than 90, 40-90 ol/(ol+opx+cpx), and both opx/(ol+cpx+opx) and cpx/(ol+cpx+opx) greater than 5.

lias   *sed.* White to bluish, compact, argillaceous limestone or cement rock, often interbedded with shale or clay; generally refers to the oldest Jurassic strata in Europe (Lias). [Armoric *liach, leach,* stone]

libollite   *sed.* Variety of asphalt resembling albertite; from Angola. [Libollo, Angola] (Gomes in 1898)

liebenerite-porphyry   *ig., altered.* Nepheline syenite porphyry whose nepheline phenocrysts are altered to a fine-grained aggregate of muscovite scales (named liebenerite under the impression that it was a new mineral); nearly identical to gieseckite-porphyry; occurs near Predazzo, Tyrol. [Leonhard Liebener, 1800-1869, of Innsbruck, Austria, a specialist on the minerals of Tyrol]

lignite   *sed.* Brownish black coal in which the

alteration of vegetal material has proceeded further than in peat but not so far as subbituminous coal; specifically, coal containing less than 8,300 BTU/lb, on a moist, mineral-matter-free basis. [L. *lignum,* wood] (Brongniart in 1807)

lignite, brown   *sed.* Variety of lignite containing less than 6,300 BTU/lb, on a moist, mineral-matter-free basis. Also named lignite-B.

lignite, woody   *sed.* Lignite still containing the fibrous structures of wood. Also named xyloid lignite.

lignite, xyloid   *sed.* = lignite, woody. [Gk. *xylon,* wood]

limburgite   *ig. extru.* Vitrophyric extrusive igneous rock, resembling basalt and containing phenocrysts of olivine and augite in an alkali-rich glassy groundmass. Also named magmabasalt. [Limburg, Kaiserstuhl, Germany] (Rosenbusch in 1872)

lime-silicate rock   *meta.* Metamorphic rock formed by thermal or contact metamorphism of impure limestones or of calcareous shales, slates, or tuffs and in which carbonate minerals are removed but calcium-silicate assemblages with quartz, pyrite, and other minerals are formed; the calcium silicates often include wollastonite, tremolite, clinopyroxene (omphacite), anorthite, grossular, epidote, zoisite, axinite, scapolite, and others.

Calc-flinta is a variety; calc-silicate rocks are essentially the same.

limepan   *sed.* Hardpan in soil in which calcium carbonate is the principal cementing agent. Also spelled lime pan.

limerock   *sed.* Unconsolidated or partly consolidated limestone, usually containing shells or shell fragments and varying amounts of quartz; hardens on exposure and is sometimes used as road metal; common in the southeastern United States.

limestone   *sed.* Coarse- to fine-grained sedimentary rocks, of various colors and textures, that are composed of calcium carbonate (more than 50% by weight), primarily in the form of calcite; they contain less than 5% dolomite and may contain many other minor components. Many of the numerous variety names are descriptive in nature, including the following: arenaceous (sand-bearing) l., argillaceous (clay-bearing) l., asphaltic l., bituminous l., carbonaceous l., cherty l., ferruginous (iron-bearing) l., fossiliferous l., glauconitic l., pyritic l., sandy l., sideritic l., siliceous l. Many additional varieties are listed below. [O. Eng. *lim,* cement]

limestone, accretionary   *sed.* Limestone formed in place by the slow accumulation of organic remains (often shells); originally these materials consist wholly or in part of aragonite that gradually

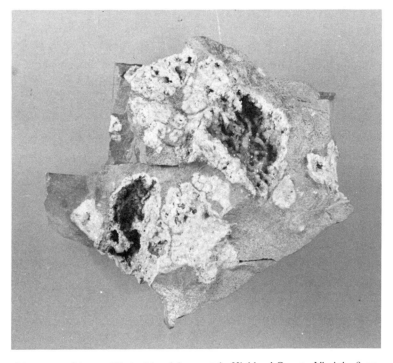

Limestone with vugs filled with calcite crystals. Highland County, Virginia. 9 cm.

inverts to calcite as the sediment ages. [L. *ad,* to + *crescere,* to grow]

limestone, algal   *sed.* Limestone composed largely of the remains of calcium-carbonate producing algae or one in which such algae serve to bind together the fragments of other fossils; these often show intricate layering of alternate algal and inorganic zones. Examples include algal biscuits, oncolites, and stromatolites.

limestone, allochemical   *sed.* General term for limestone composed of allochems, particles of calcium carbonate the size of coarse silt or larger; grains may be skeletal particles, ooliths, pellets, and intraclasts (fragments originated within the basin of deposition but transported somewhat). [Gk. *allos,* other + chemical]

limestone, allochthonous   *sed.* Clastic limestone composed of particles produced elsewhere; particles may consist of fragments of organisms, limestone clasts, ooliths, pisoliths, bahamites, and others. [Gk. *allos,* other + *chthonos,* earth]

limestone, allodapic   *sed.* Limestone believed to have been deposited by turbidity currents in relatively deep water and composed of materials derived from shallow-water reef areas. [Gk. *allos,* other + *dapedon,* level surface] (Meischner in 1964)

limestone, arkosic   *sed.* Limestone containing a relatively high proportion of grains or crystals of feldspar, either authigenic (formed in place) or detrital.

limestone, autochthonous   *sed.* Limestone formed in place by biogenic and biochemical processes; formed either by the direct constructive activity or organisms, or by the accumulation of their skeletons and tests, or by their indirect biochemical effects on the precipitation of calcium carbonate. [Gk. *autos,* self + *chthonos,* earth]

limestone, basinal   *sed.* Limestone deposited in a deep basin; generally nongraded and nontur-biditic, dark fine-grained lime mudstones and siltstones; related to allodapic limestones that occur in deep water near steep slopes. [basin]

limestone, Bedford   *sed.* Buff to gray, fine-grained, firm, compact, oolitic and fossiliferous limestone (Mississippian), quarried for dimension stone; widely use, e.g., in Washington National Cathedral, Washington, D.C. [Bedford, Lawrence County, Indiana]

limestone, bioaccumulated   *sed.* Limestone formed by the accumulation of sedentary (not colonial) organisms and their related ecological communities; the rock has a predominance of unbroken fossils, lacks sorting, has a diversity of organic components, and has a scarce fine-grained matrix; some varieties contain abundant mollusc shells, others are rich in crinoidal fragments (encrinal limestone) or foraminifers. [Gk. *bios,* life + accumulated]

limestone, biochemical   *sed.* General term for any limestone that has originated through the direct activity of organisms, e.g., bioaccumulated limestone and bioconstructed limestone. [Gk. *bios,* life + chemical]

limestone, bioclastic   *sed.* Limestone composed of fragments (clasts) derived from biochemical materials, e.g., old biochemical limestone fragments or shells or shell fragments; coquinite and encrinite and many calcarenites and calcirudites are examples. [Gk. *bios,* life + *klastos,* broken]

limestone, bioconstructed   *sed.* Limestone formed through the vital activity of colonial and sediment-binding organic communities; these rocks usually are comprised of a framework of in situ coral, algal, stromatoporoidal, and bryozoan colonies. [Gk. *bios,* life + constructed]

limestone, biohermal   *sed.* Bioconstructed limestone consisting of a bioherm reef or mound. See bioherm.

Limestone (argillaceous) (landscape marble). Bristol, England. 18 cm.

limestone, biostromal   *sed.* Biochemical limestone formed from a bedded or tabular biostrome. See biostrome.

limestone, bird's-eye   *sed.* = dismicrite

limestone, calcarenitic   *sed.* Limestone composed of considerably more than 10% calcareous-mud matrix accompanied by at least 10% clastic carbonate grains of sand or gravel size.

limestone, calcitic   *sed.* Limestone composed of essentially pure calcite.

limestone, carbonaceous   *sed.* Fine-grained, dull, black limestone containing carbonaceous matter, often producing a dusty fracture and a stain; bituminous limestone often releases a peculiar odor when it is broken; pyrite often accompanies the organic substances in these rocks.

limestone, chemical   *sed.* Limestone formed through inorganic chemical processes in which organisms played no role, e.g., oolitic limestone, many speleothems, caliche, travertine, tufa, and some evaporites and calcilutites.

limestone, cherty   *sed.* Limestone containing lenses, nodules, layers, and sometimes ooliths of chert.

limestone, clastic   *sed.* Limestone composed of fragments (clasts) derived from any earlier calcium carbonate material whether inorganic chemical (e.g., ooliths) or biochemical (shell fragments) in nature; these limestones are divided, on the basis of grain size, into calcirudites (clasts greater than 2 mm diameter), calcarenites (1/16 mm to 2 mm), calcisiltites (1/256 mm to 1/16 mm), and calcilutites (less than 1/256 mm). [Gk. *klastos,* broken]

limestone, clotted   *sed.* Recrystallized limestone with a grumous texture where patches of coarse sparry calcite irregularly invade shell fragments, ooliths, and matrix without obliterating their features and surround micritic clots or clusters of clots. Also named grumous limestone.

limestone, coquinoid   *sed.* Limestone consisting of coarse, unsorted, and usually unbroken shelly materials that have accumulated in place without subsequent transportation or agitation; like a coquina but an autochthonous limestone. [coquina]

limestone, coral   *sed.* Bioconstructed limestone consisting of the calcareous skeletons of corals accompanied by fragments of other organisms.

limestone, coral-reef   *sed.* = limestone, coral

limestone, crinoidal   *sed.* = encrinite

limestone, crystalline   *sed.* 1. Limestone composed of abundant calcite crystals as a result of diagenesis; crystals are usually larger than 20

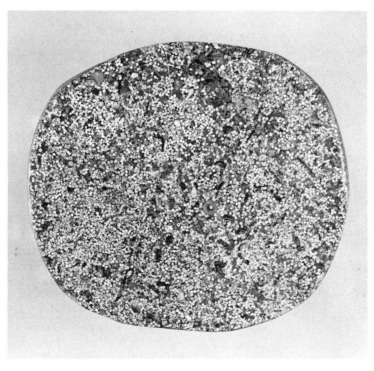

Limestone (calcitic oolite). Bristol, England. 8.5 cm.

microns; a variety of sedimentary marble. 2. Limestone (calcarenite) in which crystalline calcite cement has formed in optical continuity with the crystalline fossil fragments (clasts) by diagenesis.

limestone, crystalline   *meta.* Misleading term, commonly used in the older literature, for a metamorphosed limestone; a true metamorphic marble formed by the recrystallization of limestone as a result of metamorphism.

limestone, detrital   *sed.* = limestone, clastic [L. *deterere,* to wear down]

limestone, diatreme   *ig. extru.* Rare extrusive volcanic carbonatite rock rich in calcite. [Gk. *dia,* through + *trema,* hole; alluding to a volcanic pipe]

limestone, dispellet   *sed.* Pellet limestone with tubules or irregular patches of sparite. [Gk. *dis,* two, apart from, not + pellet] (Wolf in 1960)

limestone, dolomitic   *sed.* Limestone consisting of 50-90% calcite and 10-50% dolomite.

limestone, encrinal   *sed.* = encrinite

limestone, fenestral   *sed.* Limestone containing fenestral cavities (also referred to as bird's eyes); the fenestrals are usually filled with spar (crystalline calcite) or with sediments; essentially equivalent to dismicrite, including bird's-eye limestone. [L. *fenestra,* window]

limestone, Fontainebleau   *sed.* Calcite sand crystals that have a rhombohedral habit and contain 50-63% sand; occur at Fontainebleau and Nemours, France.

limestone, foraminiferal   *sed.* Limestone composed of the remains of foraminifers, usually without much fine-grained matrix. Varieties include fusulinid limestone and nummulitic limestone. Chalk is a related rock.

limestone, freshwater   *sed.* Limestone formed by accumulation or precipitation in a freshwater lake, a stream, or a cave, e.g., marl, travertine, tufa, and speleothems.

limestone, fusulinid   *sed.* Foraminiferal limestone composed of fusulinid tests.

limestone, grumose   *sed.* = limestone, clotted. [L. *grumus,* little pile]

limestone, high-calcium   *sed.* Limestone containing very little magnesium; the MgO equivalent is usually considered to be less than 1.1%; calcitic limestone.

limestone, hydraulic   *sed.* Impure limestone containing silica and clays in varying amounts and yielding, upon calcining, a cement that will harden under water. [Gk. *hudor,* water + *aulos,* pipe]

limestone, lithographic   *sed.* Dense, homogeneous, exceedingly fine-grained micritic limestone, having a pale creamy yellow or grayish color and often a conchoidal fracture; formerly much used in lithography and well-illustrated by the limestone from Solenhofen, Bavaria, Germany (Solenhofen stone).

limestone, madreporic   *sed.* Coral limestone, esp. one containing remains of the genus *Madrepora.*

limestone, magnesian   *sed.* Limestone consisting of 90-95% calcite and 5-10% dolomite.

limestone, marine   *sed.* Even- and fine-grained limestone, poor in fossils, and consisting of fine- to medium-grained calcite in a homogeneous mosaic or saccharoidal aggregate; it is difficult to determine whether clastic or biochemical processes predominated in its formation; some have formed by direct chemical precipitation.

limestone, micritic   *sed.* Limestone containing 90% or more micrite. Also named aphanitic limestone, matrix limestone, and calcilutite.

limestone, microclastic   *sed.* Variety of micritic limestone containing mechanical clastic, rather then chemically precipitated, grains.

limestone, microcrystalline   *sed.* Micritic limestone consisting of microscopically interlocking crystals.

limestone, mottled   *sed.* Limestone with narrow, branching, seaweedlike cylindrical masses of dolomite, often with a central tube or hole; may be inorganic or organic in origin. (Van Tuyl, 1916)

limestone, nodular   *sed.* = limestone, pellet. [L. *nodus,* knob, knot]

limestone, nummulitic   *sed.* Variety of foraminiferal limestone composed chiefly of nummulite shells, esp. of the genus *Nummulites.*

limestone, oolitic   *sed.* Calcarenitic limestone composed of abundant calcite ooliths. Also named calcitic oolite.

limestone, organic   *sed.* Limestone formed by the accumulation of minerals and materials precipitated from aqueous solutions through the action of organisms; term to be used in contrast to chemical limestone, which is formed through inorganic chemical processes.

limestone, pelagic   *sed.* Fine-grained limestone formed by the accumulation of the calcareous tests of floating organisms, esp. of foraminifers, e.g., *Globigerina.* [Gk. *pelagos,* sea]

limestone, pellet   *sed.* Limestone characterized

by having abundant pellets, e.g., limestones whose major constituents are fecal pellets (organic excrement) or carbonate muds that display rounded or ellipsoidal aggregates of grains of matrix material.

limestone, pelleted   *sed.* = limestone, pellet

limestone, phosphatic   *sed.* Limestone containing appreciable calcium phosphate minerals (mostly in the apatite group); these rocks are transitional and become phosphorite with decreasing calcite.

limestone, pisolitic   *sed.* Like an oolitic limestone (calcitic oolite) except the ooliths are greater than 2 mm in diameter and are thus named pisoliths. Also named calcitic pisolite or, more rarely, aragonitic pisolite.

limestone, quartz-flooded   *sed.* Limestone characterized by an abundance of quartz particles that were imported suddenly from an outside source by wind or water currents but that gradually die out upward and disappear within a few centimeters. (Shrock in 1948)

limestone, reef   *sed.* Bioconstructed limestone (boundstone) developed within the framework of one of the several types of reef rock.

limestone, sapropelic   *sed.* Limestone characterized by containing organic matter, either of vegetal or animal origin, in incipient stages of decomposition; rocks are black and have a dull fracture. [Gk. *sapros,* rotten + *pelos,* mud, clay]

limestone, secondary   *sed.* Limestone deposited from solution in the cracks and cavities of other rocks; esp. the limestone accompanying the salt and gypsum of the Gulf Coast, United States, salt domes.

limestone, shell   *sed.* Limestone consisting largely of fossil shells, e.g., madreporic limestone, containing corals, and encrinite, containing encrinal (crinoidal) remains.

limestone, sideritic   *sed.* Limestone in which meidum-grained, subhedral or euhedral siderite replaces calcite to a varying extent; some rocks consist of aggregates of shell fragments (unaltered or sideritized) cemented by calcite and siderite.

limestone, siliceous   *sed.* Limestone, usually thin, dark, and geosynclinal, consisting of an intimate admixture of calcite and chemically precipitated silica; term has also been applied to a limestone showing evidence of replacement of calcite by silica and to limestones that are arenaceous (containing quartz sand).

limestone, sparry   *sed.* = sparite

limestone, sparry   *meta.* Coarsely crystalline metamorphic marble.

limestone, stalactitic   *sed.* = stalactite

limestone, stylolitic   *sed.* Limestone characterized by a stylolite, a structural feature marked by an irregular and interlocking penetration of two parts, the columns, pits, and teethlike projections on one side fitting into their counterparts on the other; in cross section, resembles a suture or the tracing of a stylus. [Gk. *stylos,* pillar, column]

limestone, tuffaceous   *ig. extru., sed.* Limestone containing volcanic pyroclastic materials, e.g., sharp, angular, clear volcanic glass shards.

limestone, underclay   *sed.* Thin, dense, relatively unfossiliferous freshwater limestone underlying coal deposits; named because of its close relationship to underclay.

limestone, zebra   *sed.* Limestone banded by parallel sheet cracks filled with calcite. (Fischer in 1964)

limnogenic rock   *sed.* Any sedimentary rock formed by precipitation from fresh water, esp. that of a lake. [Gk. *limne,* pool + *genes,* born] (Grabau in 1924)

limonite rock   *sed.* Any rock whose major component is limonite (principally goethite), e.g., bog iron ore and gossan deposits.

limonite rock, oolitic   *sed.* Sedimentary rock containing closely or loosely packed, rounded to elongated and flattened ooliths of limonite, usually with a concentric structure but in some cases having a radial fibrous structure; films of chamosite may coat the ooliths, and the matrix may be sandy to clayey and also contain chamosite, siderite, dolomite, calcite, and organic debris.

limurite   *meta.* Contact metamorphic rock; formed between granite and limestone and composed of axinite (about 60%), diopside, and actinolite, with some quartz, titanite, calcite, pyrrhotite, magnetite, albite, and zoisite. [Michel Louis François Marie Limur, 1817-1901, French mineralogist of Vannes] (Frossard, before 1892)

lindinosite   *ig. pluton.* Alkali granite in which the major mafic is riebeckite (59%). [Lindinosa, Corsica] (Lacroix in 1922)

lindöite   *ig. diaschist.* Medium- to coarse-grained, light-colored dike rock, related to bostonite in texture and to solvsbergite in composition; the equivalent of an alkali granite. [Lindö island, Norway] (Brögger in 1894)

linophyre   *ig.* Any igneous rock characterized by a linophyric texture; porphyry in which the phenocrysts are arranged in lines or streaks. [L. *linea,* line] (Cross and others in 1906)

linosaite *ig. extru.* Basaltic rock either carrying a small amount of feldspathoid or showing its affinities to the alkalic rocks by the presence of sodic mafics or by its association with feldspathoidal rocks; typified by rocks from Il Fosso, Linosa, Italy. [Linosa, Italy] (Johannsen in 1938)

lintonite *ig., altered.* Agatelike variety of thomsonite that is greenish or has alternating bands of pink and green; from the north shore of Lake Superior. [Laura Alberta Linton, 1853-   ?, American chemist and physician, who analyzed the material at the University of Minnesota] (Peckham and Hall in 1880)

liparite *ig. extru.* = rhyolite. Name also has been applied to the mineral species chrysocolla, fluorite, and talc. [Lipari Islands, Tyrrhenian Sea] (Roth in 1861)

lipotexite *migmat.* Term applied to the nonliquified mafic materials occurring within a magma formed by anatexis. Also spelled lipotectite. [Gk. *lipos,* fat (like fat, insoluble in water) + *tektos,* fused]

liptobiolite *sed.* 1. Resistant plant materials that are left behind after the less resistant parts of the plant have completely decomposed, e.g., resin, gum, wax, amber, pollen, and copal. 2. = liptobiolith. [Gk. *lipos,* fat + *bios,* life]

liptobiolith *sed.* Combustible organic rock formed by an accumulation of liptobiolites, e.g., pollen peat or spore coal. [Gk. *lipos,* fat + *bios,* life]

listvenite *meta.* Greenish to yellowish contact metamorphic rock, derived from dolomite and composed of talc, quartz, and magnesite; term used locally in the gold district of Beresowsk, the Urals, Soviet Union. Also spelled listwänite. (Rose in 1842)

litchfieldite *ig. pluton.* White, somewhat schistose, plutonic nepheline syenite composed of dominant albite associated with nepheline and lepidomelane and, more rarely, cancrinite and sodalite. [Litchfield, Kennebec County, Maine] (Bayley in 1892)

litharenite *sed.* Sandstone (arenite) characterized by a rock-fragment content in excess of feldspar; wide range of composition, depending largely on types of rock fragments present, e.g., calclithite (rock fragments are chiefly limestone and/or dolomite), phyllarenite (rock fragments are chiefly shale or slate). [Gk. *lithos,* stone + arenite]

litharenite, feldspathic *sed.* Litharenite containing appreciable feldspar; specifically, sandstone (arenite) containing 10-50% feldspar, 25-90% fine-grained rock fragments, and 0-65% quartz. (McBride in 1963)

lithocalcarenite *sed.* Calcarenite composed of limestone rock fragments derived from aggregation processes or from the subaqueous reworking of biochemical, chemical, and fine-grained detrital limestones. [Gk. *lithos,* stone]

lithocalcilutite *sed.* Calcilutite of inorganic aggregation or origin, e.g., one composed of bahamite. [Gk. *lithos,* stone]

lithocalcirudite *sed.* Calcirudite composed of fragments of pre-existing limestones (often bioconstructed). [Gk. *lithos,* stone]

lithocalcisiltite *sed.* Calcisiltite composed of fragments of pre-existing limestone between lutite and arenite size. [Gk. *lithos,* stone]

lithodolarenite *sed.* Dolarenite consisting of abundant dolomite rock fragments. [Gk. *lithos,* stone]

lithodololutite *sed.* Dololutite consisting of abundant dolomite rock fragments. [Gk. *lithos,* stone]

lithodolorudite *sed.* Dolorudite consisting of abundant dolomite rock fragments. [Gk. *lithos,* stone]

lithodolosiltite *sed.* Dolosiltite consisting of abundant dolomite rock fragments. [Gk. *lithos,* stone]

lithographic stone *sed.* = limestone, lithographic

lithoidite *ig. extru.* Rhyolite without phenocrysts, made up entirely of microcrystalline felsic matter. [Gk. *lithos,* stone] (von Richthofen in 1860)

lithomarge *sed.* Smooth, compact variety of kaolin. [Gk. *lithos,* stone + L. *marga,* marl]

lithophysa *ig. extru.* Hollow, bubblelike, crystalline spherulite, often concentric in structure, and composed of quartz, alkali feldspar, and other minerals; found in pitchstone, obsidian, rhyolite, and similar glassy rocks. Lithophysae is the plural, and lithophysa is the singular. Also named stone bubble. [Gk. *lithos,* stone + *physa,* air bubble, bellows] (von Richthofen in 1860)

lithosiderite *meteor.* = meteorite, stony-iron. [Gk. *lithos,* stone + *sideros,* iron] (Brezina in 1885)

lithospor *meteor.* Type of meteorite in which silicate minerals are sporadic in metal. [Gk. *lithos,* stone (alluding to silicates) + sporadic]

lithotype *sed.* Term applied to rock species in coal petrology, e.g., clarain, durain, fusain, and vitrain. A lithotype is composed of macerals (analogous to minerals in noncarbonaceous rocks), represented by terms such as telinite (wood and bark), micrinite (fine organic debris), cutinite (cuticle), resinite (resin), and alginite (algae). [Gk. *lithos,* stone + type]

lithoxyle *sed.* Opalized wood in which the original woody structure is observable. [Gk. *lithos,* stone + *xylon,* wood] lithwacke, feldspathic   *sed.* Lithic graywacke (over 15% matrix) in which rock fragments exceed feldspar but the latter forms 10% or more of the sand fraction. [Gk. *lithos,* stone + wacke] (Pettijohn and others in 1973)

liver rock   *sed.* Freestone (sandstone) that breaks or cuts as readily in one direction as in another; its workability is not affected by stratification.

liver stone   *sed.* Bituminous limestone that gives off a bad odor upon being struck; stinkstone.

liverite   *sed.* Term sometimes applied to elaterite from Strawberry River, Utah. Also spelled liversite. [liver, alluding to its appearance] (Barb in 1944)

llanite   *ig. aschist.* Mottled, red and gray, albite-bearing rhyolite porphyry consisting of phenocrysts of red alkali feldspar and blue quartz in an aphanitic groundmass composed of quartz and feldspar (microcline and albite) with biotite, fluorite, magnetite, apatite, and zircon. Also spelled llanoite. [Llano County, Texas] (Iddings in 1904)

llanoite   *ig. aschist.* Alternate spelling of llanite; term commonly used by lapidaries.

loam   *sed.* Loose-textured soil consisting of a mixture of sand and clay containing organic matter. [O. Eng. *lam,* loam]

loam, lake   *sed.* Loess that may have been formed by deposition in lakes.

lodestone   *ig., meta.* Massive variety of magnetite showing polarity and acting like a magnet when freely suspended. Also spelled loadstone. [M. Eng. *lode,* way; apparently alluding to the guiding principle of a magnetic compass]

lodranite   *meteor.* Stony-iron meteorite consisting of a friable aggregate of granular olivine and orthopyroxene in a discontinuous aggregate of nickel-iron, the three principal phases being present in approximately equal amounts by weight. [Lodran, near Multan, Pakistan] (Meunier in 1882)

loess   *sed.* Widespread, buff to light yellow, homogeneous, commonly nonstratified, porous, friable eolian deposit, consisting essentially of a silt-sized fraction with a lesser clay-sized fraction; the silty material consists of predominant quartz, with feldspars, micas, chlorite, calcite, dolomite, volcanic shards, carbonaceous material, and other minerals, and the clayey materials are illite, montmorillonite, kaolinite, quartz, cristobalite, and dolomite. [G. *loess;* akin to dialectal Swiss *lösch,* loose; first applied in the Rhine valley]

loess, cold   *sed.* Loess derived from glacial outwash and formed in zones or garlands about the Pleistocene ice sheets, as in northern Europe and in northcentral United States.

loess, warm   *sed.* Loess derived from desert dust, such as that formed in the inland basins and steppes encircling the modern deserts of central Asia.

loessite   *sed.* Lithified loess, comparable to siltstone. [loess]

loessoide   *sed.* Dutch term for deposits believed to be derived from loess but reworked and redeposited by streams, possibly with an admixture of residual materials from in-place decomposition; deposits like this occur in southern Netherlands in Limburg province. [loess]

Lithophysae (stone bubbles) in pitchstone vitrophyre with spherulites. Yellowstone Park, Wyoming. 22 cm.

loesspuppen *sed.* Small carbonate-rich concretions occurring in loess, 1-2 cm across, and characterized by an outer dense shell and a cracked, clay-filled interior; many irregular and complex forms. Also named loessmannchen (loess manikin), loesskindchen (loess child), and loesskindel (loess small tuber). [loess + G. *Puppen,* puppet, doll]

loferite *sed.* Limestone or dolomite riddled by shrinkage pores; illustrated by the Triassic Dachstein formation (Lofer facies) in Salzburg, Austria; similar to a dismicrite. [Lofer facies] (Fischer in 1964)

loipon *sed.* General term for a residual surficial layer produced by prolonged chemical weathering and composed largely of materials derived from the original constituents of the source rock, e.g., gossan and some bauxite deposits and duricrusts. [Gk. *loipos,* remains, residue] (Shrock in 1947)

lokbatanite *sed.* Bituminous material derived from a mud volcano in the Soviet Union. [Lok Batan, Soviet Union] (Kovalevsky and Kochmarev in 1939)

lublinite *sed.* Calcitic moonmilk or moldlike encrustation on marl, composed of a microcrystalline aggregate of delicate fibrous needles. Also incorrectly spelled lubinite. [Lublin, southeastern Poland] (Morozewicz in 1907)

lucianite *sed.* Clayey material consisting of a colloidal hydrated magnesium silicate, which swells up to many times its original volume when immersed in water. [Santa Lucia, hacienda near Mexico City, Mexico] (Hilgard in 1916)

luciite *ig. hypab.* Fine-grained diorite composed essentially of plagioclase (andesine-labradorite) and blue-green hornblende, with minor quartz, orthoclase, and biotite; resembles malchite but is somewhat coarser. [Luciberg, Zwingenberg, Odenwald, Germany] (Chelius in 1892)

lucullite *sed.?* Black, carbonaceous marble, esp. that from Egypt (an island in the Nile). Also named Lucullan and Marmor Luculleum (Pliny). [Lucius Licinius Lucullus, ca. 110-ca. 56 B.C., Roman consul, the first to bring the rock to Rome from the Nile] (Agricola in 1546)

lugarite *ig. hypab.* Coarse-grained sill-rock containing shining black prisms of barkevikite (up to 3 in), small titanaugite crystals, and rectangular crystals of zoned labradorite in a groundmass of cloudy grayish analcime; resembles a labradorite-bearing ijolite in which nepheline is proxied for by analcime. [Lugar, Ayrshire, Scotland] (Tyrrell in 1912)

luhite *ig. hypab.* Calcite-rich hybrid rock in which there was assimilation of sediment; composed of olivine and titanaugite in a groundmass of clinopyroxene, melilite, hauyne, nepheline, and calcite; transitional between the polzenites and melilite-nepheline basalts. [Luh, northern Bohemia (now Czechoslovakia)] (Scheumann in 1913)

lujavrite *ig. pluton.* Coarse-grained eudialyte-bearing nepheline syenite, consisting of thin, nearly parallel, tablets of feldspar (microcline microperthite) with interstitial nepheline grains and aegirine needles. Also spelled lujauvrite, luijaurite. [Lajavr (Luijaur) Urt, Lapland, Kola Peninsula] (Ramsay in 1890)

lumachelle *sed.* = marble, fire. [It. *lumachella,* snail]

lunarite *lunar.* General term applied to the light-toned, brightly reflecting surface rocks of the lunar highlands (terrae). [L. *luna,* moon] (Spurr in 1944)

lundyite *ig. extru.* Alkali granite characterized by sodium orthoclase (72%), quartz (17%), and katophorite (an amphibole) (10%). [Lundy Island, Devonshire, England] (Hall in 1914)

lupatite *ig. extru.* Poorly described nepheline-feldspar porphyry from Zambesi; perhaps a synonym for analcime bearing kenyte. [Lupata gorge, Zambesi, Mozambique] (Mennell in 1929)

luscladite *ig. pluton.?* Olivine theralite or essexite composed of olivine and biotite imbedded in a small amount of nepheline between plagioclase (calcic ?) plates with orthoclase rims. [Ravin de Lusclade, Mont Doré, Auvergne, France] (Lacroix in 1920)

lusitanite *ig. pluton.* 1. Dark-colored albite syenite containing riebeckite, orthoclase, microcline microperthite, albite, and at times a little quartz and aegirine. [Lusitania, ancient name for Portugal] (Lacroix in 1916) 2. *IUGS,* plutonic rock in which Q is less than 5 or F is less than 10, P/(A + P) is less than 10, and M is 45-75.

lutalite *ig. extru.* Olivine-bearing leucitite composed of augite (42%), leucite (16%), olivine (6%, and mostly as phenocrysts), in a matrix of biotite, iron oxides, and glass (potentially nepheline and plagioclase). [Lutale Crater, Bufumbira lava fields, southwestern Uganda] (Holmes in 1937)

lutite *sed.* General term applied to clastic sedimentary rocks composed of silt- or clay-sized particles irrespective of composition, e.g., clay, mudstone, siltstone, shale; in each case the clast size is less than 1/16 mm in diameter. [L. *lutum,* mud, clay]

lutogenite *meta.* Rock having the same composition as kinzigite but formed by the purely isochemical metamorphism of sediments of a corresponding claylike composition. [L. *lutum,* mud, clay + *genus,* birth] (Parras in 1946)

luxulianite   *ig. pluton., autometa.* Tourmalinized granite containing orthoclase phenocrysts (filled with small tourmaline needles), quartz, mica (partially replaced by tourmaline), and cassiterite; apparently autometamorphosed granite in which replacement stopped at a halfway stage. Also spelled luxulyanite, luxuliane, and luxullianite. [Luxulyan railway station, Cornwall, England] (Pisani in 1864)

Lydian stone   *sed.* = basanite. Also named lapis Lydius and lydite. [Lydia, ancient country of western Asia Minor]

lydite   *sed.* = basanite. Also Lydian stone, lapis Lydius. [Lydian stone]

M

maberyite   *sed.* General class of North American bitumens characterized by containing a notable proportion of sulfur compounds; includes the so-called Lima oils and Canadian oils. Also incorrectly spelled mayberyite. |Charles Frederic Mabery, 1850–1927, American chemist of Cleveland, Ohio| (Peckham in 1895)

macedonite   *ig. extru.* Fine-grained, bluish black, aphanitic igneous rock composed of biotite, orthoclase, sodic-plagioclase, olivine, and other mafics; related to mugearite, a trachyandesite. |Macedon district, Victoria, Australia| (Skeats and Summers in 1912)

maculose rock   *meta.* General name applied to a group of contact metamorphic rocks with spotted, knotted or gnarled structures, like knotenschiefer, fruchtschiefer, and certain hornfels. |L. *macula,* spot, blemish| (Holmes in 1919)

macusanite   *ig. extru.* = americanite. Also spelled mancusanite. |Macusani, Peru|

madeirite   *ig. hypab.* Blackish green gabbro porphyry consisting of large black augite and green olivine crystals in a fine-grained groundmass composed chiefly of zoned plagioclase (labradorite-bytownite). |Madeira, Spain| (Gagel in 1912)

madupite   *ig. extru.* Extrusive rock containing phenocrysts of phlogopite, diopside, and perovskite in a brown glassy matrix with the composition of leucite; related to wyomingite but having more mafic minerals than leucite. |Am. Ind. *madupa,* sweet water; Sweetwater County, Wyoming| (Cross in 1897)

maenaite   *ig. hypab.* Plagioclase-bearing bostonite, differing from normal bostonite in being richer in calcium and poorer in potassium; probably a result of alteration. |Lake Maena, Gran, Norway| (Brögger in 1898)

mafite   *ig. extru.* Dark-colored aphanite with abundant mafic minerals; used in contrast with felsite. |magnesium + ferrum (mafic)|

mafraite   *ig. dike.* Theralite composed of labradorite (with sodium sanidine mantles), pyroxene, magnetite, and hornblende crystals (several cm long). |Tifâo de Mafra, Cintra, Portugal| (Lacroix in 1920)

mafurite   *ig. extru.* Rock related to olivine leucitite in which kalsilite is present instead of leucite. |Mafura, Uganda| (Holmes in 1945)

magallanite   *sed.* Variety of hard asphalt thrown up by the sea on the coast of Magallanes. |Magallanes, Argentina| (Fester, Cruellas, and Gargatagli in 1937)

magma   *ig.* Naturally occurring, high-temperature, mobile (fluid) rock material, generated within the Earth, from which igneous rocks are derived through crystallization and related processes. |traceable to Gk. *massein* to knead|

magmabasalt   *ig. extru.* = limburgite. (Boricky in 1872)

magmagranite   *ig. pluton.* Any granite whose formation is from the crystallization of a magma rather than from granitization or related processes.

magmatite   *ig.* General term for any rock formed from the solidification (usually crystallization) of a magma; synonym for igneous rock. |magma|

magnalite   *ig., altered.* Alteration product formed from basalt and containing magnesium and aluminum hydrated silicates; occurring at Oberpfalz, Bavaria, Germany. |magnesium + aluminum| (Richarz in 1920)

magnesite rock   *meta.* Magnesite-rich rock formed by the metasomatic alteration of limestone, marble, or dolomite; in addition to magnesite, other minerals might be dolomite, calcite, talc, serpentine, brucite, quartz, graphite, and pyrite; some specific varieties include listvenite and sagvandite.

magnetite rock   *ig. pluton.* = magnetitite

magnetite rock   *meta.* Banded or schistose metamorphic rock in which magnetite is the predominant mineral; other associated minerals may include hematite (including martite), quartz, grunerite, actinolite, almandine (and other garnets), hypersthene, and others. Varieties include magnetite schist, magnetite-grunerite schist, magnetite-garnet schist.

magnetitite   *ig. pluton.* Igneous rock composed essentially of magnetite; very minor apatite, hematite, siderite, pyrite, and copper sulfides may be present. Varieties, containing characteristic amounts of other minerals, include augite m., bronzite m., corundum m., diopside m., högbomite m., hypersthene m., olivine m., and spinel m. |magnetite| (Johannsen in 1938)

malaysianite   *tekt.* Name applied to tektites from Malaysia, including the Philippines, the Malay Peninsula, Indonesia, and others. (Beyer in 1933)

malchite   *ig. diaschist.* Fine-grained, black, gray, or greenish, holocrystalline lamprophyre consisting of small phenocrysts of hornblende, labradorite, and rare biotite, in a groundmass of hornblende, andesine, and minor quartz. |Malchen, local name for Melibocus, Alsbach, Germany| (Osann in 1892)

malchite, mica   *ig. diaschist.* Malchite in which biotite more or less completely proxies for hornblende, and the plagioclase is more sodic. Originally named Glimmermalchite, German for mica malchite.

malignite   *ig. pluton.* 1. Nepheline syenite having more than 5% nepheline and roughly equal amounts of pyroxene and potassium feldspar; originally a name given to a group of rocks of variable composition. |Maligna River, Ontario, Canada| (Lawson in 1896)   2. *IUGS*, plutonic rock in which F is 10-60, P/(A + P) is 10 or less, and M is 30-60.

malignite, melanite   *ig. pluton.* = ledmorite

malmstone   *sed.* Hard, grayish white, cherty sandstone whose matrix contains small opaline globules derived from sponge spicules that once filled now-empty molds; occurs in Surrey and Sussex, England; term has also been applied to chalky or marly rock. Also named malm rock. |O. Eng. *mealm,* sand|

maltha   *sed.* Thick mineral pitch, formed by drying petroleum; also a variety of ozokerite. Also named malthite. |L. from Gk. *malthe,* mixture of wax and pitch for calking ships| (Pliny, *Natural History*)

malthacite   *sed.* White to slightly yellowish, soft fuller's earth formed on basalt in Bohemia and elsewhere. |Gk. *malthakos,* soft| (Breithaupt in 1837)

mancusanite   *ig. extru.* = americanite. An alternate spelling of macusanite.

mandchurite   *ig. extru.* Glassy nepheline basanite consisting of pyroxene (32%), nepheline (11%), glassy matrix (42%), and other minor minerals. Also spelled mandschurite, manchurite. |Manchuria| (Lacroix in 1923)

manganese-silicate rocks   *meta.* Group of metamorphic rocks characterized by an abundance of various manganese silicates as essential constituents, usually well-banded with alternating quartz layers. Examples include gondite, amphibole gondite, rhodonite gondite, pyroxmangite quartzite, and manganophyllite rocks with magnetite, barite, and manganese phosphates.

manganiferous sedimentary rocks   *sed.* Sedimentary rock characterized by the presence of manganese minerals; manganese oxide minerals (pyrolusite, psilomelane, lithiophorite and others) occur as concretions, coatings, cements, and dendrites associated with sandstone, shale, vein quartz, and other rocks; rhodochrosite forms concretions in shale and may also be an important component in manganiferous carbonate and oxide-carbonate strata.

manganolite   *sed.* General name for sedimentary rocks composed of manganese minerals, esp. of the oxide minerals, pyrolusite, psilomelane, wad, lithiophorite, and others; essentially the same as manganiferous sedimentary rock. |manganese| (Wadsworth in 1891)

manganophyllite rock   *meta.* Rare metamorphic rock containing manganoan biotite (manganophyllite), manganese phosphates, magnetite, and barite.

mangerite   *ig. pluton.* Hypersthene-bearing alkali monzonite, containing microperthite and plagioclase with varying amounts of mafics, esp. hypersthene. |Manger, Norway| (Kolderup in 1903)

manjak   *sed.* Local name for a natural black variety of bitumen, having a brilliant luster and conchoidal fracture, found in the Barbados. Also spelled munjack. |Carib *manjak*| (Used earlier than 1897)

manjakite   *ig.* Equigranular rock composed of garnet, biotite, pyroxene, and variable amounts of feldspar, magnetite, hypersthene, and labradorite. |Possibly Carib *manjak;* alluding to its appearance|

marahuite   *sed.* Algal coal, similar to torbanite, and of light brown, compact, earthy appearance. Also spelled marahunite. Formerly named Turfa de Marahú. |Marahú, Bahia, Brazil| (Derby in 1907)

marble   *meta.* 1. Metamorphic rock composed essentially of fine- to coarse-grained calcite and/or dolomite, usually with a granular, saccharoidal texture; some petrologists separate the two compositions, calcite and dolomite, by using the terms marble (for calcite) and dolomite marble (for dolomite). Many varieties are recognized on the basis of characterizing accessory minerals, e.g., brucite m., diopside m., forsterite m., graphite m., grossular m., grunerite m., periclase m., phlogopite m., pyrrhotite m., serpentine m., talc m., tremolite m., wollastonite m. 2. In commerce, any carbonate rock, including true metamorphic marble as well as many types of limestone (sedimentary marble), that will take a polish and can be used as an ornamental stone. A few noncarbonate rocks, like serpentinite, have also been named marble. [L. *marmor,* marble]

marble, breccia *sed., meta.* Any marble composed of angular fragments.

marble, calc-silicate *meta.* Marble in which calcium silicate or magnesium silicate minerals are conspicuous. See marble for a list of varieties that include many of these silicates.

marble, calcite *meta.* = marble

marble, Carrara *meta.* White, fine-grained statuary marble from Carrara, Massa e Carrara province, Tuscany, Italy.

marble, dolomite *meta.* Marble in which dolomite is the essential carbonate mineral.

marble, dolomitic *meta.* 1. = marble, dolomite. 2. Calcite marble in which there is 5–50% dolomite mineral.

marble, fire *sed.* Dark-brown shell limestone (sedimentary marble), with brilliant firelike or chatoyant internal reflections proceeding from the shells; from Bleiberg in Carinthia, Austria. Also named lumachelle.

marble, flaser *meta.* Marble with a phacoidal (lenticular) texture, where ellipsoidal mineral aggregations occur in a matrix that is finer-grained. [G. *Flaser,* vein, streak]

marble, landscape *sed.* Fine-grained limestone (sedimentary marble) characterized by dark conspicuous dendritic markings that suggest natural scenery (trees, shrubbery), e.g., the argillaceous limestone from near Bristol, England.

marble, magnesian *meta.* Calcite marble containing some dolomite (usually less than 15%).

marble, onyx *sed.* Compact, translucent, parallel-banded, variety of sedimentary marble composed of calcite (or rarely aragonite) and capable of taking a good polish so it can be used as a decorative stone. Numerous names have been applied to this rock, including Algerian onyx, California onyx (aragonitic), cave onyx, Gibraltar stone, Mexican onyx, tecali.

marble, piedmontite *meta.* Marble rich in piedmontite, usually associated with manganese-silicate rocks.

marble, Potomac *sed.* Subangular, limestone-pebble conglomerate that has abundant ferruginous cement; occurs in the Triassic basins of the eastern United States, esp. in Virginia and Maryland. [Potomac River]

marble, rhodochrosite *hydrothermal?* Rare marblelike rock in which the major carbonate is rhodochrosite instead of calcite or dolomite. See rhodochrosite rock.

marble, ruin *sed.* Brecciated limestone that, when cut and polished, gives a mosaic effect suggesting ruins or ruined buildings; sedimentary marble.

marble, scapolite *meta.* Gneissoid calc-silicate rock in which the adjacent bands are markedly different, e.g., scapolite-zoisite, quartz-phlogopite, and diopside-tremolite bands.

marble, sedimentary *sed.* Any sedimentary limestone that will take a good polish and can be used as an ornamental stone; these may be banded, brecciated, fossiliferous, colorfully marked, well-crystallized, or have other interesting features.

marble, serpentine *meta.* Any serpentinite that can be cut and polished and used as a decorative stone; often applied to the variety verd antique.

marcasite rock *sed.* Any sedimentary rock in which marcasite is an important component; nodules (and crystals) of marcasite are common in limestone and also occur in shale, sandstone, and coal; marcasite has also been observed as a rare cement in feldspathic graywacke.

marchite *ig. pluton.* Pyroxenite composed of enstatite and diopside. [March River, Moravia, Czechoslovakia] (Kretschmer in 1917)

mareconite *ig. extru.* = marekanite

marekanite *ig. extru.* Black glassy obsidian balls, often showing concave indentations, that remain after perlite is broken apart at some localities. An Apache tear is an example. [Marekanka River, near Okhotsk, Siberia] (Pallas in 1793)

mareugite *ig. pluton.* Bytownite-hauyne rock occurring as fragments in ordanchite; probably hauyne gabbro. [Mareuges, Auvergne, France] (Lacroix in 1917)

marienbergite *ig. extru.* Andesine-bearing phonolite in which nepheline is entirely replaced by natrolite; the green-gray to ash-gray groundmass contains phenocrysts of sanidine, andesine, and zoned augite. [Marienberg, Aussig, Bohemia] (Johannsen in 1938)

mariposite *meta.* Dolomite marble mottled by green chromian phengite (mariposite mineral) and containing quartz and pyrite. [Mariposa County, California] (Silliman in 1868)

mariupolite *ig. pluton.* Albite-nepheline syenite that also contains acmite and biotite, with minor zircon and britholite. [Mariupol, Ukraine, Soviet Union] (Morozewicz in 1902)

markfieldite *ig. aschist.* Granite porphyry with plagioclase (labradorite) phenocrysts in a graphic granite groundmass. [Markfield, Leicestershire, England] (Hatch in 1909)

marl *sed.* Gray calcareous clay, or intimate mixture of clay and particles of calcite (often shell fragments) or dolomite; specifically, an earthy substance containing 35-65% clay and 65-35% carbonate. In Europe the term sometimes is applied to more consolidated rocks, e.g., argillaceous limestone. [Traceable to L. L. *margila,* dim. of *marga,* marl]

marl, calcareous *sed.* Soft, calcareous clay-rich earthy material, often barely consolidated, with or without distinct fragments of shells; common marl

marl, carnallitic *sed.* Marl in which carnallite has been introduced by brines; at Yorkshire, England, this material contains about 10% carnallite, 17% halite, as well as clay minerals, quartz, sylvite, anhydrite, magnesite, mica, and others.

marl, chalky *sed.* Grayish marl rich in chalk and containing up to 30 % clayey material; illustrated by the Chalk Marl near the base of the English Chalk.

marl, Folkestone *sed.* Stiff fossiliferous marl, varying in color from light gray to dark blue. Also named gault. [Possibly Folkestone, Kent, England]

marl, greensand *sed.* Marl containing glauconite sand; esp. common in the Coastal Plain areas of the southeastern United States.

marl, shell *sed.* Marl containing abundant molluscan shells, esp. in the Coastal Plain of the southeastern United States.

marlekor *sed.* Calcareous concretions in Late Glacial clays; they occur primarily in silt layers and are flattened parallel to the bedding, the bedding laminations passing through them; they often show complex bizarre forms, some with unusual bilateral symmetry. Also named Imatra stone. [Sw. *marlekor,* possibly related to Eng. marl and core] (Used as early as Linnaeus)

marlite *sed.* 1. = marlstone. 2. Somewhat indurated sheet or crust formed on the bottoms and shores of lakes by the intergrowth or cementation of a number of marl biscuits. [marl]

marloesite *ig. extru.* Fine-grained, pale gray, porphyritic extrusive rock, related to andesite, that has phenocrysts of platy albite and bronze micaceous aggregates after olivine, in a groundmass of much augite, sodic-plagioclase, and iron oxides. [Marloes, parish in Pembrokeshire, England] (Thomas in 1911)

marlstone *sed.* 1. Indurated marl; earthy or impure argillaceous limestone. 2. General term for various rocks, like calcareous mudstone, muddy limestone, shale, dolomite, oil shale and others, whose lithologic characters are not readily determined. 3. Hard ironstone worked for iron ore in England; calcareous and sideritic oolite in which the ooliths and shell chips are cemented by carbonate.

marmolite *meta.* Thin, foliated variety of serpentinite that is greenish white, bluish white, or pale asparagus green and consists of brittle, separable lamellae; from Hoboken, New Jersey. [Gk. *marmairein,* to shine; alluding to its pearly and somewhat metallic luster] (Nuttall in 1822)

Marlekor (calcareous clay concretions). Riga, Vermont. 4 cm.

marosite   *ig. pluton.* Rock intermediate between shonkinite and nepheline diorite, consisting of biotite, augite with hornblende rims, sanidine, calcic-plagioclase, nepheline, sodalite, apatite, and iron oxides. [Pic de Maros, Celebes] (Iddings in 1913)

marscoite   *ig.* Hybrid rock, formed by the partial absorption of granite material by a gabbro magma, that contains quartz and feldspar in a gabbroid matrix of abnormal composition; name intended for local use, since it did not connote a new rock type. [Marsco, Skye, Scotland] (Harker in 1904)

martinite   *ig. extru.* Fine-grained vesicular tephrite lava consisting of phenocrysts of leucite, feldspar, and augite in a dark groundmass composed of sodic labradorite, orthoclase, augite, leucite, olivine, magnetite, and apatite. [Croce di San Martino, Italy] (Johannsen in 1938)

marundite   *ig. dike.* Corundum-margarite pegmatite rock from Malelane, Transvaal; margarite (63%), corundum (33%), with minor biotite, plagioclase, apatite, and other minerals. [margarite + corundum] (Hall in 1922)

masafuerite   *ig. dike.* Dark-colored picrite, consisting of over 50% olivine and no other phenocrysts, in a groundmass of augite, calcic-plagioclase, ilmenite, and magnetite. [Masafuera Island of the Juan Fernandez group, Chile] (Johannsen in 1937)

masanite   *ig. hypab.* Quartz monzonite porphyry containing phenocrysts of zoned oligoclase and corroded quartz in a fine-grained groundmass composed of orthoclase, quartz, biotite, and others. [Masanpo, southern Korea] (Koto in 1909)

masanophyre   *ig. hypab.* Variety of masanite porphyry in which the oligoclase phenocrysts are mantled with orthoclase and in which the groundmass contains blue-green hornblende and sphene. [masanite] (Koto in 1909)

maw-sit-sit   *meta.* Bright green rock composed of chromian jadeite and albite and showing green to black spots and veins; used as a gem and ornamental stone. Also named jade-albite. [Maw-sit-sit, Tawmaw region, Burma] (Gübelin in 1965)

mayaite   *meta.* Series of jadelike rocks grading from tuxtlite (pyroxene in which diopside and jadeite molecules are about equal) to nearly pure albite; objects were made of these rocks by the ancient Mayas in Central America. [Maya peoples] (Washington in 1922)

mayberyite   *sed.* = maberyite

medfordite   *sed.* Attractive green and white jasper from Big Butte area, Jackson County, Oregon. [Medford, Jackson County, Oregon]

meerschaum   *meta.* Compact nodular to earthy-massive variety of sepiolite, often found in serpentine and magnesite where it is an alteration product; easily carved and takes a good polish; used extensively for pipe bowls. [G. *Meer,* sea + *Schaum,* foam; because of its porosity it can float]

megabreccia   *sed., meta.* Brecciated rock containing large blocks that are randomly oriented and which range from 1 m to more than 100 m in horizontal dimension; the rock may be tectonic or result from gravitational sliding on slopes. [Gk. *megas,* large + breccia]

meimechite   *ig. extru.* Ultramafic volcanic rock composed of abundant olivine (forsterite) phenocrysts (usually altered) in a serpentine-rich or glassy groundmass; the extrusive equivalent of kimberlite. [Meimecha Kotuj, Siberia] (Kotulsky in 1943)

mela-   *ig.* Common prefix meaning dark-colored or black; in many rock names the prefix indicates the rock contains more than 50% but less than 95% mafic minerals, but this is not a rigid requirement. Rock examples include the following: mela-andesite, melabasalt, meladiabase, meladiorite, melagabbro, melagranodiorite, melanorite, melasyenite, melatonalite. [Gk. *melas,* black]

melagabbro   *ig. pluton.* Gabbro whose color index exceeds 70; dark and close to ultrabasic in composition. [Gk. *melas,* black + gabbro]

melagabbroid   *ig. pluton. IUGS,* plutonic rock satisfying the definition of gabbroid, and in which pl/(pl + px + ol) is 10-35%. [Gk. *melas,* black + gabbroid]

melange   *sed.* Sizable body of rock characterized by the inclusion of fragments and blocks of all sizes, both native and exotic, embedded in a fragmented and generally sheared matrix of more tractable material. [F. *mélange,* mixture, blending] (Greenly in 1919)

melanhydrite   *ig. weathered.* = palagonite. [Gk. *melanos,* black + *hudor,* water] (Krantz in 1859)

melanophyre   *ig. hypab., extru.* Field term for any dark-colored, fine-grained igneous rock with phenocrysts. [Gk. *melanos,* black]

melanorite   *ig. pluton.* Norite whose color index exceeds 70; dark and close to ultrabasic in composition. [Gk. *melas,* black + norite]

melaphyre   *ig. hypab., extru.* Obsolete term equivalent to melanophyre; at one time limited to Carboniferous and Permian basalt. [Gk. *melas,* black] melikaria   *sed.* Boxwork or honeycomblike structure, usually consisting of quartz vein fillings, that has weathered out of a septarian concretion (septarium). |Gk. *melikeron,* honeycomb|

melilithite   *ig. extru.* = melilitholith. |melilite| (Loewinson-Lessing in 1901)

melilitholith   *ig. extru.* Light-colored rock composed almost entirely of melilite. |melilite| (Johannsen in 1938)

melilitite   *ig. extru.* Extrusive rock composed of melilite and augite but without olivine; there may be minor feldspathoids and rarely plagioclase. |melilite| (Prior in 1902)

melilitolite   *ig. pluton.* Plutonic, coarser-grained, equivalent of melilitite. |melilite|

melmafite   *ig. extru., pluton.* General name for an igneous rock composed of melilite and other mafic minerals, e.g., melilitite and melilitolite. |melilite + mafic| (Hatch, Wells, and Wells in 1961)

melteigite   *ig. pluton.* 1. Dark-colored end-member of the urtite-ijolite plutonic series that contains less than 50% nepheline and abundant mafic minerals, esp. green pyroxene. Varieties, based of characterizing accessory mafics, include: barkevikite m., biotite m., hornblende m., and melanite m. |Melteig, Fen district, Norway| (Brögger in 1921)   2. *IUGS,* plutonic rock in which F is 60-100, M is 70-90, and sodium exceeds potassium.

mengwacke   *sed.* Wacke containing from 33-90% unstable materials. |G. *mengen,* to mix, mingle + wacke| (Fischer in 1934)

mengwacke, quartz   *sed.* Wacke with 10-33% unstable materials. (Fischer in 1934)

menilite   *sed.* 1. Opaque, grayish to brown, impure opal, occurring in flattened nodular concretions in clayey shale at Ménilmontant and St. Ouen, near Paris, France. |Ménilmontant, France| (Saussure in 1797)   2. = shale, menilite.

merismite   *migmat.* Chorismite (migmatite) showing an irregular penetration of diverse rock units. |Gk. *meristos,* divided|

mesabite   *sed.* Ochreous goethite abundant at Mesabi, Minnesota. |Mesabi, Minnesota| (Winchell in 1893)

mesocoquina   *sed.* Clastic limestone composed of weakly cemented shell fragments of sand size (1/16 mm to 2 mm in diameter) or less. |Gk. *mesos,* middle + coquina|

mesopegmatophyre   *ig. aschist.* Pegmatophyre consisting of more than 5% dark minerals. |Gk. *mesos,* middle + pegmatophyre|

mesoperthite   *ig.* Perthitic feldspar consisting of an intimate mixture of about equal amounts of plagioclase (usually albite or oligoclase) and potassic feldspar; intermediate between perthite and antiperthite. |Gk. *mesos,* middle + perthite|

mesosiderite   *meteor.* Stony-iron meteorite composed of approximately equal amounts of nickel-iron and silicates; the major silicates are hypersthene and plagioclase (bytownite-anorthite). |Gk. *mesos,* middle + siderite; halfway between stony meteorite and iron meteorite| (Rose in 1864)

mesoslilexite   *ig. diaschist.* Silexite with more than 5% dark constituents. |Gk. *mesos,* middle + silexite| (Johannsen in 1932)

mesotourmalite   *ig. diaschist., autometa.* Tourmalite in which tourmaline forms 5-50% of the rock. |Gk. *mesos,* middle + tourmalite| (Johannsen in 1932)

mestigmerite   *ig. pluton.* Pyroxene malignite with large crystals of unaltered nepheline and sometimes orthoclase, accompanied by aegirine-augite, titanite, and apatite. |Mestigmer, western Oudjda, Morocco| (Duparc in 1926)

meta-   *meta.* Common prefix indicating that there has been some metamorphism of the designated rock, e.g., metabasalt, metagraywacke, metadiorite, metasediment. |Gk. *meta,* after, with|

meta-anthracite   *meta.* Coal having a fixed-carbon content of 98% or more; the highest rank of anthracite. Also named subgraphite.

meta-argillite   *meta.* Argillite that has undergone recrystallization.

meta-arkose   *meta.* Arkose that has undergone recrystallization by metamorphism so that it resembles granite or granitized sediment.

metabasalt   *meta.* Basalt that has undergone regional pressure metamorphism in which amygdaloidal, columnar, and other structures are often preserved; these may grade into greenstone (greenschist) or, at higher pressures, into blueschist.

metabasite   *meta.* General term for a metamorphosed mafic rock (basalt, diabase, and others) that has lost all traces of its original mineralogy and texture. (Hackman in 1907)

metabentonite   *meta.* Altered, somewhat indurated bentonite, characterized by clay minerals, esp. illite, that no longer have the property of adsorbing large quantities of water; they swell no more than do ordinary clays; illustrated by certain Ordovician clays of the Appalachian region. (Ross in 1928) metabitumite   *meta.* Hard, black, lustrous variety of hydrocarbon found in proximity to igneous intrusions. [Gk. *meta,* after + bitumen] (Mueller, date uncertain)

metabolite   *ig., meta.* Old term applied to an altered glassy trachyte. [Gk. *metabolos,* changeable] (Wadsworth in 1893)

metabolite   *meteor.* Iron meteorite having the same composition as an octahedrite but not showing the typical Widmanstatten structure; apparently recrystallized octahedrite. [Gk. *metabolos,* changeable]

metacarbonatite   *meta.* Metamorphosed carbonatite; term has been applied to a rutile-bearing carbonate body in Valley County, Idaho. (Heinrich in 1966)

metaconglomerate   *meta.* Deformed or otherwise altered and metamorphosed conglomerate.

metadiabase   *meta.* Rock simulating a diabase but supposed to have been formed by the metamorphism of sediments. Also named pseudodiabase, which more correctly describes the rock. [metamorphic diabase] (Dana in 1876)

metadiorite   *meta.* Rock simulating diorite but supposed to have been formed by the metamorphism of sediments. Also named pseudodiorite, which more correctly describes the rock. [metamorphic diorite]

metagabbro, glaucophane   *meta.* Metamorphosed gabbro (or diabase) in which glaucophane (with or without crossite) forms pseudomorphs after augite and labradorite has been converted to a mixture of albite and epidote (or lawsonite); other minerals that may be present include omphacite, chlorite, quartz, and magnetite.

metagraywacke   *meta.* Graywacke that may be merely cataclastically deformed, or in which quartz and potassic feldspar have been recrystallized, albite porphyroblasts may be developed, and the clayey matrix has crystallized to form a fine-grained sericite-chlorite-magnetite aggregate.

metaluminous rock   *ig.* Igneous rock whose chemistry meets the following requirement: $Al_2O_3/(K_2O + Na_2O) > 1 > Al_2O_3/(K_2O + Na_2O + CaO)$. Typical minerals of these rocks are hornblende and aluminous augite and, less commonly, melilite or combinations such as biotite and olivine. [Gk. *meta,* beside, after + aluminous] (Shand in 1943)

metamarble   *meta.* Truly metamorphic marble, used in contrast to sedimentary marble; metamorphic carbonate rock commercially valuable because it will take a polish. (Brooks in 1954)

metamorphite   *meta.* General term for any rock formed by metamorphism. [metamorphic]

metapelite   *meta.* Pelite that has undergone metamorphism of very low grade; may resemble slate but with an inferior cleavage, may be composed of montmorillonites, illites, rarely pyrophyllite, and may contain coaly streaks (up to anthracite rank).

metaquartzite   *meta.* Quartzite formed by metamorphism, as distinguished from orthoquartzite, whose nature is sedimentary (diagenetic).

metarhyolite   *meta.* Rhyolite whose mineralogy or texture has been altered by metamorphism; e.g., aporhyolite.

metasomatite   *meta.* Any rock formed by metasomatism; usually its chemical composition has been substantially changed by the alteration and replacement of its original constituents. [Gk. *meta,* beside, after + *soma,* body]

metatectite   *migmat.* Lipotexite whose mineralogy and texture have been changed mainly through metasomatism and accompanying anatexis; some use the term as a synonym for metatexite. [metasomatism + Gk. *tektos,* molten]

metatexite   *migmat.* Rock formed by metatexis, that is, the partial differential or selective anatexis of the low-melting components of a rock (generally the quartz and feldspars). [Gk. *meta,* beside, after + *tektos,* molten] (Scheumann in 1936)

metaxite   *sed.* Old term for a micaceous sandstone; originally included arkose with kaolinized feldspar. [Possibly Gk. *metaxa,* silk; alluding to its appearance] (Haüy in 1822)

meteor   *meteor.* Transitory luminous streak in the sky produced by the incandescence of an extraterrestrial solid body (a meteoroid) passing through the earth's atmosphere. [Gk. *meteoron,* meteor]

meteorite   *meteor.* Solid body that has arrived on the earth from outer space; may range in size from microscopic to a mass of many tons; essentially consisting of a nickel-iron alloy, silicate minerals (mainly olivine and orthopyroxene), or a mixture of these. Also spelled meteorolite. [Gk. *meteoron,* meteor]

meteoroid   *meteor.* Extraterrestrial solid body passing through the earth's atmosphere; when large enough to survive this passage and land on the earth's surface it becomes a meteorite. [Gk. *meteoron,* meteor]

meymechite   *ig. extru.* = meimechite

miagite   *ig. pluton.* Orbicular gabbro very similar to corsite. [Miage, glacier in the Alps] (Pinkerton in 1811) miarolite   *ig. pluton.* Miarolitic granite, that is, granite containing small irregular drusy cavities in which well-formed and terminated crystals of the constituent minerals of the rock project. [Gk. *miaros,* coarse] (Fournet in 1845)

miarolithite   *migmat.* Chorismite (migmatite) having miarolitic cavities or remnants thereof; variety of ophthalmite. [miarolitic]

miarolo   *ig. pluton.* = miarolite. [It. *miarolo,* from Gk. *miaros,* coarse]

miascite   *ig. pluton.* Biotite-bearing nepheline syenite that differs from a normal nepheline syenite in containing oligoclase in addition to microperthite, nepheline, and mica. [Miask, the Ilmen Range, Ural Mountains, Soviet Union] (Rose in 1842)

michaelite   *sed.* = siliceous sinter. [St. Michael's, Azores] (Webster in 1821)

micrite   *sed.* Consolidated or unconsolidated calcareous ooze or mud of either chemical or mechanical (clastic) origin; particles are less than approximately 0.031 mm diameter. [microcrystalline + calcite] (Folk in 1948)

microbreccia   *sed.* Sandstone containing relatively large and sharply angular particles of sand in a very fine silty or clayey matrix, e.g., graywacke. [Gk. *mikros,* small + breccia]

microclinite   *ig. pluton.* Variety of syenite composed entirely of microcline. [microcline] (Loewinson-Lessing in 1901)

microconglomerate   *sed.* Sandstone containing relatively large and rounded sand particles in a very fine silty or clayey matrix. [Gk. *mikros,* small + conglomerate]

microcoquina   *sed.* 1. Clastic limestone composed wholly or chiefly of weakly cemented shell particles of sand size (2 mm or less in diameter).   2. Variety of chalk. [Gk. *mikros,* small + coquina]

microdiabase   *ig. extru.* Aphanitic rock in the gabbro-basalt family with an ophitic fabric (once referred to as diabasic). [Gk. *mikros,* small + diabase; alluding to ophitic fabric]

microdiorite   *ig. hypab.* Fine-grained diorite porphyry. [Gk. *mikros,* small + diorite] (Lepsius in 1878)

microfoyaite   *ig. dike.* Fine-grained foyaite. [Gk. *mikros,* small + foyaite]

microgranite   *ig. hypab.* Aphanitic rock consisting of microscopic crystals of quartz and potassic feldspar with a granitoid texture. [Gk. *mikros,* small + granite]

microgranodiorite   *ig. aschist.* Granodiorite that is very fine-grained. [Gk. *mikros,* small + granodiorite]

microgranulite   *ig. hypab., extru.* Term used in connection with microgranite and rhyolite in which the groundmasses can be seen under the microscope to consist of graphic intergrowths of quartz and potassic feldspar (the crystals grew simultaneously and mutually penetrate each other). A term used in France. [Gk. *mikros,* small + granulite]

micromelteigite   *ig. hypab.* Fine-grained hypabyssal melteigite. [Gk. *mikros,* small + melteigite] (Brögger in 1921)

micrometeorite   *meteor.* Meteorite with a diameter generally less than 1 mm; because of their small size some meteoroids can enter the atmosphere without becoming intensely heated and hence without disintegration. [Gk. *mikros,* small + meteorite]

micromoldavite   *tekt.* Moldavite tektite that is very small, usually less than 1 mm in diameter. [Gk. *mikros,* small + moldavite]

micropegmatite   *ig.* Intimate and probably eutectic graphic intergrowth of quartz and potassic feldspar on a microscopic scale; resembles graphic granite but on a smaller scale. [Gk. *mikros,* small + pegmatite (term originally applied to a graphic intergrowth)] (Michel-Lévy in 1896)

microperthite   *ig.* Microscopic perthite; an intergrowth of potassic feldspar (host) and sodic-plagioclase (guest). [Gk. *mikros,* small + perthite]

microperthitite   *ig. pluton.* Rock composed entirely of microperthite. [microperthite] (Johannsen in 1937)

microquartzite   *sed., meta.* Quartzite (usually orthoquartzite) that is aphanitic and shows no individual grains; has a glassy luster on freshly broken surfaces. [Gk. *mikros,* small + quartzite]

microsparite   *sed.* Limestone whose carbonate-mud matrix has recrystallized to microspar (crystalline calcite grains, from 5 to more than 20 microns across). [Gk. *mikros,* small + sparite] (Folk in 1959)

microtektite   *tekt.* Tektite that is very small, usually less than 1 mm in diameter. A variety is micromoldavite. [Gk. *mikros,* small + tektite]

microtinite   *ig. pluton.* Light-colored, coarse-grained igneous rock characterized by vitreous plagioclase crystals (named microtine) in a matrix containing potassic feldspar and other minerals. [microtine, vitreous plagioclase, analogous to vitreous sanidine] (Lacroix in 1901)

microurtite   *ig. dike.* Fine-grained urtite dike rock. [Gk. *mikros,* small + urtite] micstone   *sed.* Fine-grained carbonate mudstone containing more than 65% by volume material less than 5 microns in diameter and consisting predominantly of carbonate minerals. [micrite + mudstone] (Lewan in 1978)

mictite   *migmat.* Coarse composite rock formed by the contamination of a magma, by the incorporation and partial or complete assimilation of country-rock fragments, under conditions of relatively low temperature and probably at relatively high levels in the crust. [Gk. *miktos,* mixed]

midalkalite   *ig. pluton.* = nepheline syenite. [Possibly middle (or medium) alkali]

middletonite   *sed.* Brown, resinous, brittle substance found between layers of coal at Middleton collieries, near Leeds, and also at Newcastle, England. [Middleton, England] (pub. by Embleton in 1878)

mienite   *ig. extru.* = dellenite. [Lake Mien, Blekinge, Sweden] (Holst in ca. 1890)

migmatite   *migmat.* Megascopically composite rock consisting of two or more petrographically different parts, one being the older country rock, in a more or less metamorphic stage, the other being newer and having a pegmatitic, aplitic, granitic, or generally plutonic appearance. [Gk. *migmatos,* mixture] (Sederholm in 1907)

miharaite   *ig. extru.* Olivine-free basalt; tholeiite or subalkaline basalt. [Miharayama Volcano, Oshima, Japan] (Tsuboi in 1918)

mijakite   *ig. extru.* Manganese-rich, red-brown basalt containing phenocrysts of augite and bytownite and occasionally biotite, hypersthene, and apatite, in a groundmass of feldspar plates, magnetite, and apparently manganese pyroxene. Also spelled miyakite. [Mijakeshima, Bonin Islands, Japan] (Petersen in 1891)

mikenite   *ig. extru.* Leucitite with much normative plagioclase. [Mikeno, Kivu area, eastern Africa] (Lacroix in 1933)

miliolite   *sed.* Foraminiferal limestone of eolian origin that is fine-grained and consists of the tests of *Miliola* and other foraminifers, cemented by calcite. [*Miliola*] (Carter in 1849)

mill-rock   *ig. extru.* 1. Any proximal, typically explosive rhyolite pyroclastic breccia. 2. Coarse acidic pyroclastic breccia found in or close to the volcanic units in which Canadian massive sulfide ore deposits occur. (Sangster in 1972)

millstone   *sed.* 1. Buhrstone formed through the silicification of limestone. 2. General term for any rock used in a mill for grinding, esp. grain.

milowite   *sed.* Trade name for a very fine-grained, white, chalklike form of quartz, occurring in large quantities on the Island of Milos, Greece. [Milos] (Barry in 1928)

mimesite   *ig. extru.* Obsolete name for a dark augite- and ilmenite-rich basalt. [Gk. *mimos,* imitation] (Cordier in 1868)

mimophyre   *meta.* Hornfels in which feldspars have developed so that the rock resembles igneous porphyry. [Gk. *mimos,* imitation + porphyry] (de Beaumont in 1841)

mimose   *ig. extru.* = mimesite. [Gk. *mimos,* imitation] (Haüy, pub. by Brongniart in 1813)

mimosite   *ig. extru.* = mimesite. [Gk. *mimos,* imitation]

minette   *ig. dike.* Gray to black lamprophyre, related to syenite, that is composed essentially of biotite crystals in an orthoclase groundmass. French miners' term. [Possibly Minkette, valley in the Vosges, or perhaps dim. of F. *mine,* ore, alluding to low content of metal in an ore] (In French literature as early as 1822)

minette   *sed.* Oolitic, brown, iron (limonite) rock from Lorraine and Luxemburg. [See minette above]

minette, natron   *ig. dike.* = minette, soda. [L. *natrium,* sodium]

minette, nepheline   *ig. dike., hypab.* Very dark, dense lamprophyre in the monchiquite-alnoite series, consisting of platy biotite and altered olivine in a dense matrix of biotite, augite, alkali feldspar, nepheline, sodalite, and iron oxides. (Pirsson in 1900)

minette, olivine   *ig. dike.* Minette in which olivine occurs as an accessory; the olivine is usually altered.

minette, pilite   *ig. dike.* Mafic-rich minette in which former olivine has been altered to actinolite (pilite). [L. *pilus,* hair]

minette, soda   *ig. dike.* Minette consisting of alkali feldspar (cryptoperthitic), dark brown mica, small amounts of aegirine, and abundant apatite and titanite. Also named natron minette. [sodium, for relatively high content] (Brögger in 1898)

minimicrite   *sed.* Dense, opaque, fine-grained micrite; generally the grains are 0.5-1.5 microns. [L. *minimus,* least + micrite]

minverite   *ig., altered.* Diabasic rock whose plagioclase is albite, in part primary and in part secondary, and that also contains brown hornblende and traces of altered olivine. [St. Minver, North Cornwall, England] (Dewey in 1910)

miskeyite   *meta.* Compact chlorite (pseudophite) used as an ornamental stone from St. Gallenkirch,

Vorarlberg, Austria. [J. von Miskey, director of the company that marketed it] (Berwerth in 1912)

missourite   *ig. pluton.* 1. Mottled, dark gray, coarse-grained rock containing leucite and 60–90% mafics, esp. augite and olivine; the plutonic equivalent of a leucite basalt. [Missouri River, near its occurrence in Highwood Mountains, Montana] (Weed and Pirsson in 1896) 2. *IUGS,* plutonic rock in which F is 60–100, M is 70–90, and potassium exceeds sodium.

mixtite   *sed.* Any coarsely mixed, nonsorted or poorly sorted, clastic sedimentary rock; the composition and origin are not considered; e.g., tillite. [L. *mixtus,* mixed] (Schermerhorn in 1966)

miyakite   *ig. extru.* = mijakite

Mocha stone   *sed.* Variety of agate consisting of white or gray chalcedony showing brown, red, or black dendritic markings resembling trees and plants. Also named moss agate. [Mocha, Yemen, original locality at entrance to Red Sea]

modlibovite   *ig. hypab.* Polzenite without monticellite; biotite (anomite variety) usually forms phenocrysts in a matrix composed of olivine, melilite, lazurite, phlogopite, biotite, and nepheline. [Modlibov, northern Bohemia, Czechoslovakia] (Scheumann in 1912)

modumite   *ig. pluton.* Calcium- and aluminum-rich facies of essexite, containing 88% bytownite and 9% pyroxene. Also named essexite anorthosite. [Modum, Oslo region, Norway] (Brögger in 1933)

molarite   *sed.* = buhrstone. [L. *mola,* mill; alluding to its use]

moldavite   *tekt.* Name applied to tektites from the valley of the Moldau River (Vltava River) of central Bohemia (Czechoslovakia). Also spelled moldauite; also named bouteillenstein, pseudochrysolite, and vltavine. [Moldau River] (Dufrénoy in 1847)

moldavite   *sed.* Variety of ozokerite from Moldavia, Rumania. [Moldavia] (Istrati in 1897)

molochite   *sed.* Ancient name for malachite, applied in recent years to a green-colored jasper rock. [malva, old name for the mallow plant, whose leaves are green] (pub. by Agricola in 1546)

monchiquite   *ig. hypab.* Lamprophyre composed of olivine, pyroxene, and usually mica (biotite) or amphibole, in an analcime matrix. The following varieties have been recognized: amphibole m., biotite m., biotite-amphibole m., hauyne m., leucite m., sodalite m. [Serra de Monchique, Portugal] (Hunter and Rosenbusch in 1890)

mondhaldeite   *ig. hypab.* Camptonitelike dike rock characterized by long hornblende needles; there are phenocrysts of hornblende, augite, bytownite, and leucite in a glassy, somewhat felty, groundmass. [Mondhalde, Kaiserstuhl, Baden, Germany] (Graeff in 1900)

monmouthite   *ig. pluton.* Urtite in which the pyroxene is substituted for by amphibole (hastingsite); essentially composed of nepheline and amphibole,

Monchiquite with pyroxene phenocrysts. Baumgarten, Kaiserstuhl, Baden, Germany. 11.5 cm.

but there are accessory cancrinite, albite, and calcite. [Monmouth Township, Ontario, Canada] (Adams and Barlow in 1910)

monomineralic rock   *ig., sed., meta.* Any rock composed entirely or nearly entirely of a single mineral, e.g., dunite (olivine), anorthosite (plagioclase), limestone (calcite), quartzite (quartz), and marble (calcite). [Gk. *monos,* single, one + mineral]

montmartrite   *sed.* Old name applied to gypsum containing calcium carbonate, from Montmartre, France. [Montmartre] (Delamétherie in 1812)

montrealite   *ig. pluton.* Ultramafic rock composed of titanaugite, less hornblende, and still less olivine, with or without plagioclase, nepheline, and minor accessories. [Montreal, Quebec, Canada] (Adams in 1913).

montroseite   *sed.* Uranium-bearing sandstone. [Possibly Montrose County, Colorado]

monzodiorite   *ig. pluton. IUGS,* plutonic rock with 0-5 Q, 65-90 P/(A + P), and plagioclase more sodic than An_{50}. [monzonitic + diorite] (Evans in 1920)

monzodiorite, foid   *ig. pluton. IUGS,* plutonic rock with 10-60 F, 50-90 P/(A + P), and plagioclase more sodic than An_{50}.

monzodiorite, quartz   *ig. pluton. IUGS,* plutonic rock with 5-20 Q, 65-90 P/(A + P), and plagioclase more sodic than An_{50}.

monzogabbro   *ig. pluton. IUGS,* plutonic rock with 0-5 Q, 65-90 P/(A + P), and plagioclase more calcic than An_{50}. [monzonitic + gabbro] (Johannsen in 1919)

monzogabbro, foid   *ig. pluton. IUGS,* plutonic rock with 10-60 F, 50-90 P/(A + P), and plagioclase more calcic than An_{50}.

monzogabbro, quartz   *ig. pluton. IUGS,* plutonic rock with 5-20 Q, 65-90 P/(A + P), and plagioclase more calcic than An_{50}.

monzonite   *ig. pluton.* 1. Granular plutonic rocks containing approximately equal amounts of potassic feldspar and sodic-plagioclase, with no essential quartz, but usually an accompanying mafic, e.g., hornblende, biotite, or augite; these are intermediate in composition between syenite and diorite; the aphanitic equivalent is latite. Some varieties, based on the major mafic, include: augite m., hornblende m., hypersthene m., olivine m. [Monzoni, Tyrolean Alps] (de Lapparent in 1864) 2. *IUGS,* plutonic rock with 0-5 Q and 35-65 P/(A + P).

monzonite, alkali   *ig. pluton.* Monzonite rich in sodium minerals; may be characterized by containing barian orthoclase, anorthoclase, albite, titanaugite, aegirine-augite, aegirine, hastingsite, and arfvedsonite.

monzonite, nepheline   *ig. pluton.* Alkali monzonite containing anorthoclase, sodic plagioclase, nepheline, biotite, diopside-aegirine-augite, barkevikite, and sometimes sodalite and olivine.

monzonite, pyroxene   *ig. pluton.* Quartz-poor member of the charnockite series, containing approximately equal amounts of microperthite and plagioclase; mangarite. (Tobi in 1971)

monzonite, quartz   *ig. pluton.* 1. Granular rocks containing approximately equal amounts of potassic feldspar and plagioclase, with essential quartz (greater than 5%); term has been variously defined in different countries. Generally synonymous with adamellite. 2. *IUGS,* plutonic rock with 5-20 Q and 35-65 P/(A + P).

monzonite, quartz-bearing   *ig. pluton.* Monzonite with too little quartz to be called a quartz monzonite or adamellite; quartz is less than 5%.

monzonorite   *ig. pluton.* Quartz-poor charnockite containing more plagioclase than microperthite. [monzonitic + norite] (Tobi in 1971)

monzosyenite   *ig. pluton.* Obsolete early term applied to the rocks later named monzonite. [Monzoni (Tyrolean Alps) + syenite] (von Buch in 1824)

monzosyenite, foid   *ig. pluton. IUGS,* plutonic rock with 10-60 F and 10-50 P/(A + P).

moonmilk   *sed.* White, cryptocrystalline substance found in caves, usually composed of carbonate minerals (calcite, aragonite, hydromagnesite, huntite, and others) and, when wet, having a texture like cream cheese, or when dry, like a very fine powder. A variety includes lublinite.

moorstone   *sed.* General term applied to the loose masses of granite found on moors in Carnwall, England.

Moriah stone   *meta.* Granular and spotted variety of serpentinite. [Possibly Mt. Moriah, Jerusalem]

moronite   *sed.* Sediment consisting of a mixture of calcium carbonate and siliceous materials formed from the remains of foraminifera and others, at Moron, Spain. [Moron, Spain] (Calderon in 1894)

morrisonite   *ig. extru., altered.* Local name for an attractively banded and colored jasper (silicified volcanic ash) occurring near the southern end of the Owyhee Reservoir in Malheur County, Oregon. [Morrison Ranch, Malheur County, Oregon] mound, algal   *sed.* Local thickening in limestone, chiefly attributed to the presence of a distinctive suite of rock types containing algae.

mountain cork   *meta.* White, gray, or yellowish massive rock composed of interlaced fibers of asbestiform tremolite-actinolite; often porous and light enough to float on water.

mountain leather   *meta.* Thin, flexible sheets composed of interlaced fibers of asbestiform tremolite-actinolite; may be white, gray, or yellowish and light enough to float on water.

mountain mahogany   *ig. extru.* Variety of banded obsidian, streaked with brown and gray and, when cut, showing a grain like that of mahogany.

mountain wood   *meta.* Compact mass of interlaced fibers of asbestiform tremolite-actinolite; gray to brown in color and resembling dry wood.

moyite   *ig. pluton.* Quartz-rich granite (quartz forms over 50% of the light-colored minerals) in which the only feldspar present is potassic. [Moyie sill, British Columbia, Canada] (Johannsen in 1919)

mozarkite   *sed.* Multi-colored Ordovician chert, used by lapidaries, from the Ozark Mountains, Missouri. [Possibly Missouri + Ozark]

muckite   *sed.* Opaque, yellow to light brownish-yellow resin from coal beds at Neudorf, Moravia; occurs as small bands and disseminated minute particles. [Fritz Muck, 1837-1891, German coal chemist who discovered it] (Schröckinger in 1878)

mud   *sed.* Term that loosely refers to a mixture of water and clay- or silt-size materials; generally sticky, plastic, and impure. [M. Eng. *mudde,* mud]

mud, black   *sed.* Black-colored mud whose dark color is usually due to organic substances that release hydrogen sulfide; developed under anaerobic conditions.

mud, blue   *sed.* Hemipelagic type of mud, whose bluish-gray color is due to iron sulfides and organic matter; occurs where the continental shelf breaks off into the deep sea.

mud, gravelly   *sed.* Unconsolidated sediment containing 5-30% gravel and having a ratio of sand to mud (silt plus clay) less than 1:1. (Folk in 1954)

mud, gray   *sed.* Marine mud intermediate in composition between globigerina ooze and red clay.

mud, green   *sed.* Marine mud whose greenish color is due to the presence of glauconite or chlorite minerals.

mud, metal-rich   *sed.* Mud rich in iron and manganese that is very common on mid-ocean ridge crests, resting immediately on basalt; ranges in thickness from a few feet to tens of feet and contains copper, nickel, zinc, and other components that may eventually become economically important.

mud, red   *sed.* Marine mud that is terrigenous and contains ferric oxide (the red color) and as much as 25% calcium carbonate.

mud, reduced   *sed.* = mud, black

mud, sandy   *sed.* Unconsolidated sediment containing 10-50% sand and having a ratio of silt to clay between 1:2 and 2:1. (Folk, 1954)

mud, sea   *sed.* = ooze, sea

mudflow, volcanic   *ig. extru., sed.* = lahar deposit

mudrock   *sed.* Broad term for rocks composed of a mixture of clay- and silt-grade materials, like mudstone, shale, argillite, and others. These compose some 45-55% of sedimentary rock sequences.

mudrock, laminated-pebbly   *sed.* Delicately laminated argillites or shales in which occur thinly scattered clasts ranging from sand- to boulder-size; laminations near the larger clasts are distorted and tend to bend down beneath them or arch over them.

mudstone   *sed.* Indurated equivalent of mud that is blocky and nonfissile (does not split easily into thin sheets).

mudstone, carbonate   *sed.* Mudstone whose cementing material is a carbonate mineral, usually calcite, but dolomite and siderite are known.

mudstone, chamosite   *sed.* Fine-grained aggregate of green, cryptocrystalline to isotropic chamosite; in some, chamosite ooliths occur in a chamositic matrix, while in others, siderite is the dominant matrix constituent.

mudstone, conglomeratic   *sed.* Mudstone with a sparse to liberal amount of pebbles or cobbles; specifically, consolidated gravelly mud containing 5-30% gravel and having a ratio of sand to mud less than 1:1. (Folk in 1954)

mudstone, dolomite   *sed.* = dolomicrite

mudstone, ferriferous   *sed.* Mudstone with iron oxide, usually more than 15%. [ferric + ferrous; alluding to iron]

mudstone, lime   *sed.* Fine-grained, nonporous, impermeable pure limestone (93-99% calcium carbonate), whose main constituent is calcite mud (micrite) (75-85%).

mudstone, pebbly   *sed.* Conglomeratic mudstone in which the larger clasts are pebble size.

mudstone, siderite   *sed.* Mudstone containing finely

disseminated siderite; common in Jurassic iron-stone sequences, associated with chamosite oolites; in color gray, bluish gray, or brown.

mudstone, tuffaceous   *ig. extru., sed. IUGS,* mudstone tuffite whose average clast size is less than 1/256 mm and by volume containing 25-75% pyroclasts, the remaining materials being sedimentary.

mugearite   *ig. extru.* Dark-colored trachyandesite consisting of oligoclase, some orthoclase, and olivine, with subordinate augite. [Mugeary, Isle of Skye, Scotland] (Harker in 1904)

muniongite   *ig. hypab.* Rock resembling tinguaite, composed of sodium orthoclase, nepheline, aegirine, and in some cases cancrinite; differs from normal tinguaite in containing a higher percentage of nepheline than orthoclase. [Muniong, New South Wales, Australia] (David and Woolnough in 1901)

munjack   *sed.* = manjak

muntenite   *sed.* Fossil resin from the Eocene beds of Oltenia, Rumania, related to copalite but different from romanite. [Georges Munteanu-Murgoci, 1872-1925, Rumanian geologist] (Istrati and Milhăilescu in ca. 1907)

murambite   *ig. extru.* Dark leucite basanite containing andesine, augite, olivine, and leucite. [Murambe Crater, north of Kigoma Crater, south-western Uganda] (Holmes in 1936)

murasakite   *meta.* Dark-violet, schistose metamor-phic rock, composed essentially of piedmontite and quartz. [Murasako, Japan] (Koto in 1887)

murite   *ig. extru.* Dark-colored, nepheline-rich phonolite, in which the mafic minerals (augite, olivine) comprise about 50% of the rock. [Cape Muri, Island of Rarotonga, Cook Islands] (Lacroix in 1927)

muscovadite   *ig. pluton.* Variety of norite charac-terized by the presence of cordierite and biotite from the partial absorption of fragments of the country rock by the magma. [Sp. *muscovado,* brown sugar; alluding to brown color of sur-faces exposed to weathering] (Winchell in 1900)

muscovado   *ig., meta.* General term for a rusty, brown, outcropping rock that resembles brown sugar; esp. used by geologists in Minnesota. It has been applied to gabbro, norite, and types of quartzite. [Sp. *muscovado,* brown sugar]

myelin   *sed.* = kaolin. [Gk. *muelos,* marrow; from its resemblance to marrow in altered rocks] (Breithaupt in ca. 1820)

mylonite   *meta.* Exceedingly fine-grained chertlike metamorphic rock, without cleavage but with a streaky or banded structure resembling flow structures in some volcanic glasses; formed by extreme milling and complete pulverization of rocks along major fault zones under strong con-fining pressure; original rocks are of varied types, including sandstones, quartzites, granites, gneisses, gabbros, peridotites, and others. [Gk. *mule,* mill; alluding to the milling of the rocks] (Lapworth in 1885)

mylonite, phyllite   *meta.* = phyllonite

myrickite   *sed.* Trade name for a gray to white chalcedony or agate containing red inclusions of cinnabar from San Bernardino County, California. Although opal containing cinnabar has also been named myrickite, the preferred name is opalite. [F. M. Myrick, Randsburg, California, who discovered the rock in 1908]

myrmekite   *ig.* Quartz intergrowth with plagio-clase (usually oligoclase) where the quartz occurs in vermicular forms or as blobs and drops; gener-ally formed during the later stages of crystalliza-tion of an igneous rock, like granite, by the replacement of potassic feldspar. [Gk. *murmekia,* ant hill, wart] (Sederholm in 1899)

N

nacrolite   *sed.* Pearly, bulbous to snail-shaped, speleothem. [F. *nacre,* mother-of-pearl] (Mowat in 1960)

nadeldiorite   *ig. pluton.* = diorite, needle. Also written nadel diorite. [G. *Nadel,* needle] (Gümbel in 1868)

nakhlite   *meteor.* = achondrite, diopside-olivine. [Nakhla, Egypt] (Prior in 1912)

napalite   *sed.* Dark reddish brown to yellow bituminous substance of the consistency of wax; shows green fluorescence. [Napa County, California] (Becker in 1888)

naphtha   *sed.* Volatile ethereal form of bitumen, rarely met with in nature; name also frequently applied to the most volatile distillate of petroleum. [Gk. *naphtha,* petroleum] (Mentioned by Strabo, who lived from 63? B.C. to 24? A.D.)

naphthode   *sed.* Concretions of bituminous limestone rich in carbonaceous matter. [Gk. *naphtha,* petroleum] (Spacek in 1927)

naphtolith   *sed.* Bituminous shale from Thelots, Saône-et-Loire, France. [Gk. *nephtha,* petroleum] (Barbier in 1911)

napoleonite   *ig. pluton.* = corsite. [Bonaparte Napoleon (Napoleon I), 1769-1821, Emperor of the French, born in Corsica, locality of the rock]

nari   *sed.* Type of caliche formed by surface or near-surface alteration of permeable calcareous rocks; occurs in the drier parts of the Mediterranean region. [Arab. *nar,* fire; alluding to its use in limekilns]

natron-   *ig.* Common prefix used with igneous rock names to indicate a high sodium content, e.g., natrongranite, natron minette, natronrhyolite, natron sussexite, natronsyenite, and others. Prefix is common in German petrology. Terms alkali and sodic are nearly equivalent. [L. *natrium,* sodium]

natroncarbonatite   *ig. extru.* Volcanic sodium-rich carbonatite composed of phenocrysts of a Ca-Na-K carbonate in a groundmass of thermonatrite; lava at Oldoinyo Lengai, Tanganyika. (Dawson in 1962)

naujaite   *ig. pluton.* Mottled very coarse-grained sodalite-rich nepheline syenite, containing micro-cline and nepheline with small amounts of albite, analcime, acmite, and sodium amphiboles. [Naujakasik, Julianehaab district, Greenland] (Ussing in 1911/12)

navite   *ig. extru.* Dark-colored aphanitic rock containing abundant phenocrysts of calcic-plagioclase, olivine (often altered), and rare augite and enstatite, in a holocrystalline groundmass consisting of feldspar and augite. [Nave, Germany] (Rosenbusch in 1887)

neapite   *ig. pluton.* Ijolite-urtite rock rich in apatite; nepheline 58%, apatite 20%, and mafics (aegirine, biotite) 15%. [nepheline + apatite] (Vlodavetz in 1930)

nebulite   *migmat.* Migmatite showing ghostlike relics of pre-existing rocks. [L. *nebula,* lacking definite form, cloudy, hazy] (Sederholm in 1923)

nectilite   *sed.* = floatstone. Also named quartz nectique and nectic quartz. [Gk. *nektos,* swimming] (Haüy, date uncertain)

neft-gil   *sed.* Mixture of brown paraffins and resin abundant in the naphtha region on Cheleken Island in the Caspian, Soviet Union. Also spelled neftdegil, naphtdachil, nephthadil, naphtagil, naphtil, and nephatil. [Russ. *nefte,* petroleum, oil + *gil,* pitch, tar] (Fritzsche in 1858; related to similar terms introduced as early as 1846)

nelsonite   *ig. dike.* Equigranular dike rock composed essentially of ilmenite and apatite, with or without rutile; apatite ilmenitite. [Nelson County, Virginia] (Watson in 1907)

nelsonite, biotite   *ig. dike.* Fine-grained, schistose dike rock composed of ilmenite or magnetite and apatite with characterizing biotite and minor titanite.

nelsonite, gabbro   *ig. dike.* Variety of nelsonite containing, in addition to ilmenite and apatite, labradorite and pyroxene.

nelsonite, hornblende   *ig. dike.* Nelsonite with large, irregular areas of pale to deep blue green, platy or fibrous hornblende, which is a dominant mineral.

nelsonite, ilmenite   *ig. dike.* Nelsonite containing about 60% ilmenite and 30% apatite, with titanite, rutile, and others.

nelsonite, magnetite   *ig. dike.* Nelsonite composed of essential magnetite and apatite, in approximately equal amounts, with brown and chloritized biotite; resembles ilmenite nelsonite.

nelsonite, rutile   *ig. dike.* Nelsonite composed of essential rutile and apatite, in approximately equal amounts; these grade into ilmenite nelsonite (normal nelsonite).

nemafite   *ig. extru.* = nephelinite. [nepheline + mafic minerals] (Hatch, Wells, and Wells in 1961)

nemite   *ig. extru.* Dark-colored leucitite. [Lake Nemi, southeast of Rome, Italy] (Lacroix in 1933)

neolite   *ig. extru.* General name for an order of volcanic rocks, embracing rhyolites and basalts, with which eruptive activity terminates in any given area. [Gk. *neos,* new, recent] (King in 1878)

nepheline-schorlomite-knopite rock   *ig. dike.* Rare dike rock that cuts the Afrikanda (Soviet Union) ring structure and is composed of nepheline, schorlomite (garnet), and perovskite (knopite).

nephelinite   *ig. extru.* Feldspar-free, fine-grained or porphyritic rock, of basaltic character, composed of nepheline and pyroxene but without olivine. [nepheline] (Roth in 1861)

nephelinite, leucite   *ig. extru.* Transitional rock between nephelinite and leucitite.

nephelinite, melilite   *ig. extru.* Transitional rock between nephelinite and melilitite.

nephelinite, olivine   *ig. extru.* Nephelinite with accessory olivine.

nephelinite, sanidine   *ig. extru.* Fine- to medium-grained, sometimes porphyritic, extrusive rock composed of pyroxene, nepheline, and sanidine, with abundant accessory magnetite accompanied by olivine; from Odenwald, Germany.

nephelinite-basanite   *ig. extru.* Extrusive igneous rock, in composition intermediate between a nephelinite and basanite; differs from a nepheline basanite in containing more nepheline than feldspar. (Johannsen in 1938)

nephelinite-phonolite   *ig. extru.* Phonolite containing more nepheline than feldspar; intermediate between nephelinite and phonolite. (Johannsen in 1938)

nephelinite-tephrite   *ig. extru.* Extrusive igneous rock intermediate in composition between nephelinite and tephrite; differs from nepheline tephrite in containing more nepheline than feldspar. (Johannsen in 1938)

nephelinitoid   *ig.* Feldspathoidal rocks whose interstitial material could not be determined optically as nepheline but could be so determined chemically, literally, nephelinelike. (Bořický in 1873)

Nelsonite (ilmenite) in which apatite is mostly weathered out. Nelson County, Virginia. 14.5 cm.

nephelinolith *ig. extru.* Extrusive rock composed almost entirely of nepheline. [nepheline] (Loewinson-Lessing in 1901)

nephrite *meta.* Jade composed of tough, compact, fine-grained, green, blue, yellow, or white amphibole (tremolite-actinolite series). [L. *lapis nephriticus,* kidney stone; worn as a remedy for diseases of the kidney] (Werner in 1780)

nephritoid *meta.* 1. Fine-grained tremolite-actinolite rock in which the fibers are parallel instead of matted as in true nephrite; occurs associated with harzburgite in the Radauthal, Harz, Germany. Also named fasernephrite. (Fromme in 1909) 2. Variety of serpentinite similar to bowenite.

neptunian dike *sed.* Sedimentary dike that cuts across sedimentary bedding and is usually filled with internal sediment; often observed in limestones and formed through small tectonic movements during sedimentation. [Neptune, in Roman mythology, god of the sea; alluding to rock formed in the sea]

neptunic rock *sed.* 1. General term for all sedimentary rocks. (Read in 1944) 2. Rock formed in the sea. [Neptune, in Roman mythology, god of the sea]

neslite *sed.* Opal similar to menilite but grayer in color; formerly popular for making sword handles. [Nesle-la-Reposte, Marne, France] (Leymérie in 1846)

neudorfite *sed.* Waxy, pale yellow variety of retinite, containing a small amount of nitrogen and occurring in a coal bed at Neudorf, Moravia. [Neudorf, Czechoslovakia] (Schröckinger in 1878)

neuquenite *sed.* Variety of asphalt from Argentina. [Neuquen Territory, Argentina] (Allen in 1932)

neurolite *meta.* = agalmatolite. [Gk. *neuron,* sinew; for fibrous structure] (Thomson in 1836)

nevadite *ig. extru.* Rhyolite consisting of large, well-formed crystals of quartz, sanidine, biotite, and hornblende in a small amount of groundmass; the phenocrysts are so abundant the rock at first appears to be phaneritic. [Nevada] (Richthofen in 1868)

névé *glacial.* French synonym for firn; term may also have a geographic meaning and refer to an area covered with perennial snow or an area of firn (a firn field). [F. dial. (Valais), *névé,* snow]

newlandite *ig. dike, pipe., meta.?* Griquaite containing pyrope (50%), chromian diopside, enstatite, and phlogopite. [Newland's diamond pipe, near Kimberley, South Africa] (Bonney in 1899)

ngavite *meteor.* Chondritic stony meteorite composed of bronzite and olivine in a friable, breccialike mass of chondrules. [Ngawi, Java]

ngurumanite *ig. dike.* Nepheline-clinopyroxene rock (the hypabyssal equivalent of ankaratrite) with iron-rich interstitial material and vuggy patches and veins of primary calcite, analcime, and zeolites. [Nguruman Escarpment, Kenya] (Saggerson and Williams in 1964)

n'hangellite *sed.* Elastic bitumen, derived from a gelatinous alga, resembling coorongite. [Lake N'hangella, Portuguese East Africa] (Boodle in 1907)

nigrite *sed.* Seldom used name for a variety of asphaltum from Utah. [L. *nigritudo,* blackness] (Peterson in 1899)

nigritite *sed.* Coalified, carbon-rich bitumen. [L. *nigritudo,* blackness]

niklesite *ig. pluton.* Pyroxenite composed of the three pyroxenes, diopside, diallage, and enstatite. [Niklesgraben, Moravia, Czechoslovakia] (Kretschmer in 1918)

niligongite *ig. pluton.* Melilite-bearing leucite ijolite, containing about equal amounts of nepheline, leucite, and titanaugite, with melilite (6%). [Niligongo, near Lake Kivu, Zaire] (Lacroix in 1933)

nitratine *sed.* Old name applied to the soda niter (saltpeter) deposits of Chile. [Possibly nitrate] (Haidinger in 1835)

nodule *ig. extru.* Knob or mass, consisting of a coarse-grained igneous rock or mineral aggregate, apparently crystallized at depth and occurring as an inclusion in an extrusive rock, e.g., olivine nodules in basalt or anorthosite nodules in diabase. [L. *nodulus,* dim. of *nodus,* knob, knot]

nodule *meta.* Knob or mass of minerals in a metamorphic rock differing in type or in proportion from the minerals of the matrix, e.g., quartz-sillimanite nodules in a micaceous sillimanite schist or leptite, or quartz-tourmaline nodules in schist. [L. *nodulus,* dim. of *nodus,* knob, knot]

nodule *sed.* Irregularly rounded knob, knot, lump, or mass of a mineral or mineral aggregate, usually without internal structure (rarely, radial fibrous) and exhibiting a contrasting composition from the enclosing sediment or rock matrix. Numerous mineralogical varieties are listed below. [L. *nodulus,* dim. of *nodus,* knob, knot]

nodule, anhydrite *sed.* Nodule composed of aphanitic or felted anhydrite occurring in sediments, esp. in dolomite rock.

nodule, barite *sed.* Spherulitic (often radial) knob of barite; usually in shale and not incorporating the sediment; not to be confused with barite concretions and barite sand crystals incorporating the sediment in which they occur.

nodule, caliche   *sed.* Calcite-rich material characteristic of regions marked by relatively high temperatures and moderate to low seasonal rainfall (caliche-forming environments); included are scattered nodules of calcite (1-2 cm across), often of diverse shapes and ranging from no structure to crudely concentric or radial structures, as well as calcite crusts on pebbles in gravels.

nodule, chert   *sed.* Small to large, subspherical to irregular masses of chert, predominantly in carbonate host rocks; sometimes coalesced to form nearly continuous layers.

nodule, ferromanganese   *sed.* Small, irregular, black to brown, friable, laminated concretionary masses, with manganese oxides alternating with iron oxides, widely distributed on the seafloor, esp. in parts of the Atlantic, Pacific, and Indian oceans; the mineralogy is complex but todorokite and goethite are relatively common constituents, and in addition to Fe and Mn they contain high concentrations of Co, Ni, Cu, Cr, and V. Earlier named pelagite.

nodule, flint   *sed.* Black to brown carbonaceous chert (flint) occurring as fine-grained nodules in chalk deposits; the nodules of the Cretaceous Chalk of England and France illustrate this type.

nodule, gypsum   *sed.* Nodule composed of gypsum often occurring as subspherical to ovoid masses (1-3 mm diameter) in dolomite; often strung out in layers parallel to the bedding.

nodule, manganese   *sed.* 1. nodule, ferromanganese.   2. Nodule composed of psilomelane (including romanechite, cryptomelane, hollandite, coronadite), as well as rhodochrosite, occurring in terrestrial sediments such as clays, shales, sandstones, and other rocks.

nodule, marcasite   *sed.* Globular and reniform marcasite masses of various sizes, occurring in some clays, shales, limestones, and coals.

nodule, phosphorite   *sed.* Phosphorite occurring as a nodule in limestone, chalk, and shale; may display colloform banding or be cryptocrystalline to granular. Nodules also are common in modern oceanic environments in many places on shallow offshore banks and plateaus (off California, New Zealand, Mexico, South Africa) and contain large tonnages of phosphate.

nodule, pyrite   *sed.* Nodule of pyrite, composed either of crystalline aggregates or radial fibers, common in dark limestones and also occurring in some shales, sandstones, and coals.

nodule, siliceous   *sed.* 1. General term for chert, flint, opal, and similar nodule composed of quartz or opaline materials.   2. Term also applied to loose fragments and lumps of chalcedony, agate, jasper, flint, chert, etc.; in most instances not true nodules, representing pebbles formed from erosion or from resistant minerals from amygdules in igneous rocks.

nonesite   *ig. extru.* Enstatite basalt porphyry composed of enstatite, labradorite, and augite phenocrysts in a groundmass of plagioclase and augite. [Nonsberg, southern Tyrol] (Lipsius in 1878)

nontectonite   *ig., sed.* General term for any rock formed by mechanical settling; fabric shows no influence of movement of adjacent grains. [L. *non,* not + tectonite]

nordfieldite   *ig.* = esmeraldite. More correctly spelled northfieldite. [Northfield, Massachusetts]

nordmarkite   *ig. pluton.* Reddish, medium- to coarse-textured, quartz-bearing alkali syenite in

Nodule (pyrite). Locality unknown. 6 cm.

which microperthite is the main constituent, accompanied by smaller amounts of oligoclase, biotite, and hornblende. [Nordmark, near Christiania, Norway] (Brögger in 1890)

nordmarkite, wollastonite   *ig. dike.* = evergreenite. (Lacroix in 1924)

nordsjöite   *ig. pluton.* Nepheline syenite composed of nepheline and orthoclase (not perthitic), with melanite, calcite, and secondary minerals. [North Sea (Nordsjö)] (Johannsen in 1938)

nordsjöite, cancrinite   *ig. pluton.* Variety of nordsjöite in which nepheline is accompanied by cancrinite. (Johannsen in 1938)

norite   *ig. pluton.* 1. Coarse-grained rock in which calcic-plagioclase (usually labradorite) and orthopyroxene (usually hypersthene) are the essential constituents; differs from gabbro by the presence of the orthopyroxene rather than clinopyroxene (augite). [Norge (Norway), original locality] (Esmark in 1823)   2. *IUGS,* plutonic rock satisfying the definition of gabbro, in which pl/(pl + px + ol) is 10-90 and the opx/(opx + cpx) is greater than 95.

norite, cordierite   *ig. pluton.* Varieties of norite containing characterizing cordierite. (Lacroix in 1898)

norite, enstatite   *ig. pluton.* Norite whose orthopyroxene is enstatite. (Teller and von John in 1882)

norite, hornblende   *ig. pluton.* Norite containing primary hornblende as an accessory dark mineral. (Cathrein in 1890)

norite, hypersthene   *ig. pluton.* Norite whose orthopyroxene is hypersthene; normal norite. (Teller and von John in 1882)

norite, ilmenite   *ig. dike.* Norite with a high percentage of ilmenite, varying from 20 to 80% and present as schlieren. (Vogt in 1893)

norite, Jotun   *ig. pluton., meta.?* = jotunite. [Jotunheim, Norway]

norite, labradorite   *ig. pluton.* Rock composed almost entirely of labradorite with a few percent of orthopyroxene and oxides; grades into an anorthosite. (Kolderup in 1896)

norite, mica   *ig. pluton.* Norite containing mica in addition to labradorite and hypersthene. (Williams in 1887)

norite, mela-quartz   Norite consisting of more than 50% mafic minerals but also containing quartz.

norite, olivine   *ig. pluton.* Norite containing accessory olivine.

norite, orbicular   *ig. pluton.* Norite in which some of the minerals are aggregated in orbs, after the manner of corsite and orbicular granite; examples occur at Romsaas, Norway, and Plumas County, California.

norite, orthoclase   *ig. pluton.* Not a norite but a hypersthene-bearing syenodiorite; porphyritic rock containing crystals of orthoclase and hypersthene in a groundmass of andesine; from near Centerville, New York. (Williams in 1887)

norite, quartz   *ig. pluton.* Norite containing accessory quartz. A similar rock is mela-quartz norite. (Teller and von John in 1882)

norite, uralite   *ig. pluton.* Norite containing secondary amphibole (uralite) formed from orthopyroxene. (Brögger in 1894)

norite-gabbro   *ig. pluton.* Unnecessary transitional name for rocks between norite and gabbro; the ratio of orthopyroxene to clinopyroxene here varies between 1:1 and 1:2.

northfieldite   *ig.* = esmeraldite. [Northfield, Massachusetts] (Emerson in 1915)

northmarkite   *ig. pluton.* = nordmarkite

noseanite   *ig. extru.* Obsolete term for nepheline basalt containing considerable nosean. [nosean] (Bořický in 1874)

noseanolith   *ig. extru.* Monomineralic extrusive rock composed entirely of nosean. [nosean] (Johannsen in 1938)

noselitite   *ig. extru.* Extrusive rock composed of nosean and pyriboles without essential olivine. [noselite (nosean)] (Johannsen in 1938)

nosykombite   *ig. pluton.* Nepheline-rich, plagioclase-bearing covite; contains nearly equal amounts of sodium sanidine, nepheline, and barkevikite, with plagioclase (8%), biotite (5%), and accessories. [Island of Nosy Komba, northwestern Madagascar] (Niggli in 1923)

notite   *ig. extru., weathered.* Variety of palagonite. [Possibly Val di Noto, Sicily] (von Waltershausen in 1846)

novaculite   *sed., meta.?* 1. Bedded variety of white or light-gray pure chert displaying under the microscope a well-developed microquartzitic texture in which grains of quartz are closely crowded together into a dense groundmass with the absence of opal and chalcedony; appears to be a thermally metamorphosed rock.   2. Term used in a more general sense for bedded chert or for certain argillaceous-siliceous rocks that can serve as whetstones. [L. *novacula,* razor] (Cordier in 1868)

O

oakstone   *hydrotherm.?* Brown, compact, banded, stalactitic barite rock that can be polished and used for ornamental purposes; occurs at Arbor Low, near Youlgreave, Derbyshire, England. [oak wood, which it resembles when polished] (Adam in 1845)

obsidian   *ig. extru.* Volcanic glass with the composition of rhyolite (and granite) that breaks with a fine conchoidal fracture and is black, gray, dark brown, or red; often banded and may show microlites and spherulites; practically free of water in contrast to pitchstone, which it resembles. [L. *obsidianus,* for rock discovered in Ethiopia by Obsidius, who served under Roman consul Laevinus, ca. 280 B.C.] (Pliny, *Natural History*)

obsidian, basalt   *ig. extru.* = hyalobasalt

obsidian, devitrified   *ig. extru., altered.* Obsidian that is actually microcrystalline, for with age all glasses eventually crystallize (devitrify); rocks older than Miocene generally contain devitrified "glass" rather than true glass.

obsidian, flowering   *ig. extru.* = obsidian, snowflake

obsidian, hydrated   *ig. extru., altered.* Obsidian containing an unusually high percentage of water; although newly formed obsidian generally contains less than 1% water by weight, it often becomes hydrated by later absorption of water, up to nearly 10%; some studies indicate that pitchstone and perlite may be derived from obsidian through hydration by meteoric water.

obsidian, mahogany   *ig. extru.* = mountain mahogany

obsidian, onyx   *ig. extru.* Obsidian showing parallel, straight bands.

obsidian, peanut   *ig. extru.* Grayish green to leek green obsidian containing brownish red spherulites vaguely resembling ordinary peanuts in size and color; used for lapidary work; occurs near Pilas, Jalisco, Mexico.

Obsidian. Lake County, Oregon. 8 cm.

obsidian, rainbow *ig. extru.* Obsidian showing iridescence; illustrated by specimens from Lake County, Oregon.

obsidian, snowflake *ig. extru.* Obsidian containing white to gray, rounded to somewhat irregular patches of crystallized, very fine-grained minerals (silica, feldspars, and others) that resemble snowflakes or flower blossoms. Also named flowering obsidian.

obsidian, trachyte *ig. extru.* Relatively rare volcanic glass with the composition of trachyte. Varieties include lassenite (fresh, unaltered) and metabolite (altered glass). Glassy trachyte pumice also occurs.

obsidianite *tekt.* Obsolete term originally applied to tektites; many stones originally described as obsidianite (tektite) were later shown to be true obsidian. [obsidian] (Walcott in 1898)

occhio di pesce *ig. extru.* Viterbite with phenocrysts of leucite (2-15 mm in diameter) in an ash gray aphanitic groundmass; the spotted appearance suggested the popular name "fish-eye." [It. *occhio di pesce,* fish eye]

ocean-picture rock *meta., altered.* Local name for a brownish, brecciated, carbonate-bearing silicified serpentinite with veinlets of pale blue to white quartz; occurs near Midway, British Columbia, Canada, near the Washington border. [Possibly named for imagined scenes exhibited on sawed surfaces]

oceanite *ig. extru.* Picritic basalt related to schoenfelsite. [Indian Ocean] (Lacroix in 1934)

ocher *sed.* Earthy mineral oxides of iron mixed with varying amounts of clay and sand, occurring in yellow, brown, and red, and sometimes used as pigments. [M.Eng. *oker,* traceable to Gk. *okhra,* derived from a term meaning yellow, pale yellow]

ocher, black *sed.* = wad

ocher, brown *sed.* Impure limonite-rich ocher. Also named ocherous limonite.

ocher, lake *sed.* Ocherous deposit formed on the bottom of a lake by bacteria capable of precipitating ferric hydroxide.

ocher, red *sed.* Earthy, soft and dull varieties of hematite mixed with clay or sand.

ocher, yellow *sed.* Yellowish limonite-rich ocher closely related to brown ocher or ocherous limonite.

ochran *sed.* Yellow, unctuous, ocherous clay. [ocher] (Breithaupt in 1832)

ocrite *sed.* Group name for the powdery ochers. [ocher]

octahedrite *meteor.* The most common type of iron meteorite, usually containing 6-14% Ni and showing an orientation of kamacite and taenite bands parallel to octahedral planes (the Widmanstatten structure) [octahedral]

odinite *ig. dike.* Greenish-gray, porphyritic lamprophyric dike rock containing phenocrysts of labradorite and pale-green augite or diallage and, less commonly, hornblende, in a groundmass of plagioclase and hornblende. [Odenwald, Frankenstein, Germany] (Chelius in 1892)

odontolite *sed.* Fossil bone or tooth colored blue by a phosphate of iron; because the material is often turquoise blue, it was confused historically with turquoise. Also named bone turquoise. [Gk. *odous, odont-,* tooth]

oil rock *sed.* 1. = shale, oil. 2. General term for any rock containing bituminous matter (pitch or asphalt) instead of coaly matter.

okaite *ig. pluton.* Hauyne melilitolite with accessory biotite, perovskite, calcite, apatite, and oxides; resembles turjaite except the feldspathoid is hauyne rather than nepheline. [Oka Hills, Quebec, Canada] (Stansfield in 1923)

okawaite *ig. extru.* Pitchstone vitrophyre with phenocrysts of anorthoclase, aegirine-augite, and magnetite in a perlitic matrix. [Okawa River, Hokkaido, Japan] (Nemoto in 1934)

oligoclasite *ig. pluton.* Light-colored diorite composed entirely of oligoclase. Also named oligosite. [oligoclase] (Kolderup in 1899)

oligomictic rock *sed.* Any clastic sedimentary rock composed of clasts of a single rock type, such as a limestone conglomerate or an orthoquartzitic breccia. [Gk. *oligos,* few + *miktos,* mixed]

oligophyre *ig. dike.* Porphyritic oligoclasite having phenocrysts of oligoclase in a groundmass of the same mineral. [oligoclase + porphyry] (Coquand in 1857)

oligosiderite *meteor.* Term applied to a meteorite containing only a small amount of metallic iron; used in contrast with holosiderite. [Gk. *oligos,* few, small + *sideros,* iron]

oligosite *ig. pluton.* = oligoclasite. [oligoclase] (Turner in 1900)

olistostrome *sed.* Sedimentary deposit consisting of a chaotic mass of intimately mixed heterogeneous materials, such as blocks and muds accumulated as a semifluid body by submarine gravity sliding or slumping of unconsolidated sediments. [Gk. *olistomai,* to slide + *stroma,* bed] (Flores in 1955)

olivine rock *ig. pluton.* = dunite

olivinite   *ig. pluton.* 1. Dunite carrying a considerable amount of pyroxene, amphibole, or more rarely, biotite.   2. *IUGS,* dunite. [olivine] (Sjögren in 1876)

olivinite, amphibole   *ig. dike.* Purple to green, fine-grained dike rock, composed of olivine, partly altered to serpentine, colorless amphibole (edenite ?), and magnetite; from Spanish Peak, California.

olivinite, hornblende-augite   *ig. hypab.* Rock composed of olivine (often altered to serpentine), hornblende (often barkevikitic), and augite.

olivinite, hornblende-enstatite   *ig. pluton.* Olivinite in which olivine is accompanied by greenish brown amphibole, enstatite, and chromite.

olivinite, magnetite   *ig. pluton.* Olivinite rich in titaniferous magnetite and containing minor biotite. (Sjögren in 1876)

olivinite, mica   *ig. dike.* Rock consisting of abundant serpentinized olivine accompanied by biotite (or phlogopite) and other minor minerals.

olivinite, titanomagnetite   *ig. pluton.* Variety of olivinite in which the olivine is intimately intergrown with magnetite (myrmekitelike); contains olivine (50%), magnetite and ilmenite (40%), and spinel (5%); occurs in Söndmöre, Norway; closely related to magnetite olivinite.

ollenite   *meta.* Hornblende schist characterized by abundant epidote, titanite, and rutile, with smaller amounts of garnet. [Col d'Ollen, Piemonte, Italy] (Cossa in 1881)

ollite   *meta.* = potstone. [L. *olla,* pot, jar]

oncolite   *sed.* Spherical to irregular algal balls, showing concentric laminations internally; smaller than a stromatolite, generally less than 10 cm in diameter. Also spelled onkolite. [Gk. *onkos,* tumor]

ongonite   *ig. extru.* Topaz-bearing keratophyre, enriched in F and Li. [Ongon region of Mongolia] (Kovalenko and others in 1971)

onkilonite   *ig. extru.* Dark-colored, feldspar-free olivine-leucite nephelinite that also contains augite and perovskite; from the island of Wilkitski, Arctic Ocean. [Onkilones, Siberian tribe inhabiting the island] (Backlund in 1915)

onkolite   *sed.* = oncolite

onychite   *sed.* Ornamental alabaster or onyx marble with yellow or brown veins, carved by the ancients into vases and other objects. [L. *onychitis,* from Gk. *onychitis lithos,* alabaster] (Theophrastus, *On Stones*)

onyx   *sed., ig.* 1. Chalcedony (agate) consisting of alternating bands of different colors that are parallel and straight.   2. Sometimes used as a qualifying adjective meaning parallel-banded, as in onyx marble and onyx obsidian.   3. Black chalcedony (agate) without banding; the black onyx of the gem trade. [Gk. *onux,* claw, fingernail]

onyx, Algerian   *sed.* Distinctly banded, stalagmitic form of onyx marble.

onyx, California   *sed.* Dark, amber-colored or brown variety of onyx marble, usually composed of aragonite; used for ornamentation.

onyx, carnelian   *sed.* Chalcedony onyx in which alternating parallel bands are carnelian; essentially the same as sardonyx but the colored bands are more reddish.

onyx, cave   *sed.* Onyx marble formed in caves as dripstone; compact, banded, composed of calcite or aragonite, and can be polished.

onyx, Mexican   *sed.* Yellowish brown or greenish-brown, translucent, onyx marble, e.g., that found in the area of Tecali, Mexico; commonly used for ornamental objects.

onyx, oriental   *sed.* Banded, mottled, or clouded onyx marble (cave onyx), usually composed of aragonite.

onyx, Yaqui   *sed.* Variety of onyx marble quarried at El Marmolito, near Rosarito, Baja California Norte, Mexico. [Yaqui Indians of Mexico, who quarried it]

ooid   *sed.* = oolith. [Gk. *oion,* egg]

oolite   *sed.* 1. Sedimentary rock made up chiefly of ooliths cemented together. Oolitic limestone (calcitic oolite) is a very common example, and other compositions include: anhydritic o., aragonitic o., chamositic o., dolomitic o., hematitic o., limonitic o., phosphatic (collophane) o., pyritic o., sideritic o., siliceous o.   2. Term often used incorrectly for oolith. [Gk. *oion,* egg]

oolith   *sed.* One of the small spherical or subspherical accretionary bodies in a sedimentary rock, resembling the roe of fish and having a diameter 0.25-2 mm; usually formed by the growth of successive layers around a nucleus in shallow, wave agitated water; many noncalcareous ooliths are formed later by the replacement of earlier calcareous types. See oolite for various compositions. [Gk. *oion,* egg]

oomicrite   *sed.* Limestone composed of ooliths in a matrix of micrite (carbonate mud). [oolith + micrite] (Folk in 1959)

oomicrudite   *sed.* Oomicrite containing ooliths more than 1 mm in diameter. [oomicrite + rudite]

oosparite   *sed.* Limestone composed of ooliths cemented by sparite (clear equant calcite). [oolith + sparite] (Folk in 1959) oosparrudite   *sed.* Oosparite containing ooliths more than 1 mm in diameter. |oosparite + rudite|

ooze   *sed.* Soft mudlike sediment covering the floor of lakes and oceans, composed chiefly of the remains of microscopic organisms (usually calcareous or siliceous) and clay minerals. |O.Eng. *wase,* mire, marsh|

ooze, abyssal   *sed.* Ooze from the deepest ocean environments. |L. L. *abyssus,* from Gk. *abussos,* bottomless|

ooze, calcareous   *sed.* Ooze containing at least 30% calcareous skeletal remains.

ooze, diatomaceous   *sed.* Siliceous ooze containing at least 30% diatom frustules.

ooze, foraminiferal   *sed.* Calcareous ooze containing the skeletal remains of foraminifera; the most widespread surface sediment type on the earth, covering one-half of the sea floor or about one-third of the planet.

ooze, globigerina   *sed.* Calcareous foraminiferal ooze containing at least 30% foraminiferal tests, predominantly of the genus *Globigerina.*

ooze, radiolarian   *sed.* Siliceous ooze covering large areas of the ocean bottom and containing at least 30% opaline silica tests of radiolarians. (Murray in 1873)

ooze, sea   *sed.* Sea mud; a rich, slimy deposit in a salt marsh or along a seashore, often used as a fertilizer.

ooze, siliceous   *sed.* General term for ooze containing opaline-silica skeletons of radiolarians, diatoms, sponge spicules, and the like.

opal rock   *sed., hydrothermal.* Term applied to the many sedimentary and hydrothermal rocks in which opal (opal-A and/or opal-CT) is the major constituent; opal is a major component of the tests of silica-secreting organisms such as diatoms, radiolarians, and sponges, and numerous rocks are composed of these materials (diatomite, radiolarite, spongolite) or are derived from them by diagenesis or low-grade metamorphism (chert, novaculite); siliceous sinters and geyserites are opaline rocks; opal is also derived from the weathering of silicate minerals and from solfataric alteration of basaltic and other rocks and then can be deposited within a variety of rocks, both sedimentary and volcanic. The variety names of opal rock are numerous and in many instances descriptive, e.g.: amber o., black o., cherry o., fire o. (yellow to red), honey o., jasper o. (jasp-opal), milk o., moss o. (dendritic), onyx o., pin fire o., precious o. (play of colors), resin o., water o. (clear), wax o. (yellow, waxy), wood o. (opalized wood). Numerous other variety names can be found in this compilation such

as cacholong, hyalite, menilite, neslite, potch, semiopal, and others.

opaline   *sed.* Rock consisting of the brecciated impure opal replacement of serpentine from the Quicksilver Region, Napa County, California. |opal| (pub. by Rogers in 1931)

opalite   *sed.* 1. General name applied to the impure colored varieties of common opal.   2. Common opal containing red inclusions of cinnabar, which has also been named myrickite, although myrickite is usually restricted to chalcedony (agate) containing cinnabar.   3. Yellow-green potch (Australian opal) showing black dendritic markings. |opal|

opdalite   *ig. pluton.* Hypersthene-biotite granodiorite. |Opdal-Inset area, Norway| (Goldschmidt in 1916)

ophicalcite   *meta.* Serpentine-calcite marble, formed from the metamorphism of a siliceous dolomite; some are highly veined and brecciated, and others consist of lobate calcite grains set in a matrix of serpentine; calcite is the major carbonate mineral, but dolomite or magnesite may be important in some specimens. |Gk. *ophis,* serpent + calcite| (Brongniart in 1813)

ophicarbonate   *meta.* Serpentine-carbonate rock in which the carbonate may be calcite (ophicalcite), dolomite, or magnesite. |Gk. *ophis,* serpent + carbonate|

ophiolite   *meta., ig.* 1. = serpentinite. (Brongniart in 1813)   2. Large group of ultramafic and mafic igneous rocks, including rocks rich in serpentine, chlorite, epidote, and albite derived from them by metamorphism, whose origin is associated with an early phase of the development of a geosyncline. |Gk. *ophis,* serpent; originally alluding to serpentine| (Steinman in 1905)

ophite   *ig. dike.* Term originally given to normal or uralitized diabases occurring in the Pyrenees; term now rarely used for the rock, but the ophitic texture is still recognized; in this texture, thin platy plagioclase crystals are partially or completely included in anhedral pyroxene (usually augite) crystals. |Gk. *ophis,* serpent; alluding to the serpentlike markings on the rock| (Palassou in 1798)

ophite   *meta.* = serpentinite. |Gk. *ophis,* serpent|

ophthalmite   *migmat.* Chorismite (migmatite) characterized by augen (e.g., feldspar augen) or lenticular aggregates of newly formed minerals. |Gk. *ophthalmos,* eye|

opoka   *sed.* Flinty, porous calcareous sedimentary rock, with a conchoidal or irregular fracture, composed of fine-grained opaline silica (up to 90%) and cemented by the presence of silica of organic origin (silicified residues of diatoms, radiolaria, sponge spicules). [Pol. *opoka,* bedrock]

oranite *ig.* Intergrowth of potassic feldspar (orthoclase or microcline) and calcic-plagioclase (near anorthite); analogous to perthite in which the plagioclase is sodic. [orthoclase + anorthite] (Alling in 1921)

orbiculite *ig. pluton.* General collective name for deep-seated igneous rocks containing laminated orbicular structures, e.g., orbicular granite, orbicular gabbro (corsite), and orbicular norite. [orbicular, from L. *orbis,* circle]

orbite *ig. aschist.* Porphyritic gabbro containing large phenocrysts of hornblende, or labradorite and hornblende, in a matrix having the composition of malchite. [Orbishöhe, Odenwald, Germany] (Chelius in 1892)

ordanchite *ig. extru.* Rock close to hauyne trachyandesite, whose phenocrysts are labradorite-andesine and hauyne, with hornblende, augite, and rarely olivine. [Banne d'Ordanche, in the Auvergne, France] (Lacroix in 1917)

ordosite *ig. pluton.* Syenite composed of aegirine, microcline, phlogopite, and apatite; the aegirine, which forms 60% of the constituents, occurs as needles, for the most part enclosed in the microcline. [Ordos Province, China] (Lacroix in 1925)

oregonite *sed.* Local name applied to an orbicular jasper found on Josephine Creek, north of Holland, Oregon; sometimes also called kinradite; not to be confused with the mineral oregonite, which is associated with josephinite (a rock) from the same area. [Oregon]

orendite *ig. extru.* Porphyritic leucite lamproite, composed of phenocrysts of phlogopite in a dull reddish-gray groundmass, consisting of leucite and sanidine in approximately equal amounts; nepheline is absent. [Orenda Butte, Leucite Hills, southwestern Wyoming] (Cross in 1897)

organic rock *sed.* General term for any sedimentary rock composed primarily of the remains of organisms (plant or animal), such as of material that originally formed part of the skeleton or tissues of the organism.

organolite *sed.* Rock composed mainly of organic material, esp. one derived from plants, e.g., bitumen, asphalt, coal, amber, and mineral resin. (pub. by Senft in 1857)

ornoite *ig. pluton.* Quartz- and microcline-bearing, hornblende-poor, diorite, that is medium-grained and in which the plagioclase is oligoclase to andesine. [Ornö Hufvud, Sweden] (Cederström in 1893)

oropion *sed.* Dark brown to black variety of bole; sometimes named mountain soap from its greasy feel. [Gk. *oros,* mountain + *pion,* fat] (Glocker in 1847)

orotvite *ig. pluton.* Diorite composed of hornblende, biotite, plagioclase (oligoclase), nepheline, and cancrinite, with minor titanite, ilmenite, and apatite. [Orotva Valley near Ditró, Rumania] (Streckeisen in 1938)

ortho- 1. Prefix often used in names of metamorphic rocks derived from igneous parent rocks; in contrast, rocks derived from sedimentary parent rocks may be designated by the prefix *para-.* 2. Prefix often used in names of sedimentary rocks formed purely by sedimentary processes; in contrast, *meta-* is used if metamorphism was involved. 3. Prefix often used in names of igneous rocks containing orthoclase. [Gk. *orthos,* straight; rarely, for orthoclase]

orthoamphibolite *meta.* Amphibolite derived from an earlier igneous parent rock; often shows zoning in plagioclase, relict ophitic textures, and remnants of pyroxene in hornblende centers.

orthoandesite *ig. extru.* Andesite containing orthopyroxene (usually bronzite); related to sanukite. [orthopyroxene + andesite] (Oebbeke in 1881)

orthoarenite *sed.* Arenite with a detrital matrix under 15%.

orthoclasite *ig. pluton.* 1. Group of orthoclase-bearing igneous rocks, including, granite, syenite, granite porphyry, and others. (Senft in 1857) 2. Igneous rock containing about 90% or more orthoclase. [orthoclase] (Merwin in 1915)

orthoconglomerate *sed.* General term for a conglomerate with an intact gravel framework, characterized by a mineral cement, and deposited by ordinary but highly turbulent water currents, either in high-velocity streams or the surf.

orthodolomite *sed.* 1. Dolomite formed by sedimentation; primary dolomite. 2. Dolomite so well-cemented that the particles are interlocking. (Tieje in 1921)

orthofelsite *ig. dike.* Rock containing phenocrysts of orthoclase in a felsitic groundmass; phenocrysts of quartz are absent. [orthoclase + felsite] (Teall in 1888)

orthogneiss *meta.* Gneiss derived from an igneous parent rock, including granite, granodiorite, rhyolite, and others; evidence for this origin might be determined by the uniform compositions of the potassium and plagioclase feldspars, the uniformity of euhedral zircons, and relict porphyritic textures. (Rosenbusch in 1891) ortholimestone   *sed.* General term for a sedimentary limestone; used in contrast with metalimestone (metamorphic), e.g., marble. (Brooks in 1954)

orthomarble   *sed.* General term for a sedimentary carbonate rock (looks like marble) that is commercially valuable because it will take a good polish; sedimentary marble; used in contrast with metamarble or marble, both metamorphic. (Brooks in 1954)

orthometamorphite   *meta.* General term for any metamorphosed igneous rock. (Philipsborn in 1930)

orthomicrite   *sed.* Unaltered or primary calcareous micrite. (Chilingar and others in 1967)

orthomicrosparite   *sed.* Microsparite that has developed by precipitation in open spaces. (Chilingar and others in 1967)

orthophyre   *ig. extru., dike.* 1. Light-colored porphyritic trachyte composed exclusively of orthoclase; occurs in dikes. 2. = porphyry, orthoclase. [orthoclase + porphyry] (Coquand in 1857)

orthopyroxenite   *ig. pluton.* 1. Perknite composed almost exclusively of orthorhombic pyroxene. [orthopyroxene] 2. *IUGS,* plutonic rock with M equal to or greater than 90, and opx/(ol + opx + cpx) greater than 90.

orthopyroxenite, olivine   *ig. pluton. IUGS,* plutonic rock with M equal to or greater than 90, 5-40 ol/(ol + opx + cpx), cpx/(ol + opx + cpx) less than 5, and opx/(ol + opx + cpx) less than 90.

orthoquartzite   *sed.* Sandstone in which the cement is quartz. Also named quartzitic sandstone, quartzose sandstone, quartzarenite, and sedimentary quartzite. [ortho- + quartzite; the derivation of a sedimentary rock name from a metamorphic name (quartzite) is awkward and should be avoided]

orthoschist   *meta.* Schist derived from an igneous parent rock, such as basalt, diabase, peridotite, rhyolite, dacite, andesite, and others.

orthoshonkinite   *ig. pluton.* Obsolete name for a shonkinite that is plagioclase-free. [orthoclase + shonkinite] (Johannsen in 1922)

orthosite   *ig. pluton, dike.* Light-colored, coarse syenite composed entirely of orthoclase. [orthoclase] (Turner in 1900)

orthosparite   *sed.* Term applied to sparite cement developed by physicochemical precipitation in open spaces. (Chilingar and others in 1967)

orthosyenite   *ig. pluton.* Syenite with orthoclase, microperthite, or anorthoclase but containing

less than 5% plagioclase in the total feldspar; orthosite is a variety. Also named kalisyenite. [orthoclase syenite] (Johannsen in 1919)

orthotarantulite   *ig. dike.* = arizonite. [orthoclase + tarantulite] (Johannsen in 1919)

orthotill   *sed.* Till formed by the immediate release from the transporting ice, as by melting and ablation. (Harland and others in 1966)

ortlerite   *ig. hypab.* Obsolete term for a greenstonelike diorite porphyry containing phenocrysts of hornblende prisms in a dark-green, fine-grained groundmass consisting of plagioclase laths. [Ortler Alps, Tyrol, Italy] (Stache and von John in 1879)

orvietite   *ig. extru.* Trachybasalt composed of approximately equal amounts of labradorite and sanidine, along with leucite, augite, and minor biotite. [Orvieto, Italy] (Niggli in 1923)

osloporphyry   *ig. hypab.* Porphyritic variety of maenaite. [Oslo, Norway] (Brögger in 1898)

ossipite   *ig. pluton.* Coarse-grained troctolite containing labradorite, olivine (mostly altered), pyroxene, biotite, apatite, and iron oxides. [Ossippe Indians, who inhabited the area of New Hampshire where the rock was found] (Hitchcock, pub. by Dana in 1872)

osteocolla   *sed.* Cellular calcareous tufa, consisting of incrusted fragments of reeds or other marsh plants. Also spelled osteocollus. [Gk. *osteon,* bone + *kolla,* glue; from supposed property of uniting broken bones] (Gesner in 1565)

ostraite   *ig. pluton.* Jacupirangite containing augite (72%), with green spinel (21%) in place of magnetite. [Ostraia Sopka, Tschistop Mountains, Urals] (Duparc in 1913)

otaylite   *sed.* Trade name for a bentonite clay. [Otay, San Diego County, California] (Spence in 1924)

ottajanite   *ig. extru.* Volcanic rock originally described as leucite tephrite with the chemical but not the mineralogical composition of sommaite; containing phenocrysts of augite and leucite in a groundmass of labradorite-bytownite, leucite, and augite, with some sanidine, nepheline, olivine, and others. [Ottajano, Vesuvius, Italy] (Lacroix in 1917)

Ouachita stone   *sed.* = Washita stone

ouachitite   *ig. hypab.* Olivine-free, dark gray or black, biotite monchiquite, with a matrix of glass or analcime. [Ouachita River, Arkansas] (Kemp in 1890)

ouenite   *ig. dike.* Fine-grained, dark diabase, composed of anorthite and chromian diopside, with less olivine and a small amount of bronzite;

occurring as dikes in peridotite. [Ouen, New Caledonia] (Lacroix in 1911)

oulopholite   *sed.* = cave flower. [Gk. *oulos,* woolly + *pholeos,* cave]

owharoite   *ig. extru.* Volcanic tuff composed of pumice and andesite fragments in a matrix of vitric and granular materials. Also named wilsonite. [Owharoa, Waihi district, southern Auckland, New Zealand] (Grange in 1934)

oxidite   *meteor.* = ball, shale. [oxidation, alluding to its formation]

oxyphyre   *ig. dike.* = leucophyre. [Gk. *oxus,* acid; alluding to high silica content] (Pirsson in 1895)

ozarkite   *ig., altered.* White, massive form of thomsonite from Magnet Cove, Ozark Mountains, Arkansas. [Ozark Mountains] (Shepard in 1846)

ozokerite   *sed.* White, greenish, or brownish, natural wax derived from paraffin oils and used for electrical insulation, waterproofing, and for wax candles; important occurrences in Galicia (eastern Europe) and Utah. Also spelled ozocerite. [Gk. *ozein,* to smell + *keros,* wax] (Glocker in 1833)

P

pacificite *ig. extru.* Extrusive basalt consisting of augite and labradorite phenocrysts in a matrix of anemousite (49%), olivine, and iron oxides; anemousite is a plagioclase whose silica percentage is lower than that required by the albite-anorthite series, and the composition has been explained as a mixture of albite-anorthite together with minor soda-anorthite (carnegieite) in the ratio 8:10:1. Also named anemousite basalt. [Pacific, for its occurrence in Hawaii] (Barth in 1930)

pacificite, olivine *ig. extru.* Variety of pacificite with phenocrysts of augite (33%), olivine (18%), and bytownite in a groundmass composed of anemousite (34%), andesine, augite, apatite, and iron oxides. Also named kaulaite. (Barth in 1930)

packsand *sed.* Very fine-grained sandstone, so loosely consolidated by a little calcareous cement that it can be easily cut by a spade.

packstone *sed.* Sedimentary carbonate rock whose granular material is arranged in a self-supporting framework but also contains some matrix of calcareous mud. (Dunham in 1962)

pagoda stone *meta., sed.* General term referring to pagodite, as well as various other rocks suitable for carving (e.g., pagodas), such as limestone, soapstone, pyrophyllite rock, and others.

pagodite *meta.* Term essentially identical to agalmatolite; relatively soft, tough rocks used in China for carving, e.g., pagodas. [pagoda] (Napione in 1798)

pahoehoe *ig. extru.* Surface structural form of basalt lava with a ropy, billowy, knotted, or mammillary form due to the solidification of surface films and progressive breaking-through of the still molten material within; usually vesicular with numerous small spherical cavities evenly scattered through the rock. Also spelled pahoihoi. Other names include ropy lava and dermolith. Descriptive variety names include: corded p., entrail p., festooned p., filamented p., sharkskin p., shelly p., slab p. [Haw. *pahoehoe*, smooth, unbroken lava] (pub. by Dutton in 1883)

paigeite *meta.* Coal black, lustrous and opaque, foliated aggregate of matted fibers and long needles, consisting of a mixture of hulsite and an iron borate; occurs in contact metamorphic marble associated with a tin deposit on Brooks Mountain, Seward Peninsula, Alaska. [Sidney Paige, 1880-1968, American geologist, U.S. Geological Survey] (Knopf and Schaller in 1908)

paisanite *ig. diaschist.* Porphyritic dike rock with rare and small phenocrysts of sanidine and quartz and gray-blue spots, which are aggregates of riebeckite, in a very fine-grained, granular to dense, nearly white, groundmass consisting of quartz and microperthite intergrown micrographically. [Paisano Pass, Texas] (Osann in 1892)

palaeophyre *ig. aschist.* Reddish diorite porphyry showing phenocrysts of zoned oligoclase, brown biotite (altered to chlorite, epidote, and titanite), brown chloritized hornblende, and corroded quartz, in a groundmass consisting of a feldspar aggregate with secondary limonite and calcite and usually some quartz; cuts across Silurian strata in the Fichtelgebirge, Germany. [Gk. *palaios*, ancient + porphyry] (Gümbel in 1874)

palaeophyrite *ig. aschist.* Obsolete collective name for diorite porphyrite, including ortlerite and suldenite; applied to the older post-Cretaceous porphyries. [Gk. *palaios*, ancient + porphyrite] (Stache and von John in 1879)

palaeopicrite *ig. extru.* Obsolete name for a picrite that occurred in Palaeozoic strata. [Gk. *palaios*, ancient + picrite] (Gümbel in 1874)

palagonite *ig. extru., weathered.* Altered and hydrated equivalent of hyalobasalt (basaltic glass), often yellow or orange, and composed of glass remnants, cryptocrystalline material, microlites of augite, olivine, and labradorite, and secondary clays, zeolites, carbonates, and chlorite. [Palagonia, Sicily] (Waltershausen in 1845)

palatinite *ig. extru.* Obsolete general term originally applied to dark extrusive rocks including melanophyre, augite porphyrite, tholeiite, and others. [Pfalz (Palatia), Germany] (Laspeyres in 1869)

paleosol *sed.* Buried soil horizon from the geologic past. [Gk. *palaios*, ancient + L. *solum*, bottom, soil]

palite *ig. pluton., meta.* Local collective name for somewhat metamorphosed diorites and quartz diorites from Germany. [Pfalz (Palatia), Germany] (Frentzel in 1911) pallasite   *ig. pluton.* Ultramafic igneous rock containing a higher percentage of iron oxides (e.g., as magnetite) than silica in its chemical composition, e.g., cumberlandite. [pallasite meteorite, alluding to similarity]

pallasite   *meteor.* Stony-iron meteorite composed of nickel-iron (both kamacite and taenite as Widmanstatten intergrowths) and olivine (rounded to angular, normally 5-10 mm across), with minor troilite, schreibersite, and farringtonite. [Peter Simon Pallas, 1741-1811, German geologist who studied the natural history of Russia and brought specimens of the meteorite to St. Petersburg in 1772] (Rose in 1862)

pantellerite   *ig. extru.* Green to black porphyritic alkali rhyolite consisting of phenocrysts of aegirine-augite or diopside, anorthoclase, and aenigmatite (cossyrite) in a fine-grained, holocrystalline groundmass, or in one consisting of aegirine microlites in a very fine-grained aggregate of feldspar, or in glass, or in pumice; the rock is without plagioclase, and quartz is rare. [Pantelleria, island south of Sicily, where it occurs at numerous volcanic vents] (Foerstner in 1881)

pantellerite, quartz   *ig. extru.* Variety of pantellerite containing some quartz phenocrysts.

pantha   *meta.* White, translucent variety of jadeite rock; term used in India. [Possibly Pantha (Pansee), Moslem tribe or sect in Yunnan province, China]

papa   *sed.* Soft and bluish clay, mudstone, siltstone, or sandstone found in North Island, New Zealand; used for whitening fireplaces. [New Zealand (Maori) *papa,* rock]

para-   Prefix often used in names of metamorphic rocks derived from sedimentary parent rocks; in contrast, rocks derived from igneous parent rocks may be designated by the prefix *ortho-.* [Gk. *para,* beside, beyond, along with]

para-amphibolite   *meta.* Amphibolite derived from earlier sedimentary parent rock.

paraconglomerate   *sed.* Conglomerate not a result of normal aqueous flow but deposited by an alternate means such as mass transport in subaqueous turbidity slides or by glacial ice; usually unstratified, and the pebbles, because of abundant matrix material, are usually not in contact. [Gk. *para,* beside + conglomerate] (Pettijohn in 1957)

parafenite   *meta.* Aegirine sanidinite formed by alkali metasomatism, similar to fenitization but in a shallower and higher-temperature environment; occurs esp. at Laacher Lake, Germany. [Gk. *para,* beside + fenite]

paraffin, native   *sed.* = ozokerite. [L. *parum,* too little + *affinis,* associated with; from its lack of affinity with other material]

paragneiss   *meta.* Gneiss derived from clastic sedimentary rocks such as conglomerate or arkose. (Rosenbusch in 1901)

pararenite   *sed.* Arenite with a detrital matrix, e.g., clays, making up 15-70% of the rock. [L. *par,* equal + arenite] (Pettijohn and others in 1973)

Palagonite. Kempenich, Lake of Laach, Germany. 11.5 cm.

paratooite   *sed.* Insoluble residue of bird guano, consisting mainly of hydrated aluminum and ferric phosphate but of variable composition. [Paratoo railway siding, South Australia] (Mawson and Cooke in 1907)

parchettite   *ig. extru.* Extrusive rock resembling leucite tephrite but containing some orthoclase and more than the usual amount of leucite. [Fosso della Parchetta, San Martina, Italy] (Johannsen in 1938)

parent rock   *ig., sed., meta.* 1. The rock formation that originally held the fragments of a rock, mineral, or ore carried elsewhere by streams and glaciers. 2. The original rock from which a given metamorphic rock was formed by metamorphism. Also named protolith.

parianite   *sed.* Complex variety of asphalt from Pitch Lake, Trinidad; so-called Trinidad pitch. [Gulf of Paria, Trinidad] (Peckham in 1895)

parophite   *meta.* Schistose, greenish, yellowish, reddish, or grayish pinite rock that carves like massive talc and resembles serpentine but is a mica; original specimens came from St. Nicholas and Famine River, Canada; a variety of agalmatolite and pinite. See these terms for other rocks similar in nature. [L. *par,* equal + Gk. *ophis,* serpent; from its resemblance to serpentine] (Hunt in 1852)

passyite   *sed.* Impure quartz occurring as white earthy masses at Contremoulins, Caux, France. [Antoine François Passy, 1792-1873, French geologist] (Marchand in 1874)

pastelite   *sed.* General term applied to a variety of jasper noted for its large wavy areas of pastel greens, pinks, reds, and browns; occurs esp. in the Western United States, e.g., Mojave desert, California, and in Arizona, Idaho, and Washington. [pastel]

pawdite   *ig. aschist.* Fine-grained, homogeneous, black or gray diabase, composed of magnetite, titanite, biotite, hornblende, zoned plagioclase (bytownite to oligoclase), and traces of quartz. [Nikolai Pawda, Urals, Soviet Union] (Duparc and Grosset in 1916)

pea grit   *sed.* = pisolite

pea ore   *sed.* Pisolitic limonite occurring as loose, small rounded grains or masses. Also named pea iron ore.

pealite   *sed.* Milk-white to greenish, vitreous variety of geyserite. [Albert Charles Peale, 1849-1913, American paleobotanist] (Endlich in 1873)

pearlite   *ig. extru.* = perlite

pearlstone   *ig. extru.* = perlite

pearlylite   *ig. extru.* Variety of obsidian used as a gem. [pearly]

peastone   *sed.* = pisolite

peat   *sed.* Dark brown or black combustible residuum produced by the partial decomposition and disintegration, in anaerobic environments, of mosses, heathers, rushes, sedges, and horsetails, and some trees such as pine, birch, and willow; usually forms in freshwater swamps and bogs and on moorland areas in temperate and polar regions. [M.Eng. *pete,* Anglo-Latin *peta,* perhaps of Celtic origin]

peat, amorphous   *sed.* Peat in which degradation of cellulose matter has destroyed the plant structures of the original materials.

peat, calcareous   *sed.* Peat containing abundant plant nutrients, such as calcium, nitrogen, potassium, and phosphorous. Also named eutropic peat.

peat, eutropic   *sed.* = peat, calcareous. [Gk. *eutrophos,* well-nourished]

peat, torch   *sed.* Resinous waxy peat derived mostly from pollen.

pebble   *sed.* Rounded, subrounded, or subangular fragment of any rock or mineral 4-64 mm in diameter. [O. Eng. *papolstan,* pebble]

pebble, algal   *sed.* Algal ball simulating an ordinary water-rounded pebble, esp. the freshwater filamentous blue-green algae.

pebble, faceted   *sed.* Pebble with faceted surfaces developed by natural agents, such as wave erosion on a beach or by the grinding action of a glacier or by wind erosion.

pebble, fine   *sed.* Pebble whose diameter is in the range of 4-8 mm.

pebble, glacial   *sed.* Pebble that is flattened and often striated by glacial action.

pebble, phosphatic   *sed.* Component of a secondary phosphorite, of either residual or transported origin, consisting of black, often varnishlike, pellets, pebbles, and nodules of phosphatic material mixed with sand and clay.

pechstein   *ig. extru.* = pitchstone. [G. *Pechstein,* pitchstone]

pectolite rock   *ig., altered.* Tough, massive, fine-grained, pale-green pectolite resembling jade; specimens occurring at Point Barrow, Alaska, were used by the Indians for implements and hammers (once named Alaska jade). Larimar is a variety of pectolite rock.

pedalfer   *sed.* General term for a soil in which there is a concentration of sesquioxides, like

Al_2O_3 and Fe_2O_3; characteristic of soil in a humid region. [Gk. *pedon,* ground + aluminum + ferrum (iron)]

pedocal   *sed.* General term for a soil in which there is an accumulation or concentration of carbonates, usually calcium carbonate; characteristic of soil in an arid or semiarid region. Also spelled pedcal. [Gk. *pedon,* ground + calcium]

pedrosite   *ig. pluton.* Perknite in which riebeckite (osannite) is the major component (97%). Also named osannite hornblendite. [Old Pedroso, Alemtejo, Portugal] (Osann in 1923)

pegmatite   *ig. diaschist.* Exceptionally coarse- and uneven-grained granite, usually found as tabular dikes, lenses, or veins, and varying from a simple composition (quartz with potassic feldspar) to one containing many accessory minerals including rare species rich in beryllium, boron, chlorine, fluorine, lithium, molybdenum, niobium, phosphorus, sulfur, tantalum, tin, tungsten, uranium, zirconium, and the rare-earth elements; pegmatites form esp. at the margins of batholiths and represent the last and most hydrous portion of a magma to crystallize. Pegmatite is a variety of granite, but pegmatitic facies of many other plutonic rocks occur. When appropriate, the term may be modified to include: alaskite p., alkali granite p., alkali syenite p., diorite p., gabbro p., hollaite p., ijolite p., kalisyenite p., larvikite p., nepheline syenite p., norite p., pulaskite p., syenite p., tonalite p. [Gk. *pegma,*

framework (term originally applied to graphic granite)] (Haüy, pub. by Brongniart in 1813)

pegmatite, corundum   *ig. diaschist.* Coarse-grained variety of kalisyenite pegmatite consisting of yellowish grains and twinned crystals of microperthitic orthoclase and long (up to 10 cm) hexagonal prisms of corundum, accompanied by very minor rutile, apatite, zircon, and other minerals; from the Ilmen Mountains, Urals, Soviet Union. (Morozewicz in 1899)

pegmatite, granite   *ig. diaschist.* = pegmatite

pegmatoid   *ig. aschist.* General term for an igneous rock with the coarse-grained texture of a pegmatite but lacking graphic intergrowths or a typically granitic composition. [pegmatite] (Evans in 1912)

pegmatophyre   *ig. aschist.* Porphyry of the granite family, whose groundmass is made up of micropegmatitic (microscopic graphic granite), semispherulitic, branching, plumose, or interpenetrating crystals of quartz and orthoclase. [pegmatite + porphyry] (Lossen in 1892)

pelagite   *sed.* Ferromanganese nodules from deep-sea environments, esp. those obtained by the Challenger Expedition from the Pacific between Japan and the Sandwich Islands (2740 fathoms). [Gk. *pelagos,* sea] (Church in 1876)

pelagosite   *sed.* Superficial thin, white, gray, or brownish, pearly crust, formed in the intertidal zone by ocean spray and evaporation, and com-

Pegmatite, small dike in pebble of biotite gneiss. Cache Poudre River, Larimer County, Colorado. 5 cm.

posed of calcium carbonate, accompanied by larger amounts of magnesium carbonate, calcium sulfate, strontium carbonate, and silica than are found in normal limy sediments. [Gk. *pelagos,* sea]

peléeite *ig. extru.* Extrusive rock in the quartz diorite family related to sakalavite and consisting of phenocrysts of calcic-plagioclase, hypersthene, and iron oxide, in a glassy groundmass whose composition indicates andesine, sanidine, and quartz. [Mt. Pelée, Martinique, West Indies] (Niggli in 1923)

Pele's hair *ig. extru.* Reddish or tobacco brown, fine capillary form of basaltic glass (hyalobasalt), produced by the blowing-out of lava during quiet fountaining of fluid lava, by cascading lava falls, or by very turbulent flows; a single strand may have a diameter less than 0.5 mm, and may be as long as 2 m. [Pele, Hawaiian goddess of fire]

Pele's tears *ig. extru.* Small, tear-shaped drops of basaltic glass (hyalobasalt) behind which trail pendants of Pele's hair. [Pele, Hawaiian goddess of fire]

Pele's tears *sed.* Name sometimes applied to clear specimens of chalcedony or opal in Hawaii. [Pele, Hawaiian goddess of fire]

pelhamite *meta.* Variety of dark gray-green serpentinite of good quality. Also spelled pelhamine. [Pelham, Massachusetts] (Shepard in 1876)

pelinite *sed.* Highly plastic and partly colloidal secondary (transported) clay. [Gk. *pelinos,* made of clay] (Searle in 1912)

pelionite *sed.* Bituminous coal from Tasmania, resembling English cannel coal. [Monte Pelion, Tasmania] (Petterd in 1894)

pelite *sed.* General term for a clastic sedimentary rock composed of particles less than 1/16 mm (including silt and clay sizes) and consisting primarily of clay minerals with minute particles of quartz and other minerals; equivalent to lutite. [Gk. *pelos,* mud, clay] (Naumann in 1849)

pellodite *sed.* = pelodite

pelmicrite *sed.* Limestone composed of peloids (spherical, cylindrical, or angular grains, composed of microcrystalline carbonate, without internal structure) in a matrix of micrite (carbonate mud). [pellet + micrite] (Folk in 1959)

pelodite *sed.* Lithified equivalent of varved clay; term is sometimes extended to include any lithified glacial rock flour, composed of glacial pebbles in a silty of clayey matrix. Also spelled pellodite. Varvite is considered a synonym by some

petrologists. [Possibly Gk. *pelos,* mud, clay] (Woodworth in 1912)

pelsparite *sed.* Limestone composed of peloids (spherical, cylindrical, or angular grains, composed of microcrystalline carbonate, without internal structure) cemented by sparite, clear equant calcite. [pellet + sparite] (Folk in 1959)

pencatite *meta.* Marble in which calcite is nearly equal to periclase or brucite in abundance; related to predazzite. [Giuseppe Marzari-Pencati, 1779-1836, Italian geologist of Tyrol] (Roth in 1851)

pencil stone *meta.* Compact pyrophyllite rock, once used for making slate pencils.

penikkavaarite *ig. hypab.* Rock similar to essexite, composed essentially of augite, barkevikite, and green hornblende, in a feldspar (chiefly andesine, some orthoclase) matrix. [crest of Penikkavaara, Kuusamo parish, Finland] (Johannsen in 1938)

peperin *ig. extru.* Ash gray to yellow gray, soft, unconsolidated tuff, containing numerous dark inclusions in an earthy cement of lighter color, so that it resembles grains of pepper; in composition it includes augite, olivine, leucite, mica, magnetite, sanidine, glass shards, and earthy materials; from Albano Hills, near Rome, Italy. [It. *peperino,* pepper]

peperite *ig., sed.* Shallow-water or marine hybrid rock consisting of a mixture of lava with sediment; usually consists of somewhat globular, but not water worn, particles of volcanic glass (0.5 mm to 1 cm and larger) and a siliceous-carbonate or argillaceous matrix. [Gk. *peperi,* pepper; alluding to its appearance]

peracidite *ig. diaschist.* = silexite. [L. *per,* through (in reference to large amount) + acid] (Rinne in 1921)

peralboranite *ig. extru.* Theoretical, light-colored, alboranite with less than 12.5 volume percent mafics and 0-12.5 volume percent silica surplus. [L. *per,* through, by + alboranite] (Burri in 1937)

peralkaline rock *ig.* Igneous rock whose chemistry meets the following requirement: $Al_2O_3/(K_2O + Na_2O)$ is less than 1; in these rocks the typical mafic minerals are sodic pyroxenes (aegirine, aegirine-augite) and sodic amphiboles (riebeckite, arfvedsonite); in some the place of Al_2O_3 is taken by Fe_2O_3, ZrO_2, or TiO_2. Examples include p. foid syenite, p. granite, p. rhyolite, p. syenite, and p. trachyte. [L. *per,* through (in reference to large amount) + alkaline] (Shand in 1943)

peraluminous rock *ig.* Igneous rock whose chemistry meets the following requirement: $Al_2O_3/(K_2O + Na_2O + CaO)$ is greater than 1; excess Al_2O_3 is accommodated by the micas and by minor amounts of corundum, tourmaline, topaz, garnet, and more rarely, cordierite and sillimanite. Examples include p. foid syenite, p. granite, p. syenite. [L. *per,* through (in reference to large amount) + aluminous] (Shand in 1943)

peridotite   *ig. pluton.* 1. Feldspar-free, phaneritic igneous rock composed of more than 90% mafic minerals of which olivine is essential (over 50%); there are many varieties including, among others, dunite, cortlandite, harzburgite, lherzolite, saxonite, wehrlite. [F. *péridot,* olivine] (Cordier in 1868)   2. *IUGS,* plutonic rock with M equal to or greater than 90, and ol/(ol + opx + cpx) greater than 40.

peridotite, hornblende   *ig. pluton.* 1. Peridotite containing characterizing hornblende, e.g., cortlandite and schriesheimite.   2. *IUGS,* plutonic rock with M equal to or greater than 90, 40-90 ol/(ol + hbl + px), and px/(ol + hbl + px) less than 5.

peridotite, mica   *ig. pluton.* Peridotite containing characterizing biotite; because there is usually much more olivine than biotite, the rock also has been named mica olivinite.

peridotite, pyroxene   *ig. pluton.* 1. Peridotite containing characterizing pyroxene (orthopyroxene or clinopyroxene), e.g., harzburgite and lherzolite.   2. *IUGS,* plutonic rock with M equal to or greater than 90, 40-90 ol/(ol + hbl + px), and hbl/(ol + hbl + px) less than 5.

peridotite, pyroxene-hornblende   *ig. pluton.* 1. Peridotite containing characterizing pyroxene and hornblende, as illustrated by weigelith.   2. *IUGS,* plutonic rock with M equal to or greater than 90, 40-90 ol/(ol + hbl + px), px/(ol + hbl + px) greater than 5, and hbl/(ol + hbl + px) greater than 5.

peridotite, serpentinized   *meta.* Rock often altered or partly altered, because of the unstable nature of olivine and the other mafics that compose peridotite, to serpentinite and containing serpentine, uralite, chlorite, talc, phlogopite, magnetite, and magnesium-rich carbonates.

peridotite, vitrophyric   *ig. extru.* Glassy, sometimes vesicular, vitrophyre, showing olivine phenocrysts, that represents the extrusive or hypabyssal counterpart of kimberlite; occurs esp. in Tanganyika and northern Siberia and is illustrated by meimechite.

perilith   *ig. extru.* = bomb, cored. [Gk. *peri,* around + *lithos,* stone]

perknide   *ig. pluton.* Field term for any holocrystalline igneous rock composed almost entirely of mafic minerals, e.g., pyroxenite, hornblendite, peridotite. [Gk. *perknos,* dark, and relationship to perknite] (Johannsen in 1931)

perknite   *ig. pluton.* Collective name for igneous rocks containing as their main constituent a pyroxene or amphibole, including, e.g., enstatolite,

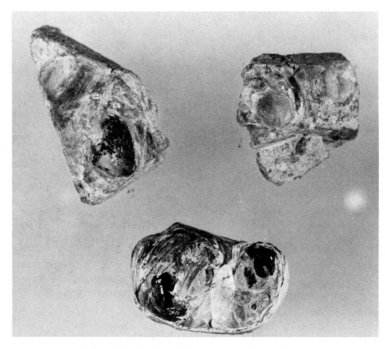

Perlite with obsidian (black) nuclei. Arizona. 3.5 cm.

bronzitite, hypersthenite, websterite, and hornblendite. [Gk. *perknos,* dark; alluding to the dark colors] (Turner in 1901)

perlite   *ig. extru.* Colorless, grayish, bluish, greenish, reddish, or brownish volcanic glass having the composition of rhyolite and generally a higher water content (less than 4%) than obsidian; often showing numerous concentric cracks showing onionlike partings and imparting a pearly luster; some perlites may have been derived from obsidian through hydration by meteoric water. [F. *perle,* pearl; alluding to luster] (Beudant in 1822)

perthite   *ig.* Regular to semiregular exsolution intergrowth consisting of a single-crystal host of potassic feldspar and thin plates, spindles, blebs, or crystals of guest albite or oligoclase, oriented or partly oriented according to the host structure; intergrowths vary from ones undetectable microscopically (cryptoperthite) to ones that can be seen with the unaided eye. [Perth, Ontario, Canada] (Thomson in 1843)

perthite, myrmekite   *ig.* Myrmekitelike intergrowth of predominant microcline (or other potassic feldspar) and vermicular plagioclase, resembling quartz-plagioclase myrmekite.

perthite, pyroxene   *ig.* Lamellar intergrowth of two or more pyroxenes in a manner similar to a perthite composed of two feldspars.

perthitophyre   *ig. pluton.* Light-colored syeno-

gabbro containing microperthite; from Volyn, Russia. [perthite] (von Chrustschoff in 1888)

perthosite   *ig. pluton.* Light-colored monzonite with approximately equal amounts of cryptoperthite and antiperthite and less than 3% aegirine-augite. [perthite] (Phemister in 1926)

Petoskey stone   *sed.* Limestone composed of masses of fossil colony coral (*Hexagonaria percarinata*) occurring abundantly in Devonian reef limestones (Traverse Group) in several Michigan counties (Charlevoix, Emmet, Cheboygan, Presque Isle, Alpena); widely scattered and often found as rounded pebbles and cobbles on beaches of Lake Michigan. [Petoskey, Emmet County, Michigan]

petrified wood   *sed.* Rock material formed by the mineralization of wood in such a manner that the original forms and structures of the wood are preserved; most fossil wood of this type is replaced by quartz or opal (e.g., agatized wood, jasperized wood, opalized wood), but other minerals may be involved, including azurite, barite, bornite, brucite, calcite, carnotite, chalcocite, cinnabar, coffinite, dolomite, hematite, limonite, malachite, siderite, sphalerite, uraninite, and others. See also silicified wood. [L. *petra,* rock]

petroleum   *sed.* Natural substance consisting of crude oils, chiefly short and long-chain hydrocarbons, and gas, mainly methane, which in most cases has migrated into porous reservoir

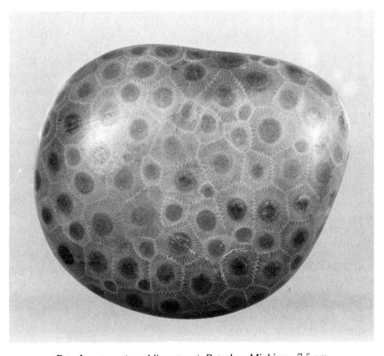

Petoskey stone (coral limestone). Petoskey, Michigan. 7.5 cm.

rocks from source rocks; several hundred organic compounds have been recorded from natural oil. [L. *petra* (Gk. *petros*), rock + *oleum,* oil] (Agricola in 1546)

petrosilex *ig. extru.* Old general term for aphanitic crystalline rhyolites and trachytes and devitrified glasses; essentially the same as amausite and felsite. [L. *petra,* rock + *silex,* flint]

petrosilex *sed., ig., meta.* 1. Chert (esp. flint) occurring as a rock mass, or as part of a mass, rather than as detached nodules. 2. Broad term including various flinty rocks, e.g., felsite, phonolite (clinkstone), compact feldspar rocks, hornstone, and hälleflinta. [L. *petra,* rock + *silex,* flint]

phacoidal rock *meta.* General term for metamorphic rock, usually formed by mechanical deformation or cataclastic processes, consisting of ellipsoidal or lensoid units in a finer-grained matrix that is brecciated and sheared; certain augen gneisses belong in this category. [Gk. *phakos,* lentil seed]

phanerite *ig. pluton.* General term for any igneous rock having grains of its essential minerals large enough to be seen without the aid of a microscope. [Gk. *phaneros,* visible]

philippinite *tekt.* = rizalite. [Philippines]

phlebite *migmat.* Migmatite with a veined appearance (e.g., veined gneiss); term used when it is impossible to discriminate between an arterite and venite. [Gk. *phlebos,* vein] (Scheumann in 1936)

phonolite *ig. extru.* 1. Aphanitic extrusive rock composed of alkali feldspar (esp. anorthoclase or sanidine), nepheline, and mafics; the extrusive equivalent of nepheline syenite. Also named clinkstone and echodolite. [Gk. *phone,* sound; alluding to the alleged ringing sound emitted when the rock is struck] (Klaproth in 1801) 2. In a broad sense, any extrusive rock composed of alkali feldspar, any feldspathoid, and mafics, giving rise to the following varieties: analcime p., hauyne p., leucite p., nepheline p. (phonolite), nosean p., sodalite p. Many petrologists call them trachyte (see trachyte, foid).

phonolite, latite *ig. extru.* Dark-gray to black porphyritic rock, without nepheline, consisting of small phenocrysts of pyroxene, feldspar, and rare biotite, in a rather fine-grained groundmass composed of nearly equal amounts of alkali feldspar and oligoclase accompanied by sodalite, nosean, analcime, pyroxene, and others; from the Cripple Creek district, Colorado. (Graton in 1906)

phonolite, natrolite *ig. extru.* = marienbergite. (Hibsch in 1904)

phoscorite *ig. pluton.* Iron-oxide-rich ultramafic peridotite, composed of abundant magnetite with much olivine, apatite, and phlogopite; first noted in Phalaborwa, South Africa. [Possibly phosphate core; alluding to its composition and geological structure] (Russell and others in 1954)

phosphate, blue-rock *sed.* Name applied to the Ordovician bedded phosphorite of Tennessee.

phosphate, white-bedded *sed.* Name applied to phosphatic limestone characterized by partial replacement of calcite by calcium phosphate and by a matrix consisting of cryptocrystalline quartz; occurs in Tennessee.

phosphate rock *sed.* = phosphorite

phosphorite *sed.* Sedimentary white, yellow, brown, or gray rock, consisting mainly of phosphate minerals, the most abundant of which are members of the apatite group esp. carbonate-fluorapatite (francolite) and carbonate-hydroxylapatite (dahllite); these minerals are often cryptocrystalline and essentially isotropic (called colophane) and may be accompanied by calcite, chalcedony, pyrite, anhydrite, gypsum, and various detrital and organic and fossiliferous materials. Also named phospholite and phosphate rock. Many of the varieties are descriptive in nature, e.g., arenaceous p. (sandy, grading into sandstone), argillaceous p. (clayey, grading into shale), bituminous p., carbonaceous p., coprolitic p., coquinoid p. (replaced mollusc shells), crinoidal p., fossiliferous p., glauconitic p., nodular p., oolitic p., ovulitic p. (ovoidal pellets), pebbly p., pellet p., pisolitic p., shaly p., stalagmitic p. [phosphorous, alluding to phosphate composition]

phosphorite, bone *sed.* = bone bed

phosphorite, organic *sed.* Phosphorite owing its origin to organisms and often containing one or more of the following materials: bitumen, coprolites, guano, shell and bone fragments, sponge spicules, diatoms, radiolaria, mollusc tests, crinoids, fossil fish, sharks' teeth.

phosphorite, primary bedded *sed.* Marine phosphorite formed by the slow precipitation of colloidal phosphate, under anaerobic conditions with close control of the water temperature and acidity.

phosphorite, secondary residual *sed.* Phosphorite derived from a sediment, originally lean in phosphate, that is enriched in phosphate through the leaching out of the nonphosphatic constituents, primarily calcite; illustrated by the land-pebble phosphate deposits of Florida.

phosphorite, secondary transported *sed.* Secondary phosphorite formed from a lean phosphatic sediment through the leaching out of the nonphosphatic constituents and then partly transported; illustrated by the river-pebble phosphate deposits of Florida.

phosphorolite *sed.* General term including phosphorite, guano, apatite-rich rocks, and the like. [phosphorus] (Wadsworth in 1891)

phthanite *sed., meta.* Old term variously defined and applied to several kinds of rocks including the following: chert, hornstone, basanite, lydite, compact micaceous or talcose quartz grit, siliceous schists, silicified shale. Also spelled phtanite. [Gk. *phthanein,* to come first] (Haüy in ca. 1822)

phyllarenite *sed.* Litharenite composed chiefly of foliated, phyllosilicate-rich, metamorphic rock fragments (schist, phyllite, slate). [Gk. *phullon,* leaf; alluding to phyllosilicate minerals + arenite] (Folk in 1968)

phyllite *meta.* Fine-grained, micaceous rock with a foliation intermediate in perfection between slaty and schistose and characterized by having a satiny or glossy luster; intermediate in grain size between slate and mica schist and essentially equal to a sericite schist. Metacrysts are common and may include chloritoid, biotite, garnet, magnetite, and pyrite. Major mineralogical varieties include chlorite p., paragonite p., sericite p., and sericite-chlorite p. [Gk. *phullon,* leaf; alluding to platy cleavage] (Naumann, 1849)

phyllonite *meta.* Fine-grained, highly schistose mylonite whose fine texture results from the destruction of coarser-grained units and whose foliation in many cases results from closely spaced shear planes; resembles phyllite and commonly contains low-grade metamorphic minerals. [Gk. *phullon,* leaf] (Sander in 1911)

-phyre *ig.* Suffix commonly used in igneous rock names to signify porphyry. [porphyry]

phytocollite *sed.* Black gelatinous nitrogenous humic substance occurring beneath or within peat deposits. [Gk. *phuton,* plant + *kolla,* jelly] (Lewis in 1881)

phytogenic rock *sed.* Biogenic sedimentary rock either produced by plants or directly attributable to the presence or activities of plants, e.g., peat, coal, diatomite, and algal deposits. [Gk. *phuton,* plant + *genes,* born]

piauzite *sed.* Slaty, brownish or greenish-black, asphaltlike substance, with a high melting point; occurs in Carniola and Styria. [Piauze, near Neustadt, Carniola, Yugoslavia] (Haidinger in 1844)

picotitfels *ig. pluton.* Monomineralic igneous rock whose only essential component is picotite (chromian spinel); original specimens, from New South Wales, contain picotite (85%) and serpentine (15%). Also named picotitite. (Judd in 1895)

picotitite *ig. pluton.* = picotitfels

picrite *ig. extru.* 1. Extrusive equivalent of the peridotites; dark-colored and containing olivine, pyroxene, and hornblende, with ilmenite and magnetite. (Tschermak in 1866) 2. Very dark olivine diabase (in the peridotite family), composed chiefly of olivine and augite, with or without hornblende. [Gk. *pikros,* bitter; on account

Phyllite. Dixon, Taos County, New Mexico. 8.5 cm.

of high magnesia, whose salts are bitter] (Rosenbusch in 1877)

picrite, hornblende   *ig. pluton.* Obsolete name for certain uralitized olivine-rich rocks from Anglesey, Wales. (Bonney in 1881)

picrolite   *meta.* Green, gray, or brown serpentinite, often having a long, splintery fracture; originally found at Taberg, Sweden. [Gk. *pikros*, bitter; on account of high magnesia, whose salts are bitter] (Hausmann in 1808)

picture rock   *sed.* Sandstone showing bands of strongly contrasting colors due to successive waves of infiltration of water solutions containing oxides of iron; when cut, planar surfaces show patterns which resemble landscapes; term esp. applied to rock from Kanab, Kane County, Utah.

picture stone   *sed.* Sandstone, similar to picture rock but finer-grained; from St. George, Washington County, Utah.

picurite   *sed.* = coal, pitch. [L. *piceus*, made of pitch]

pienaarite   *ig. pluton.* Titanite-rich (17%) malignite whose feldspar is anorthoclase. [Pienaar Creek, Pretoria, South Africa] (Brouwer in 1910)

pietricikite   *sed.* Brown substance, very much like ozokerite in most physical characteristics but almost completely insoluble in ether and with a higher melting point. Incorrectly spelled zietrisikite. [Mt. Pietricica, Moldavia] (Istrati in 1897)

pikeite   *ig. pipe.* Dense, porphyritic, greenish-black, phlogopite peridotite (kimberlite) with olivine phenocrysts and also containing augite, chromite, magnetite, secondary serpentine, and rarely diamonds. [Pike County, Arkansas] (Johannsen in 1938)

pilandite   *ig. pluton., dike.* Name originally applied to the dike and border facies of hatherlite (leeuwfonteinite). [Pilansberg, West Transvaal] (Henderson in 1898)

pinite   *meta., ig. altered.* Massive, very fine-grained, impure muscovite-rich rock, variously colored and often containing admixed chlorite, clays, and other minerals; formed by the alteration of cordierite, feldspars, nepheline, scapolite, spodumene, and similar silicates. Numerous other names have been applied to these materials, esp. with their discovery in different localities, e.g., agalmatolite, gieseckite, liebenerite, pagodite, parophite, and others. [Pini mine at Aue, near Schneeberg, eastern Germany] (Karsten in 1800)

pinolite   *meta.* Metamorphic rock containing ferroan magnesite (breunnerite) as crystals and granular aggregates in a matrix of phyllite or talc schist. [L. *pinus*, pine; from resemblance of crystal clusters to pine cones] (Rumpf in 1873)

pipestone   *sed.* Argillaceous sedimentary rock carved by American Indians into tobacco pipes; term usually used in reference to catlinite.

pisolite   *sed.* 1. Sedimentary rock made up chiefly of pisoliths cemented together; pisolitic limestone (calcitic pisolite) is the most common example, but other compositions have been noted (see oolite); essentially identical to an oolite, but grains are greater than 2 mm in diameter. Also named pea grit and peastone.   2. Term often used incorrectly for pisolith. [Gk. *pisos*, pea]

pisolith   *sed.* One of the spherical or subspherical accretionary pealike bodies that occurs in some chemical sedimentary rocks and has a diameter greater than 2 mm; like an oolith but larger; calcite and aragonite are the usual components, but other minerals occur (see oolite). [Gk. *pisos*, pea]

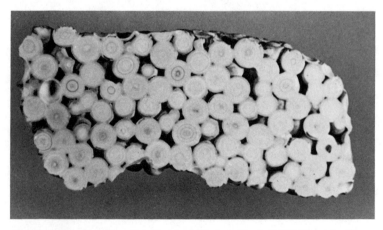

Pisolite (aragonitic) (ktypeite). Karlsbad (Karlovy Vary), Czechoslovakia. 12 cm.

pisosparite *sed.* Limestone containing at least 25% pisoliths and no more than 25% intraclasts and in which the sparry-calcite cement is more abundant than the carbonate-mud matrix (micrite); like an oosparite. |pisolith + sparite| (Folk in 1959)

pitch *sed.* = asphalt. |O.Eng. *pic,* pitch|

pitch, iron *sed.* At the pitch lake in Trinidad, land asphalt that has overflowed the lake onto the land and hardened to such an extent it resembles refined lake asphalt.

pitchstone *ig. extru.* Black, gray, olivine green, brown, or red extrusive igneous glass with a conchoidal fracture and exhibiting a pitchy luster; chemically different from obsidian (rhyolite glass) only in that it contains 4-10% water, while obsidian usually contains less than 1%. |pitch, for its luster| (Used in Germany as early as 1759)

plädorite *ig. pluton.* = pläthorite

plaffeiite *sed.* Amber yellow fossil resin found in the Flysch (Lower Tertiary) shales in Switzerland. |Plaffein, Switzerland| (Tschirsch and Kato in 1926)

plagiaplite *ig. diaschist.* Diorite aplite composed essentially of sodic-plagioclase (oligoclase to andesine), with or without green hornblende, and accessory quartz and mica. |plagioclase + aplite| (Duparc and Jerchoff in 1902)

plagifoyaite *ig. pluton.* Plutonic rock containing 10-60% feldspathoid and a ratio of plagioclase to total feldspar of 10-50% (by volume); these unusual rocks grade into essexite. |plagioclase + foyaite|

plagioclasite *ig. pluton.* = anorthosite. |plagioclase|

plagiogranite *ig. pluton.* General term used by Russian petrologists for igneous rocks having a low potassium content, including rocks ranging from quartz diorite to trondhjemite; these rocks contain plagioclase rather than potassic feldspar, which is essential to a true granite. |plagioclase + granite| (Kruschov in 1931)

plagiophyre *ig.* Rock resembling orthophyre (orthoclase porphyry) in texture but containing plagioclase rather than orthoclase; type specimens contained platy andesine with interstitial chloritic minerals, iron oxides, and rare orthoclase. |plagioclase + porphyry| (Tyrrell in 1912)

plagisyenite, foid *ig. pluton.* = monzosyenite, foid. |plagioclase + syenite|

plasma *sed.* Dull, dark to leek green, subtranslucent or opaque, fine-grained quartz rock, approaching jasper in texture and appearance; color possibly due to the presence of chlorite, celadonite, microfibrous amphibole, or other green silicate. |Gk. *plasma,* form, image; alluding to the carved seals and signets made of the rock|

plaster stone *sed.* = gypsum rock

pläthorite *ig. pluton.* Obsolete name for a mafic-rich hornblende-bearing biotite granite or a biotite-bearing hornblende granite. Also spelled plädorite. |Gk. *plethore,* fullness; alluding to abundant mafic constituents| (Lang in 1877)

plauenite *ig. pluton.* Unnecessary name for a plagioclase-bearing syenite. |Plauenscher Grund, near Dresden, Germany| (Brögger in 1895)

pleonaste-pyrope rock *ig.* Rare ultramafic rock in which the major minerals are ferran spinel (pleonaste) and pyrope; occurs with dunite and ariegite at Manget Heights northeast of Spitskop, eastern Transvaal. (Strauss and Truter in 1950)

pleonastite *meta.?* Rock composed of ferroan spinel (pleonaste) and chlinochlore surrounding corundum crystals. |pleonaste|

plinthite *sed.* Material in soil consisting of a mixture of clay and quartz and also rich in iron and aluminum sesquioxides but poor in humus; highly weathered and with time becomes a hardpan (ironpan). |Gk. *plinthos,* tile, brick|

plumasite *ig. hypab.* Unusual rock related to diorite and composed of coarse, white, oligoclase (84%) with imbedded corundum crystals (16%) that are partly altered to margarite. |Plumas County, California| (Lawson in 1901)

plutonite *ig. pluton.* General name for any deep-seated (plutonic) igneous rock. |Gk. Plouton, Pluto and the lower world| (Scheerer in 1862)

pollenite *ig. extru.* Olivine-bearing phonolite, containing phenocrysts of sanidine, plagioclase (labradorite), nepheline, clinopyroxene, amphibole, and biotite, in a glassy groundmass; related to tautirite but containing olivine and more glass. |Vallone di Pollena, Monte Somma, Italy| (Lacroix in 1907)

polyhalite rock *sed.* Sedimentary evaporite rock in which polyhalite is an essential constituent; associated minerals include accessory dolomite, halite, anhydrite, and kieserite.

polyhalite-anhydrite rock *sed.* Sedimentary evaporite rock containing about equal amounts of polyhalite and anhydrite with accessory magnesite, langbeinite, kieserite, leonite, and pyrite.

polyhalite-halite rock *sed.* Sedimentary evaporite rock containing about equal amounts of polyhalite and halite with accessory magnesite, dolomite, kieserite, langbeinite, and bischofite.

polylitharenite   *sed.* Litharenite containing a diversity of sand-sized rock clasts, e.g., volcanic, sedimentary, and metamorphic. [Gk. *polus,* many + litharenite] (Folk in 1968)

polylitharenite, feldspathic   *sed.* Polylitharenite containing more than 10% feldspar. (Folk in 1968)

polynigritite   *sed.* Type of nigritite occurring finely dispersed in argillaceous rocks. [Gk. *polus,* many; alluding to dispersed condition + nigritite]

polzenite   *ig. hypab.* Group of lamprophyres containing olivine and melilite; included in the group are modlibovite, luhite, and vesecite. [Polzen, Bohemia, Czechoslovakia] (Scheumann in 1912)

ponzaite   *ig. extru.* = trachyte, Ponza. [Ponza Islands, Italy] (Reinisch in 1912)

ponzite   *ig. extru.* Alkali trachyte composed of potassic feldspar or sodium-bearing potassic feldspar, some sodic-plagioclase, aegirine or other sodium pyroxene or sodium amphibole, and usually showing phenocrysts of augite and magnetite (representing resorbed biotite); the feldspathoid-free member of the Ponza trachyte (ponzaite) group. [Ponza Islands, Italy] (Washington in 1913)

poppy stone   *sed.* Jasper consisting of red orbicular areas set in a yellow matrix, esp. the type from California, used for cutting cabochons. Also named poppy jasper.

porcellanite   *sed., meta.* 1. Sedimentary rock composed of clay and silt with large amounts of chalcedony or opal so that it is intermediate between a siliceous shale and chert; has the texture, luster, hardness, and conchoidal fracture of unglazed porcelain. 2. Fine-grained silicified tuff. 3. Indurated or baked clay or shale, occurring on the roof or floor of a burned-out coal seam. Also spelled porcelainite and porcelanite. [It. *porcellana,* porcelain]

porcellophite   *meta.* Variety of serpentinite with a smooth, porcelainlike fracture. [It. *porcellana,* porcelain + Gk. *ophis,* serpent]

porfido rosso antico   *ig. aschist.* Altered andesite porphyry showing small (1-3 mm) white phenocrysts of altered plagioclase (andesine) in a dense aphanitic dark purple red groundmass, whose color is due to red epidote (withamite); much used in ancient times by both the Egyptians and Romans for statues, columns, vases, and sarcophaguses; quarried at Djebel Dokhan, Egypt. [It. for red antique porphyry]

porfido verde antico   *ig. aschist.* Diabase porphyry composed of phenocrysts of augite and labradorite in a groundmass of the same minerals, used for statues and as a decorative stone by the ancient Greeks and Romans; occurs south of Levetsova (formerly Croceae) on the old road from Sparta to Gythion in Laconia (Peloponnesus), southern Greece. [It. for green antique porphyry]

porodine   *sed.* Term applied to amorphous rock solidified from a colloidal condition, such as an opal rock derived from gelatinous silica. [Gk. *porodes,* like tufa] (Breithaupt in 1841)

porodite   *ig. extru.* General name for all the altered, fragmental forms of extrusive igneous rocks, e.g., some tuffs and schalstein. [Gk. *porodes,* like tufa]

poros   *sed.* Variously defined limestones, occurring in Egypt and the Peloponnesus and extensively used as building materials by the ancient Greeks; apparently included several materials, e.g., tufa, travertine, fossiliferous limestone, and onyx marble. [Gk. *poros,* pore] (Theophrastus, *On Stones*)

porphyrite   *ig. hypab., extru.* Obsolete term now a synonym for porphyry; term one time used in reference to plagioclase-bearing rocks containing phenocrysts, whereas porphyry was used in reference to potassic feldspar (orthoclase)-bearing rocks with phenocrysts. The following porphyries were once named porphyrites: dacite p., diabase p., diorite p., gabbro p., norite p., tonalite p. In a more restricted sense porphyrite was also a synonym for andesite porphyry. [porphyry]

porphyrite, albite   *ig. dike.* = albitophyre

porphyrite, augite   *ig. dike.* = augitophyre

porphyrite, hamrongite   *ig. dike.* Syenodiorite porphyry with well-defined white calcic-oligoclase phenocrysts in a grayish-green groundmass of sodic-andesine, orthoclase, titanaugite, and secondary chlorite, quartz, and serpentine. [Hamrånge parish, Sweden] (von Eckermann in 1928)

porphyrite, labradorite   *ig. pluton.* = labradophyre

porphyrite, quartz   *ig. extru.* Obsolete name for dacite porphyry.

porphyroid   *meta.* Metamorphic rock with a texture resembling igneous porphyry, where phenocrysts of feldspar or other minerals occur in a finer groundmass and in which there is evidence that the rock was formed from clastic sediment. [porphyry] (Lossen in 1869)

porphyry   *ig. hypab., extru.* 1. Term applied to any igneous rock showing larger crystals (phenocrysts) in a finer groundmass, which may be glassy, microcrystalline, finely crystalline, or coarsely crystalline. When the term is modified by a rock name the inference is that a rock of that composition shows phenocrysts. The number of possibilities is very large, and the following are common examples: andesite p., dacite p., diabase p., diorite p., essexite p., gabbro p., granite p., granodiorite p., larvikite p., latite p., maenaite p., monzonite p., nepheline syenite p., norite p., phonolite p., pitchstone p., syenite p., syenodiorite p., tonalite p., trachyte p. [Gk. *porphyra,* purple; term applied to a purplish rock, mottled with white phenocrysts, from Djebel Dokhan, Egypt (porfido rosso antico)] (Pliny, *Natural History*)   2. Obsolete synonym for trachyte porphyry.   3. Term used colloquially in the western United States for almost any igneous dike or sheet rock associated with an ore body.

porphyry, augite   *ig. dike.* = augitophyre

porphyry, claystone   *ig., altered.* Porphyry whose naturally fine-grained groundmass is more or less kaolinized, so as to be soft and earthy, suggesting a hardened clay rock with phenocrysts.

porphyry, felsite   *ig. hypab.* Rhyolite containing phenocrysts of quartz, orthoclase, usually some plagioclase, and biotite, in an aphanitic, occasionally porous, rarely vesicular groundmass of quartz and orthoclase. (Naumann in 1849)

porphyry, foyaite   *ig. pluton., dike.* Nepheline syenite porphyry showing phenocrysts of orthoclase and nepheline in a matrix of the same minerals with pyroxene (aegirine, aegirine-augite, diopside), sodium amphiboles, biotite, and others.

porphyry, granophyric granite   *ig. aschist.* Granite porphyry in which the groundmass is composed of quartz and potassic feldspar intergrown after the manner of graphic granite.

porphyry, labradorite   *ig. pluton.* = labradophyre. (Zirkel in 1894)

porphyry, Lenne   *ig. extru.* General term applied to the keratophyres and associated crystal tuffs of the Lenne Valley, Westphalia, western Germany. [Lenne Valley, Germany] (von Dechen in 1845)

porphyry, leucite   *ig. hypab.* Dike rock whose composition corresponds to a leucite phonolite and containing leucite phenocrysts; in most of the known examples the leucite has been replaced by mixtures of orthoclase and nepheline (pseudoleucite).

porphyry, leucite-syenite   *ig. pluton.* = arkite

porphyry, nepheline   *ig. hypab.* Seldom used synonym for nepheline syenite porphyry.

porphyry, orthoclase   *ig. hypab.* Porphyritic hypabyssal rock of the syenite family composed of sharp phenocrysts of orthoclase in a ground-

mass made up of the same feldspar with very minor plagioclase, quartz, and mafics. Other names, essentially synonyms, include orthophyre and orthosite.

porphyry, quartz   *ig. extru.* Old name applied to rhyolites of pre-Tertiary age, esp. by European geologists.

porphyry, quartz-free   *ig. extru., hypab.* = porphyry, orthoclase

porphyry, rhombic   *ig. hypab.* = rhomb-porphyry

posepnyte   *sed.?* Oxygenated hydrocarbon from the Great Western mercury mine, Lake County, California; occurs in green to brown plates and nodules that are often hard and brittle. [Franz Posepny, 1836-1895, Austrian geologist of Pribram, Bohemia, who discovered it] (Schröckinger in 1877)

potch   *sed.* Variety of opal that does not exhibit a play of colors but occurs with precious opal in the Australian opal deposits; may be colorless or any color and varies from opaque to transparent. Some varieties include: agaty-p. (colors arranged in layers or bands), beer-bottle p. (amber color), blue-bottle p. (transparent blue, from Queensland), crockery p. (white), p.-and-color (with a little precious opal mixed in), snide (large pieces of potch, in Queensland). [Possibly Eng. potch, related to poach, cook]

potstone   *sed.* Large concretion or boulder in the roof of a coal seam, having the rounded appearance of the bottom of an iron pot, and easily detached. Also named pot bottom or pot.

potstone   *meta.* Impure soapstone or massive talc rock used in prehistoric times to make cooking pots and vessels.

pozzolan   *sed., ig. extru.* Siliceous materials such as diatomaceous earth, opaline chert, and certain tuffs that can be finely ground and combined with Portland cement (in proportion of 15-40% by weight); these cements are highly resistant to the penetration and corrosion of salt water. Originally the term was applied to a leucite tuff from near Pozzuoli, Italy. Also spelled pozzolana, pozzuolane, puzzuolana, and puzzolane. [Pozzuoli, Italy]

prase   *sed.* Translucent, light or grayish yellow green, macrocrystalline or microcrystalline quartz (usually chalcedony); the color has been attributed to the presence of silicate minerals like actinolite, hornblende, or chlorite. Plasma is similar but is opaque. [Gk. *prason,* leek; alluding to green color] (Theophrastus, *On Stones*)

prasinite   *meta.* Variety of greenschist in which hornblende, chlorite, and epidote are present in approximately equal amounts. [Gk. *prason,* leek; for green color] (Kalkowsky in 1886)

predazzite *meta.* Brucite marble in which brucite is usually pseudomorphous after periclase; calcite is greater than brucite (and periclase), and there may also be some forsterite. [Predazzo, Italy] (Petzholdt in 1843)

prehnite rock *ig. extru., altered.* Any rock in which prehnite is a major component; pink prehnite, superficially resembling thomsonite (lintonite), occurs as amygdules in lava of Keweenanwan age in the Lake Superior region; other prehnite-rich rocks, also containing diopside, epidote, quartz, and chlorite, are known to have formed by calcium metasomatism of serpentized ultramafic igneous rocks. Uigite is also a variety of prehnite rock.

previtrain *sed.* Woody lenses in lignite that are the equivalent of vitrain in coal of higher rank. [L. *pre,* before + vitrain]

prilepite *sed.* Resinous substance occurring as reniform crusts on coal shales at Prilep, Bohemia. [Přilep] (Bořický in 1873)

propylite *ig. extru., altered.* Greenstonelike rock originally thought to be the oldest of the Tertiary extrusives but now known to be an altered andesite containing chlorite, epidote, serpentine, calcite, quartz, pyrite, and iron oxides. [Gk. *propulon,* before the gate; alluding to its existence at the beginning of the Tertiary] (Richthofen in 1867)

proteolite *meta.* Old name for contact metamorphic rocks produced by granite intrusions in slate in Cornwall, England; specifically, term has been applied to andalusite hornfels, consisting of quartz, mica, and andalusite. [Gk. Proteus, sea god, fabled to assume various shapes, thus, changing, varying] (Boase in 1832)

proterobase *ig. aschist.* Diabase whose mafic mineral is primary hornblende; if the hornblende is secondary, the rock is a uralite diabase (uralitite), although some early workers also included this rock under proterobase. [Gk. *proteros,* earlier (since the original specimens were older than the accompanying normal diabases) + diabase] (Gümbel in 1874)

protogine *ig. pluton.* Obsolete name applied to certain sheared granites of the Alps in the belief that they represented a part of the original crust of the earth, including rocks from the Aar massif, the Gotthard massif, and Mont Blanc. [Gk. *protos,* first + *genes,* born] (Jurine in 1806)

protolith *ig., sed., meta.* Parent rock (usually unmetamorphosed) from which a given metamorphic rock was formed by metamorphic processes. Also named parent rock. [Gk. *protos,* first]

protomylonite *meta.* 1. Low-grade crush breccia formed before the development of a mylonite; the rock faintly retains the original rock structures. 2. Mylonite formed from the granulation and flowage of a recently formed contact metamorphic rock due to overthrusts along the contact surfaces between the intrusion and the country rock. [Gk. *protos,* first + mylonite] (Backlund in 1918)

protoquartzite *sed.* Lithic sandstone intermediate in composition between subgraywacke and orthoquartzite, containing 75-95% quartz, less than 15% clay matrix, and 5-25% unstable materials in which rock fragments exceed feldspar grains. [Gk. *protos,* first + quartzite] (Krynine in 1951)

provitrain *sed.* Vitrain in which some plant structures are visible under the microscope. [L. *pro,* for + vitrain (meaning before vitrain)]

prowersite *ig. diaschist.* Potassium-rich, sodium-poor, fine-grained, greenish-gray minette composed of orthoclase, biotite, and diopside. Also named prowersose. [Prowers County, Colorado] (Cross in 1906; Rosenbusch in 1908)

psammite *sed.* General term for a clastic sedimentary rock composed of particles between 1/16 and 4 mm in diameter (sand and granule or coarse sand); equivalent to arenite. [Gk. *psammos,* sand] (Haüy, pub. by Brongniart in 1807)

psephite *sed.* General term for a clastic sedimentary rock composed of particles greater than 4 mm in diameter (pebbles, cobbles, boulders); equivalent to rudite. [Gk. *psephos,* pebble] (Brongniart in 1813)

pseudo- Common prefix used with rock names to indicate they either resemble or mimic the rock in some way. [Gk. *pseudes,* false]

pseudobreccia *sed.* 1. Sedimentary rock consisting of angular fragments that appear as if they could be put together like the pieces of a jigsaw puzzle, bound by a homogeneous nonclastic cement; formed by the disruption of an original sediment, followed by a certain dispersal of the fragments with a filling of the interstitial voids with cement. 2. Partially and irregularly replaced (dolomitized) limestone, characterized by an appearance that gives the rock a texture mimicking that of a breccia. (Tiddeman, pub. by Strahan in 1907)

pseudochrysolite *tekt.* = moldavite. [pseudo- + chrysolite (old name for olivine)]

pseudoconcretion *sed.* General term for sedimentary objects resembling concretions but not formed by diagenesis, e.g., mud balls, armored mud balls, oncolites, and lake balls.

pseudoconglomerate *sed.* Any rock that can be mistaken for normal sedimentary conglomerate, e.g., crush conglomerate, sandstone filled with numerous rounded concretions, or pseudobreccia in which the fragments are rounded. (Van Hise in 1896)

pseudodiabase *meta.* = metadiabase. (Becker in 1888)

pseudodiorite *meta.* = metadiorite. (Becker in 1888)

pseudojade *meta.* General term applied to numerous rocks other than nephrite and jadeite rock that resemble jade in appearance. Examples include: agalmatolite, bowenite, californite, grossular, hydrogrossular, pectolite rock (larimar), prehnite rock, pseudophite, pinite, serpentinite, steatite, verdite, and others. (McMahon in 1890)

pseudojadeite *meta.* Green jadelike albite feldspar from a jadeite quarry at Tawmaw, Upper Burma. (Clarke in 1906)

pseudomicrite *sed.* In limestone, calcareous micrite formed by secondary changes such as "degenerative" recrystallization (crystal diminution) of floral or faunal materials. (Chilingar and others in 1967)

pseudomicrosparite *sed.* In limestone, microsparite that has developed by recrystallization or by grain growth. (Chilingar and others in 1967)

pseudonodule *sed.* Rock consisting of a ball-like, hassocklike, or pillowlike mass of sandstone enclosed in shale or mudstone, characterized by a rounded base with upturned or inrolled edges, and resulting from the settling of sand into underlying clay or mud that welled up between isolated sand masses. Also named hassock or flow-roll. (Macar in 1948)

pseudo-ozokerite *sed.* Ozokerite from Central Persia differing in some respects from Galician ozokerite. (Förtsch in 1898)

pseudophite *meta.* Green, compact, massive chlorite rock resembling serpentinite and composed in part of clinochlore and in part of peninite. [pseudo- + ophite (old name for serpentine)] (Kenngott in 1855)

pseudosparite *sed.* Limestone consisting of relatively large, clear calcite grains (sparite) that have developed by recrystallization. (Folk in 1959)

pseudostalactite *sed.* Club-shaped or stalactitelike projectile attached to the ceiling of a submarine cave and composed of aggregates of serpulid tube worms (small polychaete worms) that are heavily coated by magnesian calcite cement precipitated from the sea water; observed at Carrie Bow Cay, south of Tobacco Reef, Caribbean Sea. (Macintyre and Videtich in 1982)

pseudosteatite *sed.* Impure, dark green variety of halloysite clay. (Thomson and Binney, date uncertain)

pseudotachylyte *meta.* Black cataclastic metamorphic rock, occurring as lenticles in mylonites and locally along overthrust belts and consisting of vitreous material that under the microscope appears as a nonpolarizing glassy substance (actually cryptocrystalline); nearly identical to flinty crush-rock and ultramylonite. [pseudo- + tachylyte (igneous basalt glass also named hyalobasalt)] (Shand in 1914)

pseudotill *sed.* Any nonglacial sedimentary deposit resembling glacial till.

pseudotillite *sed.* Indurated pseudotill; nonglacial tillitelike rock formed on land by flow of nonglacial mud or deposited by a subaqueous turbidity flow, e.g., conglomeratic mudstone; nearly equivalent to tilloid. (Schwarzbach in 1961)

psilomelane *sed.* Black and smooth, nodular, botryoidal, reniform, or stalactitic masses composed of one or more hard hydrated manganese oxides (including minerals such as romanechite, cryptomelane, coronadite, or hollandite) and often quartz, iron oxides, and other species. Also rarely spelled psilomelanite. [Gk. *psilos,* bare, smooth + *melas,* black] (Haidinger in 1828)

psilomelane-chalcedony *sed.* Black-banded quartz rock consisting of chalcedony and black hydrated manganese oxides (collectively named psilomelane); some of these materials from Mexico are used for gems. Also named psilomelane-quartz.

pterolite *ig.* Fan-shaped, massive rock, consisting of ferrian biotite (lepidomelane) with barkevikite and aegirine; from the Brevik region, Norway. [Gk. *pteron,* feather, plume] (Breithaupt in 1865)

puddingstone *sed.* Conglomerate, esp. one consisting of well-rounded pebbles set in an abundant matrix; an attractive example from near Fairburn, South Dakota, consists of red, yellow, tan, orange, and lavender jasper pebbles in a very hard orthoquartzite cement. [plum pudding]

puglianite *ig. pluton.* Coarse-grained foid monzogabbro occurring as fragments in the lavas of Monte Somma, Italy, and composed of augite, leucite, and anorthite, with minor sanidine, hornblende, apatite, and magnetite. [Pugliani, Monte Somma, Italy] (Lacroix in 1917)

pulaskite *ig. pluton.* Granular to trachytoid alkali syenite consisting essentially of potassic feldspar (micro- or cryptoperthitic orthoclase), and a small amount of nepheline and mafics (biotite, amphibole, pyroxene). [Pulaski County, Arkansas] (Williams in 1891)

pulverite   *sed.* Sedimentary rock composed of silt- or clay-sized aggregates of nonclastic origin, simulating in texture clastic lutite, e.g., rock formed of diatom frustules (diatomaceous earth or diatomite). [L. *pulveris,* dust]

pulverulite   *ig. extru.* Variety of ignimbrite with a pulverulent or powdery, texture. [L. *pulveris,* dust] (Marshall in 1935)

pumice   *ig. extru.* Light-colored rock froth, usually with a composition near rhyolite, that forms as crusts on more compacted lava or occurs in the form of volcanic ejectamenta; the glassy material is so filled with small air bubbles that the pore space may be much greater than the glass. [L. *pumicis,* perhaps from *spuma,* foam] (Pliny, *Natural History*)

pumice, trachyte   *ig. extru.* Pumice whose composition is that of trachyte rather than rhyolite.

pumpelleyite rock   *meta.* Rock containing essential amounts of pumpelleyite; formed by the low-grade hydrothermal alteration of a variety of rocks including andesite, basalt, spilite, skarn, and hornfels; pumpelleyite may be an abundant constituent of chlorite schist, antinolite schist, and glaucophane schist. Chlorastrolite is a variety of nearly pure pumpelleyite rock that occurs in altered amygdaloidal trap rock.

pungernite   *sed.* Variety of oil shale, rich in kerogen, from Ordovician rocks in Estonia. [Gross-Pungern, Estonia] (Bulgarine in 1851)

puzzolane   *sed., ig. extru.* = pozzolan. Also spelled puzzuolana.

pyrite rock   *sed.* A rare sedimentary rock in which pyrite may appear as a major component; e.g., certain pyritic limestones grade into beds of granular pyrite in the Tully (Devonian) of New York and the Greenhorn (Upper Cretaceous) of Wyoming; oolitic pyrite occurs in a fine-grained shale matrix in the Wabana iron ores of Bell Island, Newfoundland, with barite in sediments at Meggen Lenne River, Westphalia, Germany, and in quartz-rich sandstone near East Stone Gap, Virginia. See also iron sulfide rock and pyrite nodule.

pyritosalite   *ig. pluton.* Igneous silexite, associated with essexite from Tofteholmen, Hurum, Oslo, Norway, containing quartz (86%), pyrite (7%), muscovite (4%), and minor accessories. [pyrite + possibly L. *salire,* to spring forth] (Brögger in 1931)

pyrobiolite   *sed.* Sedimentary rock containing

Puddingstone (sandy conglomerate). Schirmeck, France. 10 cm.

organic remains altered by volcanic action. [Gk. *pur,* fire + *bios,* life]

pyrobitumen   *sed.* Any of the dark-colored, fairly hard, nonvolatile native hydrocarbon complexes, which may or may not contain oxygenated substances and are often associated with mineral matter; on heating they generally yield bitumens rather than melt. [Gk. *pur,* fire + bitumen]

pyroclasite   *sed.* Hard, massive, reniform to tuberose, variety of guano, from Monks Island (Los Monjes) off the coast of Venezuela in the Caribbean. [Gk. *pur,* fire + *klasis,* fracture; because it flies to pieces when heated] (Shepard in 1856)

pyroclast   *ig. extru. IUGS,* individual crystal, crystal fragment, glass fragment, or rock fragment generated by disruption as a direct result of volcanic action; shapes assumed during disruption or during subsequent transport must not have been altered by later redeposition processes. [Gk. *pur,* fire + *klastos,* broken]

pyroclastic deposits   *ig. extru. IUGS,* either consolidated (including coherent, cemented, indurated) or unconsolidated assemblages of pyroclasts; they must contain more than 75% pyroclasts by volume. [pyroclast]

pyroclastic rock   *ig. extru.* 1. General term for igneous rock composed of consolidated pyroclasts ejected by volcanic explosions or aerial expulsion from a volcanic vent; they are transported aerially and deposited upon land surfaces, in lakes, or in marine waters; examples include such materials as agglomerate, volcanic breccia, lapilli tuff, and tuff.   2. *IUGS,* predominantly consolidated pyroclastic deposits. [pyroclast]

pyroguanite   *sed.* Massive hard guano from Monks Island (Los Monjes) off the coast of Venezuela in the Caribbean. [Gk. *pur,* fire + guano; from erroneous impression it was hardened by heat] (Shepard in 1856)

pyrolite   *ig. synthetic.* Synthetic variety of peridotite composed primarily of pyroxene and olivine. With variations in pressure, the following varieties have been observed: plagioclase p. (low pressure) composed of olivine, aluminum-poor pyroxene, and plagioclase; spinel p. (higher pressure) composed of olivine, aluminum-rich pyroxene, and spinel; garnet p. (much higher pressure) composed of olivine, aluminium-poor pyroxene, and garnet. [pyroxene + olivine] (Ringwood in 1963)

pyrolith   *ig.* Obsolete synonym for igneous rock. [Gk. *pur,* fire + *lithos,* stone] (Grabau in 1904)

pyromeride   *ig. extru.* Quartz-rich felsite or devitrified rhyolitic glass, characterized by a conspicuous spherulitic or lithophysa structure and having a nodular appearance. [Gk. *pur,* fire + *meros,* part] (Haüy, pub. by Monteiro in 1814)

pyrophane   *sed.* 1. Variety of opal that becomes translucent by the absorption of melted wax when hot but again becomes opaque on cooling.   2. = fire opal. [Gk. *pur,* fire + *phanes,* appearing]

pyrophyllite rock   *meta.* Metamorphic rock in which pyrophyllite is an essential component; may be a major component of some schists, e.g., pyrophyllite-kyanite schist; also occurs in massive stellate aggregations and as fine-grained compact forms, the latter being used for making slate pencils (pencil stone) and ornamental carvings (named agalmatolite at times).

pyropissite   *sed.* Grayish-brown, earthy, friable, coaly substance, made up primarily of water, humic acid, wax, and silica; frequently found associated with brown coal and massive limestone. [Gk. *pur,* fire + *pissa,* pitch] (Kenngott in 1853)

pyroretinite   *sed., meta.?* Type of retinite found in the brown coals near Aussig in Bohemia, Czechoslovakia; perhaps formed by the intrusion of a basaltic dike in the coal. [Gk. *pur,* fire + retinite] (Reuss in 1854)

pyroschist   *sed.* General term applied to sedimentary rocks that are impregnated with combustible, bituminous materials, e.g., oil shale. [Gk. *pur,* fire + *schistos,* cleft (not a metamorphic schist)] (Hunt in 1863)

pyroxenide   *ig. pluton.* Field term for any holocrystalline, medium- to coarse-grained igneous rock composed chiefly of pyroxene, e.g., pyroxenite. [pyroxene]

pyroxenite   *ig. pluton.* 1. Medium- to coarse-grained, plutonic igneous rock, composed essentially of pyroxene, regardless of its species. Also named pyroxenolite. [pyroxene] (Senft in 1857; Coquand in 1857).   2. *IUGS,* plutonic igneous rock with M equal to or greater than 90, and ol/(ol + opx + cpx) less than 40.

pyroxenite, biotite   *ig. pluton.* Rare variety of pyroxenite found in the ejectamenta of Vesuvius, Italy, and consisting of about 60% augite, 37% biotite, and interstitial glass.

pyroxenite, hornblende   *ig. pluton. IUGS,* plutonic rock with M equal to or greater than 90, ol/(ol + hbl + px) less than 5, and 50-90 px/(px + hbl).

pyroxenite, olivine   *ig. pluton. IUGS,* plutonic rock with M equal to or greater than 90, 5-40 ol/(ol + hbl + px), hbl/(ol + hbl + px) less than 5, and px/(ol + hbl + px) less than 90.

pyroxenite, olivine-hornblende   *ig. pluton. IUGS,* plutonic rock with M equal to or greater than 90, 5-40 ol/(ol + hbl + px), and pyroxene more abundant than amphibole (hornblende).

pyroxenolite   *ig. pluton.* = pyroxenite. (Lacroix in 1895)

pyterlite   *ig. pluton.* Rapakivi without oligoclase (or other plagioclase) mantles on the potassic feldspar; consisting of ellipsoidal microcline perthite (40%), quartz (38%), oligoclase (14%), with lesser lepidomelane and other accessories. [Pyterlahti, Wiborg, southern Finland] (Wahl in 1925)

Q

quarfeloid   *ig.* In Johannsen's classification of igneous rocks, rock with either a combination of quartz and feldspar or of feldspathoid and feldspar. [quartz + feldspar + feldspathoid] (Johannsen in 1917)

quartz-barite rock   *ig. hypab.* Rock composed of about 70% quartz and 30% barite, occurring as a network of pegmatitelike veins in the Salem district of Madras, India; believed to be of magmatic (igneous) origin. Also spelled quartz-barytes rock. (Holland in 1897)

quartz diorite   *ig. pluton. See* diorite, quartz

quartz porphyry   *ig. extru. See* porphyry, quartz

quartz sillimanitisé   *meta.* Quartz containing needles of sillimanite, often crowded in matted aggregates. Also named faserkiesel.

quartzarenite   *sed.* Sandstone composed of 95% or more quartz grains. Also named orthoquartzite. Also written quartz arenite.

quartzfels   *ig., sed., meta.* Poorly defined term for quartz-rich rocks, including orthoquartzite, quartzite, silexite, and others.

quartzite   *sed.* Sandstone consisting chiefly of quartz grains cemented by quartz, the cement often having crystallographic continuity around each grain. The preferred names are orthoquartzite or quartzarenite. Also named quartzitic sandstone, quartzose sandstone, sedimentary quartzite, and siliceous sandstone. [quartz]

quartzite   *meta.* Granular, nonfoliated (granoblastic) metamorphic rock, consisting mainly of quartz and formed by the recrystallization of quartz-rich sedimentary rocks (sandstone, conglomerate, chert, and others) by either regional or contact metamorphism. Also named metaquartzite. Variations in colors, in textures, and in impurities (up to 40%) give rise to numerous varieties, e.g.: white q., orange q., red q., black q., brown q., coarse-grained q., schistose q., conglomeratic q., micaceous q., feldspathic q., aluminous q., garnetiferous q., chloritic q., actinolitic q., magnetitic q., hematitic q. [quartz]

quartzite, aluminous   *meta.* Mineralogical variety of quartzite containing one or more minerals rich in aluminum derived from earlier clays, micas, feldspars, and similar materials. Important varieties include: andalusite q., andalusite-sericite q., kyanite q., corundum q., alunite q., and diaspore-sericite q.

quartzite, amygdaloidal   *meta.* Quartzite in which larger lenticular quartz elements are elongated along the bedding planes and are set in a fine-grained groundmass; may represent relict pebbles. (Cayeux in 1929)

quartzite, aphanitic   *sed., meta.* = microquartzite

quartzite, arkose   *sed.* Orthoquartzite with a notable amount of feldspar, e.g., well-indurated arkosic sandstone or arkose. Also named arkosite and quartzitic arkose.

quartzite, carbonaceous   *meta.* Gray to black quartzite containing carbon, usually graphite, derived from original organic matter in the parent sandstone.

quartzite, conglomeratic   *meta.* Quartzite derived from a quartz-rich conglomerate; often the former quartz pebbles are rotated so their long axes lie within the foliation planes, or the pebbles are deformed and elongated to augen or thin lenses.

quartzite, feldspathic   *meta.* Quartzite containing microcline, orthoclase, or sodic-plagioclase; the feldspar may have been in the original sediment or, in some cases, introduced metasomatically.

quartzite, flexible micaceous   *meta.* = itacolumite

quartzite, fuchsite   *meta.* Pale to deep emerald-green, micaceous quartzite whose color is due to chromian muscovite (fuchsite); illustrated by portions of the Medicine Peak quartzite in the Medicine Bow Mountains, Wyoming.

quartzite, martite   *meta.* Magnetite-bearing quartzite (magnetitic quartzite) in which the magnetite has been oxidized to hematite (martite). [martite (hematite pseudomorph after magnetite)]

quartzite, metamorphic   *meta.* = quartzite. Also named metaquartzite.

quartzite, micaceous   *meta.* Quartzite distinguished by containing mica, usually muscovite (sericite) or biotite, but other types are known, e.g., green chromian muscovite (fuchsite).

quartzite, oily *sed., meta.* = microquartzite. [For greasy luster]

quartzite, phyllitic *meta.* Micaceous quartzite in which the mica flakes are present along the bedding planes in discontinuous layers; these are transitional to schistose quartzites and mica schist.

quartzite, porphyritic *meta.* Quartzite displaying a fine-grained groundmass in which larger quartz elements are scattered, as if the rock resulted from two phases of crystallization. (Cayeux in 1929)

quartzite, porphyroblastic *meta.* Quartzite displaying a fine-grained groundmass in which there are large, irregularly lobated or sometimes subglobular individuals; larger units are usually relics of early pebbles.

quartzite, pyroxmangite *meta.* Manganese silicate rock containing, in addition to quartz, pyroxmangite and rhodonite; occurs at Simsiö, Finland.

quartzite, schistose *meta.* Transitional rock between schist and quartzite in which quartz dominates.

quartzite, secondary *meta.* Quartzite apparently formed by large-scale metasomatic replacement of mainly acidic to intermediate igneous rocks, esp. volcanic and hypabyssal types; rock common in the Soviet Union.

quartzite, sedimentary *sed.* = orthoquartzite

quartzite, specularite *meta.* = itabirite

quartzite, spherulitic *meta.* Quartzite in which larger quartz units have affected the crystallization of the surrounding groundmass over a radial distance that may be equal to their own diameter; very elongated quartz granules occur in radial disposition around the larger elements, which act as nuclei.

quartzolite *ig. pluton.* 1. Igneous rock composed nearly entirely of primary quartz. Essentially the same as peracidite and silexite. 2. *IUGS*, plutonic rock with Q greater than or equal to 90.

quartzwacke *sed.* = wacke, quartz

queenstownite *impact., uncertain.* = glass, Darwin. [Queenstown, western Tasmania] (pub. by Suess after 1914)

queluzite *meta.* Spessartine garnetite at times containing amphiboles, pyroxenes, and micas, with or without secondary black manganese oxides; weathered residual deposits derived from this rock are mined for manganese. [Queluz, Minas Gerais, Brazil] (Derby in 1901)

quercyite *sed.* Phosphorite consisting of an intimate interbanded mixture of amorphous and finely fibrous components. [Quercy, ancient district in France] (Lacroix in 1910)

quernstone *sed.* Sandstone used for grinding in a quern (hand-operated mill); essentially the same as a carstone (in Norfolk, England) and certain types of millstone. [O. Eng. *cwyrn, cweorn*, millstone]

quetzalztli *meta.* Term used in Mexico for a translucent, green jade. [Aztec, from *quetzalli*, large, green, rich feather]

quicksand *sed.* Mass of fine sand consisting of smooth rounded grains with little tendency for mutual adherence and usually completely saturated with water flowing upward through the voids, forming a soft, shifting, highly mobile mass that yields easily to pressure and tends to suck down and readily swallow heavy objects touching its surface.

quickstone *sed.* Consolidated rock that flowed when wet under the influence of gravity before lithification.

quinzite *sed.* Rose-colored common opal. Also spelled quincite but not to be confused with the variety of the mineral sepiolite named quincite (or quincyite) from the same locality. [Quincy, near Mehun, France]

quisqueite *sed.* Lustrous, black, brittle asphaltite, composed of carbon and sulfur and accompanying the vanadium ores of Peru. [Quisque district, near Mina Ragra, Pasco, Peru] (Hewett in 1907)

quoiceneck *sed.* Grayish black clay with streaked shining surfaces; term used in Shropshire, England. [quoice, a local variation of queest, ring-dove or wood pigeon + neck]

R

raabsite *ig. diaschit.* Lamprophyre composed of alkali amphibole, microcline, biotite, and accessories including apatite. [Raabs, Nieder Austria] (Waldmann in 1935)

radiolarite *sed.* 1. Consolidated, earthy material composed of the complex opaline skeletons or tests of radiolaria, which are marine protozoans; other components may include, calcite, diatom shells, clay, pumice fragments, and others. Radiolarian earth is similar but unconsolidated. 2. Hard chert derived from radiolaria.

radiolarite, arenaceous *sed.* Somewhat hard, siliceous rock, often with zones of cherty development, with abundant detrital material and glauconite often replacing radiolaria.

radiolarite, carbonaceous *sed.* Gray to black radiolarian chert, containing carbonaceous pigments and microgranular quartz, with clays and phosphates.

radiolarite, ferruginous *sed.* Red radiolarian chert, containing iron oxide pigments, fossil radiolaria, opaline to holocrystalline quartz, and usually some clay.

radiumite *hydrotherm., altered.* Mixture of black uraninite (pitchblende), yellow uranophane, and orange gummite, formerly occasionally used as a gem material. [radium]

rafaelite *ig. hypab.* Analcime syenite containing sodium orthoclase, labradorite, analcime, and others but no nepheline. [San Rafael Swell, Emery, Utah] (Johannsen in 1938)

rafaelite *sed.* Vanadiferous asphalt occurring in Argentina. [San Rafael, Argentina] (Windhausen and Vignau in 1912)

raglanite *ig. pluton.* Gray corundum-rich (5%) nepheline diorite, containing oligoclase (69%), nepheline, and other minor minerals. [Raglan, Craigmont Mountain, Ontario, Canada] (Adams and Barlow in 1910)

rammell *sed.* Mixed argillaceous and sandy rocks. [M. Eng. *ramel,* rubbish, from O.F. *ramaille,* branches, twigs]

ramosite *ig. extru.* Old term for scoria. [Possibly ramose; or from Ramos, San Luis Potosi, Mexico]

randanite *sed.* Dark diatomaceous earth, containing humic materials, from France. Also spelled randannite and also named ceyssatite. [Puy-de-Dôme (Randan) region, France] (Solvétat in 1848)

rapakivi *ig. pluton.* Variety of porphyritic granite (sometimes quartz monzonite) whose phenocrysts consist of ellipsoidal orthoclase crystals that are usually surrounded by a mantle of plagioclase (oligoclase, more rarely albite or andesine); hornblende and biotite are often present. [Finn. *rapakivi,* rotten stone; because of its property of breaking easily] (Hjarne in 1694)

raphaelite *sed.* Carbonized asphalt resembling coal, from Argentina. Not to be confused with rafaelite. (Longobardi in 1922)

raqqaite *ig. extru.* Lava with the composition of a pyroxenite. [Possibly Raqqa, Syria] (Eigenfeld in 1965)

rash *sed.* Impure and unusable coal; the coal is mixed with clay, shale, or other foreign materials taken from the top or bottom of the seam; a substance intermediate in character between coal and shale. [Eng. dial. *rash,* rough, crisp, brittle]

rashing *sed.* Soft scaly earth or shale immediately beneath a coal seam, often containing much carbonaceous matter. [Eng. dial. *rash,* rough, crisp, brittle]

rattle box *sed.* Limonite klapperstein from Chester County, Pennsylvania.

rattlestone *sed.* = klapperstein. Also named rattle rock.

rauhaugite *ig. hypab.* Dolomite carbonatite, with approximately 90% dolomite, 8% apatite, and additional alkali feldspar, barite, and pyrite. [Rauhaug, Fen complex, Norway] (Brögger in 1921)

recomposed rock *sed.* Sedimentary rock produced in place by the cementation of the fragmental products of surface weathering; e.g., arkose consisting of feldspar and quartz that has been so little reworked and so little decomposed it looks very much like the original granite, thus recomposed granite.

red beds *sed.* Terrigenous clastic sediments (sandstone, siltstone, and shale), colored by hematite, as very thin coatings around grains and infiltration of clays; deposited in rivers, floodplains, alluvial fans, and the like. Examples are the Triassic and Permian sedimentary rocks of the western United States.

reddle   *sed.* = ruddle

redstone   *sed.* Reddish arkosic wacke, representing a reconstructed granite, with a greater percentage of quartz than the original granite and a clay-rich matrix that is commonly stained red.

redwitzite   *ig.?* Collective name for a coarse- to medium-grained, streaked rock, with alternating feldspar-poor to feldspar-rich aplite zones. [Marktredwitz, Oberpfalz, Germany] (Wilmann in 1919)

reef milk   *sed.* Matrix material in reef rock consisting of white, opaque microcrystalline calcite or aragonite derived from abrasion of the reef core and flank.

reef rock   *sed.* Calcium carbonate framework rock; one with a loosely knit reticulating network with large voids and coated cavities, which makes up reefs and is usually formed through the growth of coral colonies, encrusting algae, fenestrate bryozoans or other forms. [M. Eng. *riff,* from O.N. *rif,* ridge]

reef rock, algal   *sed.* Organic reef rock in which algae are or were the principal organisms secreting calcium carbonate from which the reef is built.

reef rock, coral   *sed.* Organic calcium carbonate reef rock built up esp. of in-place coral colonies but often including algae, accumulated skeletal fragments, sands, and other materials.

regalite   *sed.?* Green variety of quartz, or white quartz with green veins; local term used for lapidary materials in Utah. [Possibly regal]

regolith   *sed.* General term for the loose materials that mantle the land areas of the Earth and rest on the solid rocks; materials are derived from the decay of rocks, accumulations of vegetation, talus, sediments, wind-blown sands, and glacial deposits. [Gk. *rhegos,* blanket] (Merrill in 1897)

reibungsbreccia   *sed.* = breccia, fold. [G. *Reibung,* rubbing, friction + breccia]

rensselaerite   *meta.* White to dark green, waxlike, talcose rock, from St. Lawrence County, New York. [Stephen van Rensselaer, 1765-1839, American army officer and politician of Albany, New York] (Emmons in 1837)

resin, fossil   *sed.* Ancient solidified and mainly amorphous products of secretions and excretions of certain trees. There are two major kinds:

the amber group resins, usually containing succinic acids and found in clay and other sediments, and the retinite group resins, usually devoid of succinic acid and found in brown coal or peat. [Gk. *rhetine,* L. *resina,* resin]

resin, mineral   *sed.* General term for any of a group of resinous, usually fossilized, natural hydrocarbons, e.g., fossil resin, asphalt, bitumen. [Gk. *rhetine,* L. *resina,* resin]

resinite   *sed.* = retinasphalt. [Gk. *rhetine,* L. *resina,* resin] (Haüy in 1822)

restite   *meta., granitization.* Dark rock composed of mafic minerals left behind after the mobilization and removal of the quartz-feldspar components during granitization. Also named rest rock. [rest]

restormelite   *meta. ?* Massive, grayish green agalmatolitelike material from Cornwall, England. [Restormel mine, Cornwall] (Church in 1870)

reticulite   *ig. extru.* Foamlike, brown or black basaltic glass (hyalobasalt) in which the vesicles are enclosed by paper-thin walls. [L. *reticulum,* little net]

retinalite   *meta.* Serpentinite that is massive, honey yellow to light oil green, and waxy or resinous. [Gk. *rhetine,* resin] (Thomson in 1836)

retinasphalt   *sed.* Light brown variety of retinite usually found with lignite. Also named resinite. [Gk. *rhetine,* resin + asphalt] (Hatchett in 1804)

retinellite   *sed.* Light brown, resinlike substance, from the Tertiary coal of Bovey in Devonshire, England. [dim. of Gk. *rhetine,* resin] (Dana in 1868)

retinite   *sed.* General name applied to various resins, esp. those from beds of brown coal; like amber in appearance but containing little or no succinic acid. [Gk. *rhetine,* resin] (Zinken in 1884)

rheoignimbrite   *ig. extru.* Ignimbrite, on the slope of a volcanic crater, that developed secondary flowage because of high temperatures. [Gk. *rheos,* flow + ignimbrite]

rhizoconcretion   *sed.* Thin concretionlike structure in a sedimentary rock, cylindrical or conical, usually forked or branching, and resembling a root or root system; usually composed of calcium carbonate or, more rarely, of chert. [Gk. *rhiza,* root + concretion]

rhizocretion   *sed.* Hollow rhizoconcretion formed around the root of a living plant. [Gk. *rhiza,* root + L. *crescere,* to grow] (Kindle in 1923)

rhodochrosite rock   *hydrotherm. ?* Massive, fine- to coarse-grained rhodochrosite, often handsomely banded as in botryoidal cavity linings and stalactitic formations and occurring in some ore deposits; stalactitic and thick banded crusts occur in galena veins in the Catamarca Province, Argentina, and have been worked as an ornamental stone; rhodochrosite also occurs as nodules in some sedimentary rocks. The variety from Argentina has been named rosinca at times.

rhodonite rock   *meta.* Manganese-silicate rock in which rhodonite is an essential constituent, esp. rhodonite gondite, a rock composed of rhodonite, spessartine, and quartz; some of these rocks consist only of rhodonite and quartz, and others are nearly pure rhodonite.

rhomb-porphyry   *ig. hypab.* Porphyritic rock, variously described as syenite, trachyte, or monzonite, in which rhomb-shaped feldspar phenocrysts are characteristic; the phenocrysts have been described as sodium orthoclase, sodium microcline, or andesine. Also spelled rhomben-porphyry. (Buch in 1810)

rhomb-porphyry, analcime   *ig. extru.* = shackanite. (Daly in 1912)

rhomb-porphyry, nepheline   *ig. dike.* Nepheline syenite porphyry, containing greenish gray phenocrysts of sodium orthoclase and microperthite in a fine-grained groundmass; occurs as dikes in larvikite in southern Norway. (Brögger in 1890)

rhyoandesite   *ig. extru.* Aphanitic extrusive equivalent of granodiorite. [rhyolite + andesite]

rhyobasalt   *ig. extru.* Aphanitic extrusive equivalent of granogabbro. [rhyolite + basalt] (Johannsen in 1919)

rhyodacite   *ig. extru.* Aphanitic extrusive equivalent of granodiorite; essentially equivalent to rhyoandesite. [rhyolite + dacite] (Winchell in 1913)

rhyolite   *ig. extru., hypab.* Generally, light-colored (white, yellow, pink, reddish), aphanitic rock belonging to the granite family; often porphyritic with quartz and sanidine phenocrysts in a groundmass of the same composition, which may be holocrystalline, hypocrystalline, or vitreous. With a decrease in quartz, the rock grades into trachyte. [Gk. *rhyax*, stream, torrent; alluding to lava flow] (Richthofen in 1860)

rhyolite, alkali   *ig. extru., hypab.* Sodium-rich rhyolite in which the chief feldspars may be anorthoclase or strongly perthitic orthoclase and minor albite; the major accessory amphiboles are usually sodium rich (riebeckite, arfvedsonite, or hastingsite).

rhyolite, bird's-eye   *ig. extru.* = rhyolite, spherulitic

rhyolite, cristobalite   *ig. extru.* Rhyolite in which the major silica mineral is cristobalite rather than quartz, e.g., some of the rhyolite encasing thunder eggs in the Western United States.

rhyolite, potassic   *ig. extru., hypab.* The more common rhyolite in which the feldspar is sanidine (usually clear), and, if present, plagioclase is normally oligoclase; accessory mafics include biotite and more rarely hornblende.

Rhyolite porphyry. Roosevelt Mountain, Deadwood, South Dakota. 12.5 cm.

rhyolite, sodaclase   *ig. extru.* Aphanitic extrusive equivalent of albite granite (sodaclase granite).

rhyolite, sodic   *ig. extru., hypab.* = rhyolite, alkali

rhyolite, spherulitic   *ig. extru.* Aphanitic rhyolite containing very symmetrical spherulites of feldspars (usually potassic); some of these rocks may represent devitrified obsidian. Also named bird's-eye rhyolite, esp. in Juab County, Utah, and other Western localities.

rhythmically banded rock   *meta.* = wrigglite. (Shabynin in 1977)

riband   *sed.* Variety of jasper with broad, ribbonlike stripes of alternating colors; similar to jasp-onyx, except the bands are usually much wider. [riband, for its decorative ribbon appearance]

ribbon rock   *meta.* = wrigglite. Also named ribbon-rock tactite or ribbon-rock skarn, as well as rhythmically banded rock. (Jahns in 1944)

ricolettaite   *ig. pluton.* Dark-colored, coarse-grained rock containing anorthite and pyroxene, with minor amounts of olivine, orthoclase, biotite, and iron oxides. [Ricoletta Mountain, Traversellit Valley, Tyrol] (Johannsen in 1920)

ricolite   *meta.* Variety of serpentine marble having parallel bands of alternating green and white colors; the type material occurs near Redrock, New Mexico. [Sp. *rico,* rich; alluding to the rich green colors in the stone] (pub. by Merrill in 1897)

riedenite   *ig. pluton.* Melteigite composed of large biotite tablets in a granular aggregate of white nosean, biotite, and pyroxene with small amounts of titanite and apatite. [Rieden in the Laacher Lake region, Germany] (Brauns in 1921)

rimstone   *sed.* Calcium carbonate deposit that forms a rim around cave pools. [O. Eng. *rima,* rim]

ringite   *ig. hypab.* Feldspar-bearing, aegirine carbonatite, composed of calcite (70%), aegirine (20%), alkali feldspar, and apatite, titanite, and others. Varieties include: aegirine r. (more than 30% aegirine), albite r. (more than 40% albite), r. pegmatite (an intergrowth of calcite and aegirine). [Ringe, Fen complex, Norway] (Brögger in 1921)

rischorrite   *ig. pluton.* Lepidomelane-nepheline syenite in which the feldspar (microcline perthite) is a eutectic intergrowth with nepheline. [Rischorr, Khibina complex, Kola Peninsula, Soviet Union] (Kupletsky in 1932)

rizalite   *tekt.* General term for tektite from the Philippines. Also named philippinite. [Rizal Province, Philippines] (Beyer in 1928)

rizzonite   *ig. extru.* Local term for a variety of limburgite from Monte Rizzoni, Italy. [Monte Rizzoni, Tyrol] (Doelter in 1902)

rochlederite   *sed.* Reddish brown fossil resin found in brown coal in Bohemia. [Friedrich Rochleder, 1819-1874, Austrian chemist of Vienna who described it in 1851] (Dana in 1868)

rock   *ig., sed., meta., etc.* Naturally formed aggregate composed of one or more minerals and/or mineraloids, including loose incoherent masses as well as firm solid masses, and constituting an essential or appreciable part of the earth or other celestial body.

rock anhydrite   *sed.* = anhydrite rock

rock borate   *sed.* = borate rock

rock flour   *sed.* See flour, rock

rock graphite   *meta.* = graphite rock

rock gypsum   *sed.* = gypsum rock

rock meal   *sed.* White, calcite-rich efflorescence that is light, like cotton, and becomes a powder with the slightest pressure; common near Paris at quarries of Nanterre.

rock phosphate   *sed.* = phosphorite

rock rot   *sed.* General term for products of decomposition (composed esp. of clays); used in contrast to rock flour, which is pulverized rock matter with very little decomposition.

rock salt   *sed.* = halite rock

rock sulfur   *sed.* = sulfur rock

rockallite   *ig. dike.* Rock composed of quartz, albite, and aegirine-acmite, occurring as segregations in a coarse-grained light-colored aegirine-riebeckite granite off the coast of Ireland. [Rockall, island off Ireland] (Judd in 1897)

rodingite   *ig. pluton.* Coarse- to medium-grained gabbroic rock associated with dunite, commonly enriched in calcium, barium, and strontium, and containing clinopyroxene and grossular; altered varieties contain serpentine or prehnite. [Roding River, Dun Mountain, New Zealand] (Bell in 1911)

rodite   *meteor.* Olivine-hypersthene chondrite meteorite; the presence of chondrules was not recognized at first; essentially the same as amphoterite, a name also now considered unnecessary. [Roda, Spain]

Rofna-porphyry   *ig. pluton.* = taspinite. [Rofna Valley, Upper Rhine, Switzerland]

rogenstein   *sed.* Oolite in which the ooliths are cemented by an argillaceous material. [G. *Rogen,* roe + *Stein,* stone] rogueite   *sed.* Variety of milk-white agate showing bright-green markings, patterns, and designs; esp. applied to specimens from the Rogue River valley, southwestern Oregon. [Rogue River, Oregon]

romanite   *sed.* = rumanite (the resin)

rongstockite   *ig. pluton.* Medium- to fine-grained, dark gray rock consisting of plagioclase (zonal with labradorite cores and oligoclase rims), orthoclase, some cancrinite, augite, biotite, brown hornblende, and others; resembles essexite but is poorer in nepheline and contains sodic- instead of calcic-plagioclase. [Rongstock, Czechoslovakia] (Tröger in 1935)

rose, barite   *sed.* Clusters of tabular barite crystals, somewhat symmetrically disposed so they resemble a full-blown rose; usually these are barite sand crystals in which the included sand equals or exceeds the barite; in the United States, common in the Permian red Garber sandstone near Norman, Oklahoma. Also named barite rosette.

rose, desert   *sed.* Somewhat flat, rose-shaped or rounded, chalcedony nodule that is often slightly pinkish; examples occur in Riverside County, California, and in Graham County, Arizona.

rose-garnet   *meta.* Trade name for a decorative stone consisting chiefly of pink grossular (rosolite), vesuvianite, and wollastonite, from Xalostoc, Morelos, Mexico.

rosette rock   *sed.* Sedimentary rock resembling a full-blown or double rose; may be sand crystals (e.g., gypsum or barite), concretions, or nodules (e.g., chalcedony desert rose). See barite rose and desert rose.

rosinca   *hydrotherm.?* Trade name for the banded rhodochrosite rock occurring in Argentina. Also named rhodochrosite rock and Inca rose. [Sp. *rosinca,* from *rosa,* rose]

rosso di Levanto marble   *meta., altered.* Reddish, altered serpentinite used as a decorative stone. [It. *rosso,* red + Levanto, Liguria, Italy]

rosthornite   *sed.* Brown, greasy retinite with a low oxygen content (4.57%); found as lenticular masses in coal at Sonnberge, Carinthia. [Franz Edler von Rosthorn, 1796-1877, Austrian geologist of Carinthia] (Höfer in 1871)

rottenstone   *sed.* Soft, friable rock, consisting largely of siliceous particles resulting from the decomposition of siliceous limestone whose calcareous material has been removed; essentially the same as tripoli. [rotten]

rougemontite   *ig. pluton.* Medium- to fine-grained, gray gabbro, consisting of anorthite (46%) and titanaugite (35%) with lesser amounts of olivine, iron oxides, and others. [Rougemont Mountain, Quebec, Canada] (O'Neill in 1914)

routivarite   *ig. pluton.* Anorthosite composed of labradorite (96%) and almandine garnet (3%). [Routivare, Norrbotten, Swedish Lapland] (Sjögren in 1893)

Rose (chalcedony) (desert rose). Graham County, Arizona. 5 cm.

rouvillite   *ig. pluton.* Medium- to coarse-grained, light gray, mottled black and white, theralite consisting predominantly of labradorite or bytownite and nepheline, with small amounts of titanaugite, hornblende, pyrite, and apatite. [Rouville County, Quebec, Canada] (O'Neill in 1914)

rubasse   *meta.?* Quartz rock colored red by included abundant small scales or flecks of iron oxide, usually hematite; occurs in Brazil and other places and may be used as a gem. Also spelled rubace and rubacelle. [Fr. *rubace,* from L. *rubeus,* red] (Trevoux in 1752)

rubble   *sed.* Loose mass of rough, irregular, angular rock detritus, commonly overlying outcropping rock; unconsolidated equivalent of breccia. Fragments are coarser than sand size and include the following: granule r., pebble r., cobble r., boulder r. [O.F. *robel,* rubbish]

rubblerock   *sed.* = breccia

rubblestone   *sed.* 1. = rubblerock.   2. = rubble. 3. = graywacke.

rubolite   *sed.* Red-colored common opal. [L. *rubere,* to be red]

rudaceous rock   *sed.* = rudite. [L. *rudus,* crushed stone, rubble, debris] (Grabau in 1904)

ruddle   *sed.* Red variety of ocher. Also spelled reddle and also named red chalk. [O. Eng. *rudu,* redness]

rudite   *sed.* Sedimentary rock composed of a significant amount of fragments coarser than sand (greater than 2 mm in diameter); there are no special size, shape, or roundness qualifications for the fragments, which may be granules, pebbles, cobbles, or boulders. Examples include conglomerate and breccia. [L. *rudus,* crushed stone, rubble, debris] (Grabau in 1904)

rumanite   *sed.* 1. Brownish yellow to brown, amberlike resin, obtained from different places in Rumania, e.g., from sandstone in the Buseo district. Also written rumänite, roumänite, and romanite. [Rumania] (Helm in 1891)   2. Term also applied to a variety of opal from Rumania.

runite   *ig. pluton., dike.* = graphic granite. [rune, for similarity to characters of a runic inscription] (Pinkerton in 1811)

rutterite   *ig. pluton.* Equigranular, medium-grained, dark pink syenite, composed of microperthite, microcline, and albite, with small amounts of nepheline, biotite, amphibole, graphite, and magnetite. [Rutter, Sudbury district, Ontario, Canada] (Quirke in 1936)

S

sabalite   *sed.* Trade name for a yellowish to greenish, banded, phosphatic material, similar to, and including some, variscite; from Clay Canyon, Fairfield, Utah. [Possibly Sabal Palmetto, for its similarity in color and banding to the leaves] (pub. by Sterrett in 1914)

sabarovite   *ig., meta.?* Rock related to charnockite and composed of antiperthitic oligoclase (58%), quartz (32%), hypersthene (7%), and accessories. [Sabarovo, near Bug, Ukraine, Soviet Union] (Bezborodko in 1931)

saccharite   *meta?* Massive, granular rock composed chiefly of feldspar (andesine), occurring in veins in serpentinite at the chrysoprase mines near Frankenstein, in Silesia. [Gk. *sakcharon,* sugar; alluding to its texture] (Glocker in 1845)

saernaite   *ig. pluton.* = syenite, cancrinite. Also spelled särnaite. (Brögger in 1883)

sagenite   *ig., sed.* Crystalline or chalcedony quartz containing numerous needlelike crystals of other minerals such as rutile, tourmaline, actinolite, or goethite. [Gk. *sagene,* net; alluding to the reticulated structure often exhibited; originally applied to prismatic twinned rutile without quartz] (Saussure in 1796)

sagger   *sed.* Fireclay, often forming the floor of coal seams, used for making saggers, ceramic boxes in which pieces are placed while being baked. Also spelled saggard, seggar, and sagre. [safeguard]

sagvandite   *ig., meta.?* Redish brown, massive rock, consisting of large euhedral crystals of bronzite in a groundmass of bronzite and magnesite, with minor picotite (chromian spinel), magnetite, pyrite, and talc; possibly a hybrid rock. [Sagvand, Balsfjord, Tromsö, Norway] (Pettersen in 1883)

Saint Stephen stone   *sed.* Translucent, white or grayish chalcedony containing round, blood-red spots uniformly distributed throughout it; spots are generally no larger than dots, so the stone often appears to be of a uniform rose-red color. [Saint Stephen, the stone formerly venerated as having been stained during his martyrdom]

sakalavite   *ig. extru.* Glassy quartz basalt consisting of phenocrysts of labradorite, augite, and iron oxides, in a glass matrix that potentially contains quartz, labradorite, augite, and sanidine.

[Port Bergé, Sakalava district, Madagascar] (Lacroix in 1923)

sakenite   *meta.* Sapphirine-bearing metamorphic rock mainly composed of anorthite, augite, spinel, corundum, diaspore, and sapphirine. [Sakena, Madagascar] (Lacroix in 1940)

salcrete   *sed.* Thin, hard crust of salt-cemented sand grains, occurring on a marine beach that is periodically saturated by saline water. [L. *sal,* salt + *crescere,* to grow] (Yasso in 1966)

salinastone   *sed.* General term for a sedimentary rock composed dominantly of saline minerals (often precipitated, but may be fragmental), e.g., halite rock, anhydrite rock, and gypsum rock. [salina, from L. *sal,* salt] (Shrock in 1948)

salitrite   *ig. diaschist.* Lamprophyre consisting predominantly of an aegirine-bearing diopside, in part with aegirine-rich mantles, and of abundant titanite; minor accessories include apatite, microcline, anorthoclase, and baddeleyite. [Salitre Mountains, Minas Geraes, Brazil] (Tröger in 1928)

salt, blebby   *sed.* Miners' term for saline rock consisting of coarse-grained polyhalite aggregates in a halite matrix.

salt, rock   *sed.* = halite rock

saltpeter   *sed.* 1. Naturally occurring niter (potassium nitrate) bodies. 2. General term for nitrate-rich deposits, e.g., earthy cave deposits, Chile saltpeter, Peru saltpeter. [L. *sal petrae,* rock salt]

saltpeter, Chile   *sed.* Sodium nitrate (nitratite) occurring in beds in a desert region near the boundary of Chile and Peru but chiefly in Chile.

saltpeter, Peru   *sed.* Same as Chile saltpeter but occurs in Peru.

sancyite   *ig. extru.* Rhyolite containing numerous large sodium sanidine phenocrysts and smaller andesine phenocrysts associated with augite, biotite, and accessories in a glassy matrix. [Aiguilles du Sancy, Mont Doré, France] (Lacroix in 1923)

sand   *sed.* 1. Rounded, subrounded, or subangular fragment of any mineral or rock between 1/16 mm and 2 mm in diameter; coarse sand, also named granule, is 2-4 mm in size. 2. Loose aggregate of unlithified particles of sand size. 3. Quartz fragments or aggregates; when other minerals are involved the term is usually modified by using a color or mineralogical adjective, e.g., black sand, ruby sand, feldspathic sand, glauconitic sand. [O.Eng. *sand*]

sand, black *sed.* Black-colored sands in beach and alluvial placers, rich in magnetite or ilmenite, or in some situations contain other black minerals such as cassiterite, rutile, thorite, betafite, euxenite, columbite, and schorl.

sand, carbonate *sed.* Sand-size detritus composed principally of calcite clasts of biochemical (skeletal) or inorganic chemical (ooliths) origin; when cemented they give rise to the calcarenites.

sand, coarse *sed.* = granule

sand, cover *sed.* Wind-blown deposit of fine to very fine sand, usually containing more than 90% quartz and believed to have been deposited by heavy snowstorms during the glacial epoch.

sand, dune *sed.* Type of wind-blown sand piled up into a sand dune, usually consisting of rounded quartz grains having diameters slightly smaller (0.1-1 mm) than beach sands.

sand, dust *sed.* Medium to coarse silt, having diameters 0.025-0.04 mm and washed out by a stream having a velocity of 1.5 mm/sec. (Searle in 1923)

sand, fine *sed.* Variously defined term, but in geology usually refers to a sand particle size with a diameter 0.125-0.25 mm or to a loose aggregate of particles of this size.

sand, gas *sed.* Sand or sandstone body containing a large quantity of natural gas.

sand, gravelly *sed.* 1. Unconsolidated sediment containing 5-30% gravel and having a ratio of sand to mud (silt + clay) greater than 9:1. (Folk in 1954) 2. Unconsolidated sediment containing more particles of sand size than of gravel size, more than 10% gravel, and less than 10% of all other finer sizes. (Wentworth in 1922)

sand, green *sed.* = greensand

sand, gypsum *sed.* Sand composed of gypsum, as at White Sands National Monument, New Mexico.

sand, oil *sed.* Unconsolidated sand or sandstone (more rarely porous limestone and dolomite) from which oil is obtained by drilled wells.

sand, olivine *sed.* Relatively rare sands rich in green olivine occurring in some volcanic regions, e.g., on the beach at Hanauma Bay, Koko Head area, Oahu, Hawaiian Islands.

sand, pyroclastic *ig. extru., sed.* Sandy deposits composed of pyroclastic materials from volcanic explosions.

sand, ruby *sed.* Red-colored beach sand containing much garnet, as at Nome, Alaska.

sand, shell *sed.* 1. Marine sand containing up to 5% shell fragments. 2. Sand-sized calcium-carbonate detritus composed essentially of broken shells.

sand, silica *sed.* Industrial term for a sand or an easily disaggregated sandstone with a very high percentage of quartz (silica); a source for glass and silicon.

sand, silty *sed.* Unconsolidated sand containing appreciable silt; specifically, may contain 50-90% sand and have a ratio of silt to clay greater than 2:1. (Folk in 1954)

sand, singing *sed.* = sand, sounding

sand, sounding *sed.* Sand, usually clean and dry, that emits a musical, crunching, or humming sound when disturbed by sliding down a dune or when it is stirred or walked over. Also named musical sand, singing sand, roaring sand, and whistling sand.

sand, tar *sed.* Variety of oil sand in which asphalt fills the interstices.

sand, volcanic *ig. extru., sed.* = sand, pyroclastic

sand, white *sed.* 1. sand, gypsum. 2. Pure quartz sand.

sand-calcite *sed.* = sand crystal, calcite

sand crystal *sed.* Large euhedral or subhedral crystal (usually a carbonate or sulfate mineral) filled with detrital sand inclusions (up to 60%), and developed by crystal growth in an unconsolidated sand deposit; consists of sandstone in which the cementing material is a single crystal or aggregate of attached single crystals.

sand crystal, barite *sed.* Barite crystals filled with detrital sand; well-known are the reddish-stained barite sand crystals (barite roses) from the areas around Norman, Oklahoma.

sand crystal, calcite *sed.* Euhedral calcite crystals, as large as 10 cm long, that may incorporate much sand during their crystallization; noted localities are in the Badlands of western South Dakota and at Fontainebleau near Paris, France (Fontainebleau sandstone).

sand crystal, celestite *sed.* Relatively rare crystals of celestite filled with detrital sand; specimens occur near Weatherford, Oklahoma.

sand crystal, gypsum *sed.* Single crystals and rosettes of gypsum commonly filled with detrital sand; in some deposits the sand may form hour-glass inclusions, e.g., near Jet, Alfalfa County, Oklahoma.

sand spike   *sed.* Sandy concretion whose form is spikelike in that it has a long conic shape with a globular termination at the base of the cone; examples occur at Mt. Signal, Imperial County, California.

sandrock   *sed.* General term applied to sandstone easily crumbled between the fingers.

sandshale   *sed.* Sedimentary deposit consisting of thin alternating beds of sandstone and shale.

sandstone   *sed.* 1. Arenite composed of rounded or angular clasts whose average size ranges from 1/16 mm to 2 mm, often set in a fine-grained matrix (silt or clay) and usually more or less firmly united by a cementing material; the name

Sand crystals (barite rose). Norman, Oklahoma. 8.5 cm.

for consolidated sand.   2. Cemented clastic sand rock containing about 85% or more of quartz. The many varieties may be named according to the presence of nonquartz clasts (feldspathic s., glauconitic s.), to the presence of nonsand particles (argillaceous s., pebbly s.), or to the presence of various cementing materials (anhydrite, ankerite, barite, calcite, celestite, chalcedony, dolomite, gypsum, halite, hematite, limonite, opal, quartz, siderite, or organic matter). Arenite is essentially identical. Numerous varieties are listed below.

sandstone, argillaceous   *sed.* Sandstone containing an indefinite amount of fine silt or clay and usually too weak for use as a building stone. [argil, white clay, from Gk. *argillos,* white]

sandstone, arkosic   *sed.* = arkose

sandstone, beach   *sed.* = beachrock

sandstone, Berea   *sed.* Fine-grained, blue-gray to buff, pure sandstone (95% quartz) of Mississippian age; widely used building stone, quarried esp. at Amherst, Ohio. [Berea, Cuyahoga County, Ohio] (Newberry in 1870)

sandstone, calcarenaceous   *sed.* Transitional rock between sandstone composed of detrital quartz and calcarenite composed of detrital carbonate debris (skeletal and oolitic).

sandstone, calcareous   *sed.* Sandstone whose precipitated cement is calcite or, more rarely, aragonite or dolomite.

sandstone, clayey   *sed.* Argillaceous sandstone containing more than 20% clay. (Krynine in 1948)

sandstone, conglomeratic   *sed.* 1. Sandstone containing 5-30% gravel and having a ratio of sand to mud (silt + clay) greater than 9:1. (Folk in 1954)   2. Sandstone containing more than 20% pebbles. (Krynine in 1948)

Sand crystals (calcite) (Fontainebleau sandstone). Fontainebleau, near Paris, France. 9 cm.

Sand crystals (gypsum rose). Corpus Christi, Texas. 11 cm.

sandstone, crystal   *sed.* 1. Sandstone in which the quartz clasts have been enlarged by the authigenic crystallization of silica so the grains show regenerated crystal faces and at times nearly perfect quartz euhedra; sparkles in bright light.   2. Sandstone in which the cement (usually calcite) has been deposited as large units with crystallographic continuity; if the sand is incompletely cemented and the calcite is euhedral, a calcite sand crystal may result.

sandstone, dirty   *sed.* Sandstone containing much matrix, esp. clay; quartz particles are held together by the claylike material.

sandstone, eolian   *sed.* Sandstone composed of wind-blown sand, e.g., dune sand; usually well-sorted but consists of grains slightly finer than those formed by water-deposited sands.

sandstone, feldspathic   *sed.* Feldspar-rich sandstone; specifically, sandstone intermediate in composition between an arkose and a quartz sandstone, containing 10-25% feldspar and less than 20% matrix material of clay, sericite, and chlorite. (Pettijohn in 1957)

sandstone, ferruginous   *sed.* Sandstone containing detrital or authigenic cement composed of one or more of the iron oxides, limonite (goethite) and hematite; brownstone is an example.

sandstone, flexible   *meta., sed.?* = itacolumite

sandstone, Fontainebleau   *sed.* Calcite sand crystals from Fontainebleau, near Paris, France.

sandstone, fritted   *meta.* Metamorphic rock formed chiefly by subjecting sandstone to extreme heat; usually consists of quartz grains, rounded by melting, in a network matrix composed of silica glass (lechatlierite) or of glass and tridymite (often inverted to quartz), and, if the sandstone was argillaceous, there also may be feldspars, cordierite, sillimanite, or mullite. [F. *fritte,* from L. *frigere,* to roast]

sandstone, glauconitic   *sed.* Sandstone containing sufficient glauconite sand to impart a greenish color to the rock; greensand is a related rock.

sandstone, graywacke   *sed.* = graywacke

sandstone, lithic   *sed.* Sandstone characterized by an excess of rock fragments over feldspar particles and by a predominance of secondary precipitated cement over primary detrital matrix; subgraywacke is an example. Also named litharenite.

sandstone, micaceous   *sed.* Sandstone containing muscovite (including sericite); usually present as a clastic cement.

sandstone, molasse   *sed.* Sandstone (arkose or a lithic arenite) containing poorly sorted, coarse sand, rich in rock fragments and generally calcareous; characteristic of the molasse facies of sedimentation. [F. *mollasse,* soft]

sandstone, normal   *sed.* Sandstone (arenite) composed almost exclusively of quartz, with only subordinate amounts of other minerals.

sandstone, pebbly   *sed.* Sandstone containing 10-20% pebbles. (Krynine in 1948)

sandstone, phosphatic   *sed.* Sandstone in which collophane, a poorly organized carbonate-fluorapatite or carbonate-hydroxylapatite, forms the cement; in some sandy rocks, phosphatic granules or ooliths may be mixed in all proportions with detrital quartz.

sandstone, quartz   *sed.* = orthoquartzite

sandstone, quartzitic   *sed.* = orthoquartzite. (Krynine in 1940)

sandstone, quartzose   *sed.* 1. = orthoquartzite. (Shrock in 1948)   2. Sandstone containing at least 95% quartz but one that is not cemented with silica; to be used in contrast with orthoquartzite. (Krynine in 1940)

sandstone, sericitic   *sed.* Micaceous sandstone in which sericite (derived from the decomposition of feldspars) intermingles with finely divided quartz and fills spaces between the quartz clasts.

sandstone, siliceous   *sed.* = orthoquartzite

sandstone, silty   *sed.* Sandstone containing more than 20% silt. (Krynine in 1948)

sandstone, stylolitic   *sed.* Sandstone characterized by the presence of stylolitic seams. See stylolitic limestone for discussion of this feature.

sandstone, tuffaceous   *ig. extru., sed.* 1. Tuffs that have been eroded, retransported, and redeposited, mixed with sands and grading from arenaceous tuffs to tuffaceous sandstones.   2. *IUGS,* tuffite whose average clast size varies between 1/16 mm and 2 mm and which by volume contains 25-75% pyroclasts, the remaining materials being sedimentary.

sandstone, vitrified   *meta.* = sandstone, fritted. [L. *vitrum,* glass]

sandstone, volcanic   *ig. extru., sed.* Indurated deposit of rounded, water-worn pyroclastic fragments and a subordinate amount of nonvolcanic detritus; tuffaceous sandstone is an example.

sandstone dike   *sed.* Clastic or sedimentary dike formed by the injection or intrusion of sand into openings discordant to the structure of the country rock; often associated with events like earthquakes. See also dike, clastic and dike, sedimentary.

sandstone sill   *sed.* Same as a sandstone dike but having a concordant position as a tabular body parallel to older bedding; commonly mistaken for a sandstone bed, since it is parallel to bedding.

sanidinite   *ig. pluton.* Igneous plutonic rock composed almost entirely of sanidine. [sanidine] (Nose in 1808)

sanidinite   *meta.* Buchite containing, through metasomatism, sanidine and plagioclase with various amounts of glass, cordierite, sillimanite, hypersthene, corundum, and others; rock often considered to represent metamorphism at maximum temperature and minimum pressure. [sanidine]

sannaite   *ig. hypab.* Gray green porphyritic rock containing phenocrysts of barkevikite, less pyroxene, and still less biotite, in a fine-grained to dense groundmass of alkali feldspar, aegirine, chlorite, calcite, and mica pseudomorphs after nepheline. [Sannavand, Fen region, Telemark, Norway] (Brögger in 1921)

sansicl   *sed.* Unconsolidated sediment consisting of a mixture of sand, silt, and clay, in which no component forms 50% or more of the mixture. [sand + silt + clay]

santorinite   *ig. extru.* Sodium-rich variety of hypersthene andesite containing plagioclase (zoned, labradorite cores and oligoclase rims) with a groundmass of microlites of sodic oligoclase. [Santorini (Thira), Greece] (Becke in 1899)

sanukite   *ig. extru.* Gray or black, glassy, hornstonelike bronzite andesite, consisting of orthopyroxene needles lying in a groundmass of clear glass with abundant magnetite grains. [Sanuki, Shikoku Island, Japan] (Weinschenk in 1891)

sanukitoid   *ig. extru.* General term for all textural modifications of rocks derived from orthopyroxene-andesite magma, e.g., sanukite. [Sanuki, Shikoku Island, Japan] (Koto in 1916)

sapphirine rock   *meta.* Rock in which sapphirine is an important component; sapphirine, in association with spinel, cordierite, corundum, sillimanite, plagioclase, amphiboles, and pyroxenes, occurs in certain highly metamorphosed rocks, esp. those derived from ultramafic igneous types, e.g., in gneiss, granulite, and schist. Important varieties include sakenite and sapphirinite.

sapphirinite   *meta.* Sapphirine rock whose major components are sapphirine and spinel; occurs in Madagascar. (Lacroix in 1940)

saprodil   *sed.* Sapropelic coal of Tertiary age. [Gk. *sapros,* rotten + *odme,* stink]

saprolite   *sed.* Superficial, soft, earthy, clay-rich, decomposed rock, formed in place by chemical weathering of igneous, sedimentary, and metamorphic rocks; characterized by the preservation of the structures in the unweathered rock. [Gk. *sapros,* rotten] (Becker in 1895)

sapromyxite   *sed.* = coal, boghead. Also named tomite. [Gk. *sapros,* rotten + *muxa,* slime] (Zalessky in 1915)

sapropel   *sed.* Jellylike organic ooze or sludge that accumulates subaqueously in shallow to deep marine basins, lagoons, and lakes and is composed of plant remains (algae and fragments of higher plants) putrefying in an anaerobic environment; potential source material for petroleum and natural gas. [Gk. *sapros,* rotten + *pelagos,* sea]

sapropel-calc   *sed.* Sediment in which the amount of remains of calcareous-algae exceeds that of sapropel.

sapropel-clay   *sed.* Sediment in which the amount of clay exceeds that of sapropel.

sapropelite   *sed.* = coal, sapropelic

sapropsammite   *sed.* Sapropel rich in sand. [sapropel + Gk. *psammos,* sand]

sard   *sed.* Translucent reddish brown to brown chalcedony (agate); very close to carnelian, but the term sard is applied to the types that are more brown than red. [Sardis, on the Hermus River, capitol of the ancient kingdom of Lydia] (Theophrastus, *On Stones*)

sardoine   *sed.* = sard; not to be confused with sarduine, an artificially colored sard.

sardonyx   *sed.* Banded quartz rock consisting of straight parallel layers of sard (if brownish) or carnelian (if reddish) alternating with colorless or white chalcedony. [sard + onyx]

särnaite   *ig. pluton.* = syenite, cancrinite. [Särna, Sweden] (Brögger in 1883)

satelite   *meta.* Grayish to greenish blue serpentinite with a fibrous structure and showing slight chatoyant effect; occurs in Tulare County, California; used for cat's-eye gems. Also named California cat's-eye and serpentine cat's-eye. [Possibly satin, alluding to satinlike luster] (pub. by Sterrett in 1909)

satin spar   *sed.* 1. Fibrous, chatoyant, variety of gypsum with a silky luster; necklaces and trinkets are cut and polished from the material; in the Niagara Falls area, New York, the rock is called Niagara spar or Falls spar. 2. Less commonly applied to chatoyant and fibrous varieties of aragonite (Karlsbad spring stone), calcite, celestite, and other minerals. [satin, for its luster] saussurite rock   *meta.* Tough, compact, white, green, or gray aggregate composed of a mixture of albite (or other sodic-plagioclase) and epidote, clinozoisite, or zoisite, together with variable amounts of sericite, prehnite, calcite, and calcium-aluminum silicates; formed from the alteration of calcic-plagioclase and often occurs as zones in anorthosite (esp. labradoritite) rocks. [Horace Benedict de Saussure, 1740-1799, Swiss geologist; named by his son, Nicholas Théodore de Saussure] (Saussure in 1806)

saxonite   *ig. pluton.* Peridotite composed of olivine and orthopyroxene (enstatite); similar to harzburgite but contains less than 5% iron oxide minerals, while harzburgite contains more. [Saxony, Germany] (Wadsworth in 1884)

scanoite   *ig. extru.* Extrusive rock related to ghizite but containing normative nepheline and no biotite. [Scano flow, Monte Ferru, Sardinia, Italy] (Lacroix in 1924)

schalstein   *ig. extru., altered.* Term applied to altered basaltic and spilitic rocks and tuffs; shearing structures, with incipient schistosity and cleavage, and partial replacement of the rock by calcite are often characteristic features. Miners' term. [G. *Schale,* shell, pan, basin + *Stein,* stone] (pub. by Becher in 1789)

scharizerite   *sed.* Nitrogenous hydrocarbon occurring as black patches in phosphatic earth in a cave in Styria. [Rudolf Scharizer, 1859-1935, Austrian mineralogist of Graz] (Schadler in 1926)

schaumopal   *sed.* Porous variety of opal from the Virunga district, Tanganyika Territory, East Africa. [G. *Schaum,* foam, froth + opal] (Hauser in 1911)

scheibeite   *sed.* Fossil resin found in brown coal; not to be confused with the mineral of the same name. [Robert Scheibe, 1859-1923, German-Colombian mineralogist] (von Linstow in 1912)

schillerfels   *ig.* Igneous rocks containing orthopyroxenes, e.g., the peridotites saxonite and harzburgite, which exhibit a play of colors due to the arrangement of minute inclusions in the pyroxenes. [G. *Schiller,* iridescence, play of colors + fels]

schist   *meta.* Metamorphic rock readily split into thin flakes or slabs and composed of 50% or more of phaneritic minerals of micaceous or prismatic habit. Numerous varieties are recognized, and most of them are based on the mineralogy of the rock, either the major micaceous or prismatic mineral, or the major metacrysts set in these as a matrix. The following varieties have been observed: actinolite s., andalusite s., anthophyllite s., anthophyllite-plagioclase s., antigorite s., biotite s., biotite-zoisite s., chlorite

Schist (actinolite). Chester, Vermont. 15 cm.

s., chlorite-muscovite s., chloritoid s., clinozoisite s., corundum-biotite s., crossite s., cummingtonite s., damourite (hydromica) s., diopside s., epidote s., fuchsite (chromian muscovite) s., garnet s., glaucophane s., graphite s., grunerite s., hematite s., hornblende s., kyanite s., lawsonite-chlorite s., magnetite s., mica s., muscovite s., ottrelite s., paragonite s., piedmontite s., pyrophyllite s., quartz-albite s., quartz-cummingtonite s., quartzofeldspathic s., quartzose s., sericite s., serpentine s., sillimanite s., specularite (hematite) s., staurolite s., stilpnomelane s., talc s., tourmaline s., tremolite s. Also rarely spelled shist. [Gk. *schistos,* cleft; alluding to the good cleavage of most varieties]

schist, amphibole *meta.* Schist in which one or more of the amphibole-group minerals is the principal component: the major amphiboles forming schists include actinolite, anthophyllite, crossite, cummingtonite, glaucophane, grunerite, hornblende, and tremolite.

schist, augen *meta.* Schist containing crystalline aggregates of lenticular shape (augen) parallel to and alternating with the schistose foliation. [G. *Augen,* eyes]

schist, calc *meta.* Foliated rocks containing calcite as an important or dominant constituent along with quartz, magnetite, and various silicates, e.g., sericite, chlorite, antigorite, talc, tremolite, actinolite, biotite, hornblende, diopside, grossular, and others.

schist, calc-silicate *meta.* Regional metamorphic foliated or schistose rock composed predominantly of one or more of the various Ca, Ca-Mg, Ca-Al, or Mg silicates, usually with some calcite; grade into calc-silicate gneiss; a variety of calc-silicate rock.

schist, chlorite *meta.* Schist containing a chlorite (commonly clinochlore) as the principal micaceous mineral, usually with one or more other micaceous mineral (biotite, muscovite) and quartz, and with or without albite; metacrysts are commonly garnet, magnetite, and pyrite.

schist, curly *meta.* Schist crinkled or folded in such a way as to show an undulating or wavy surface.

schist, lime-silicate *meta.* = schist, calc-silicate

schist, mica *meta.* Schist in which one or more of the mica-group minerals is the principal component; minerals usually include biotite, muscovite (including sericite and fuchsite), paragonite, and hydromica.

schist, mylonite *meta.* Mylonite displaying a strong foliation resembling a schist; term is superflu-

ous since mylonite by nature shows a banded structure.

schist, ottrelite *meta.* Schistose rock containing small, rather euhedral, metacrysts of ottrelite, a manganiferous chloritoid, in a matrix of muscovite or chlorite often with chloritoid. [Ottré, village in Belgium on the border of Luxemburg]

schist, quartz *meta.* Schist composed of quartz whose foliation is due to streaks and lenticles of very fine-grained quartz; micas may be present but in lesser quantities than in mica schist. Also named quartzose schist; it grades into a schistose quartzite.

schist, specular *meta.* Schist composed principally of hematite whose foliation is due to aligned flakes or platy crystals (specularite) of the mineral. Also named specularite schist.

schoenfelsite *ig. extru.* Dark-colored, mafic-rich basalt, containing bytownite (28%), olivine (53%), augite, iron oxides, and minor devitrified glass. [Altschönfels, Vogtland, Germany] (Uhlemann in 1909)

schonite *artificial.* Glass from Skåne, Sweden; once incorrectly regarded as a tektite but now considered to be an artificial substance. [Possibly G. *schön,* beautiful] (pub. by Suess in ca. 1908)

schorenbergite *ig. hypab.* Nepheline leucitite with phenocrysts of nosean, and sometimes of leucite, in a groundmass of leucite, nepheline, and aegirine. [Schorenberg, Rhine Province, Germany] (Brauns in 1921)

schorl rock *ig. diaschist., autometa.* = tourmalite

schörlquartzit *ig. diaschist., autometa.* = tourmalite

schörlschiefer *ig. diaschist., autometa.* = tourmalite

schraufite *sed.* Hyacinth- to blood red resin occurring as small masses and in layers in a schistose sandstone (Carpathian sandstone) near Wamma, in Bukovina, eastern Europe. [Albrecht Schrauf, 1837-1897, Austrian mineralogist of Vienna] (Schröckinger in 1875)

schriesheimite *ig. pluton.* Hornblende peridotite in which the major mineral hornblende encloses rounded grains of olivine (poikilitic texture); diopside and phlogopite are constant accessories. [Schriesheimer Valley, north of Heidelberg, Germany] (Salomon and Nowomejsky in 1904)

schungite *meta.?* = shungite

scintillite *sed.* Quartz helictitic speleothem consisting of a microcrystalline silica core and a sparkling euhedral quartz covering. [L. *scintillare,* to sparkle, to shine] (Deal in 1964)

scleretinite *sed.* Hard, black to brown, fossil resin. [Gk. *skleros,* hard + *rhetine,* resin] (Mallet in 1852)

scoop stone *sed.* Term applied to sea amber dredged (scooped) from the Baltic Sea.

scoria *ig. extru.* 1. Irregular, rough, clinkerlike, more or less vesicular to frothy fragment of dark lava, thrown out in an explosive eruption or formed by the breaking up of the first-cooled crust of a lava flow; usually of basalt composition, and thus heavier, darker, and less cellular, and more crystalline than pumice. 2. Bomb-size (greater than 32 mm in diameter) pyroclast irregular in form and very vesicular. [Gk. *skoria,* dross, that which is thrown off]

scoria, basalt *ig. extru.* Light-weight, very cellular to frothy mass of basalt lava; common scoria.

scorilite *ig. extru.* Volcanic glass. [Gk. *skoria,* dross, that which is thrown off]

scree *sed.* Rock waste at the base of a cliff, or a sheet of rock waste covering a slope below a cliff; essentially the same as talus. [O.Eng. *scridan,* to glide]

scyelite *ig. pluton.* Coarse-grained, somewhat altered, hornblende peridotite, consisting of hornblende filled with serpentine inclusions derived from olivine, and conspicuous mica with magnetite. [Loch Scye, Reay Parish, West Caithness, Scotland] (Judd in 1885)

seastone *sed.* = amber, sea

seatearth *sed.* Clayey soil occurring beneath coal seams, representing the soil that supported the vegetation from which the coal was formed; typically massive with rootlets and siderite nodules. Also named underclay.

sebastianite *ig. pluton.* Gabbro composed of anorthite, biotite, and a little augite and apatite, but without feldspathoids; in fragments found among ejectamenta of Monte Somma Volcano. [San Sebastian, Monte Somma, Italy] (Lacroix in 1917)

sedarenite *sed.* Litharenite composed chiefly of sedimentary rock fragments and having any clay content, sorting, or rounding; examples are sandstone arenite and shale arenite. [sedimentary arenite] (Folk in 1968)

sediment *sed.* Solid fragmental material originating from the weathering of rocks and transported or deposited by wind, water, or ice, or accumulating by other natural agents, such as chemical precipitation from solution or secretion by organisms, and forming at the Earth's surface at ordinary temperatures. [L. *sedimentum,* settling]

sediment, volcaniclastic *ig. extru., sed.* General term for all fragmental volcanic rocks formed by any mechanism or origin, emplaced in any physiographic environment (on land, under water, or under ice), or mixed with any nonvolcanic fragments in any proportion. [volcanic + Gk. *klastos,* broken] (Fisher in 1961)

sedimentary dike *sed.* See dike, sedimentary

seebenite *meta.* Variety of hornfels containing feldspar and cordierite as the dominant minerals. [Seeben near Klausen (possibly in Tyrol)] (Salomon in 1898)

selagite *ig. extru.* Somewhat decomposed, dull, earthy, fine-grained, gray-green trachyte showing abundant biotite crystals in a groundmass of orthoclase and diopside with a little quartz and sporadic olivine. [Gk. *selagein,* to beam brightly; alluding to the glistening of mica] (Haüy in 1822)

selbergite *ig. hypab.* Hypabyssal phonolite consisting of abundant phenocrysts of leucite, nosean, sanidine, aegirine-augite, and biotite in a fine-grained matrix of nepheline, alkali feldspar, and aegirine. [Selberg, near Rieden, Germany] (Brauns in 1921)

selenolite *sed.* General term for a sedimentary rock composed of gypsum or anhydrite. [Possibly selenite, crystallized gypsum] (Wadsworth in 1891)

semianthracite *sed.* Coal between bituminous and anthracite in rank, having a fixed-carbon content of 86–92%; closely resembles anthracite in physical appearance. [L. *semi,* half + anthracite]

semiopal *sed.* Opaline rock, nearly opaque, ranging from white to various colors; specimens often have a lower water content than most opal and are composed essentially of a cristobalitelike phase. [G. *Halbopal,* half (semi-) opal] (Werner in 1789)

semischist *meta.* General term applied to somewhat schistose rocks that develop in the course of transition from massive rocks (such as a graywacke) to schists in the earlier stages of progressive regional metamorphism; these transitional rocks have been observed in New Zealand, California, and elsewhere. [L. *semi,* half + schist]

Seneca oil *sed.* Type of petroleum, esp. that from Seneca Lake (Seneca Oil Spring) near Cuba, Allegany County, New York. [Seneca Indians of the region]

sepiolite rock *meta.* = meerschaum

septaria *sed.* Plural of septarium.

septarian boulder *sed.* = septarium

septarian concretion *sed.* = septarium

septarian nodule *sed.* = septarium

septarium *sed.* Argillaceous, spherical to sub-spherical, concretion, traversed by cracks that radiate and widen toward the center and intersect a series of cracks concentric with the margins; the cracks are often filled or partly filled by crystalline minerals (calcite, barite, dolomite, siderite, sphalerite, and others); formed by the dehydration and shrinkage cracking of a concretion whose exterior was case-hardened. Septaria is the plural. Also named septarian boulder, septarian concretion, septarian nodule, and petrified turtle. [L. *septa,* partitions]

serpent stone *sed., meta.* Highly absorbent aluminous stone, once believed to be formed by snakes and to be efficacious in drawing out poison; also called adder stone. Some serpentinites have also been considered to have the same power because of their fancied resemblance to a serpent's skin.

serpentine rock *meta.* = serpentinite

serpentinite *meta.* Rock consisting almost wholly of one or more of the serpentine minerals (antigorite, chrysotile, lizardite), and at times with various amounts of calcite, magnesite, dolomite, talc, chromite, magnetite, and others; in color they are usually yellow to green to greenish black, although some red varieties are known, and they may be dull to waxy in luster and vary from homogeneous to banded, streaked, and spotted structures. Examples of the numerous varieties include: ophicalcite, ricolite, rosse di Levanto marble (red), verd antique. Polished serpentinite is often named serpentine marble. [serpentine]

serpophite *meta.* General term for compact, seemingly amorphous, varieties of serpentine. Also named serpentinophite. [serpentine + ophite] (Lodochnikov in 1933)

sesseralith *ig. pluton.* Corundum-bearing hornblende gabbro, consisting of labradorite (55%), brown hornblende and augite (41%), corundum (12%), with some iron oxides, apatite, and secondary actinolite and chlorite. [Sessera Valley, Piemonte, Italy] (Millosevich in 1927)

settlingite *sed.?* Hard, brittle, pale yellow to deep red mineral resin, having the form of rounded, flattened drops; encrusting vein walls in an old mine at Northumberland, England. [Settling Stones mine, Northumberland, England] (Des Cloizeaux in 1874)

severite *sed.* = lenzinite. [St. Sever, France] (Beudant in 1824)

shackanite *ig. extru.* Dark-gray analcime trachyte, showing small glassy rhombs of zoned anorthoclase with orthoclase rims and a few augite prisms

Septarium. Coal Brook, Dale, England. 12 cm.

in a dense groundmass composed of analcime, anorthoclase, and orthoclase. [Shackan, between Midway and Osoyoos lakes, British Columbia, Canada] (Daly in 1912)

shale   *sed.* Fine-grained, argillaceous sedimentary rock, laminated or fissile parallel to the bedding and in which there has been no substantial change in mineral composition since its accumulation other than changes resulting from compaction or diagenesis; the clay-mineral content varies from less than 40% to almost 100%, consists primarily of illite with varying amounts of chlorite, kaolinite, and montmorillonite, and is accompanied by quartz, micas, calcite, feldspars, pyrite, organic matter, and others. The variety names are based primarily on colors, on mineral impurities, and on chemical compositions. Some of the important varieties are listed below. [O.Eng. *scealu,* scale, shell]

shale, alum   *sed.* Shale containing one or more of the alum minerals (hydrous alkali aluminium sulfates, like kalinite); these sulfates form in pyrite-rich shale from reactions between the clay minerals and sulfuric acid that forms from the decomposition of the pyrite. Also named ampelite.

shale, bituminous   *sed.* Shale of marine or freshwater origin, usually gray to dark brown or black and containing organic matter, either as completely structureless material or as whole or fragmentary organisms. Oil shale is a variety.

shale, black   *sed.* Synonym for carbonaceous shale; however, the black color of some shales is due to abundant, fine-grained pyrite, rather than organic substances, although they usually occur together.

shale, calcareous   *sed.* Shale usually containing more than 6% calcite, although many shales contain up to this amount; the carbonate is present as finely precipitated cement or as small fossiliferous materials; these rocks grade into shaly limestone.

shale, cannel   *sed.* Carbonaceous shale formed by the accumulation of sapropelic sediments, accompanied by an approximately equal amount of silt and clay.

shale, carbonaceous   *sed.* Black shale composed of illite and other clay minerals closely associated with abundant organic materials, quartz, and major amounts of pyrite; the organic materials do not occur in a pure state but are intimately associated with the clays and other components.

shale, chloritic   *sed.* Poorly laminated shale, containing some silt-size particles, including quartz, feldspar, and unstable minerals, and having abundant chlorite, esp. in the finer matrix; often associated with graywacke beds.

shale, clay   *sed.* 1. Shale consisting chiefly of clay minerals.   2. Shale consisting of no more than 10% sand and having a silt to clay ratio of less than 1:2. (Folk in 1954)

shale, diatomaceous   *sed.* Clay-rich diatomite showing shaly partings.

shale, feldspathic   *sed.* Kaolinitic shale characterized by a feldspar content greater than 10% in its silt-sized particles; commonly associated with coarser arkosic debris.

shale, ferruginous   *sed.* Red or rusty-colored shale containing ferric oxide; since an average shale

Shale (oil shale). Anvil Points, Rifle, Garfield County, Colorado. 7.5 cm.

contains 6.47% iron oxides, the iron oxide in this variety should be more than 15%. (Pettijohn in 1975)

shale, iron *meteor., altered.* Laminated material consisting of iron oxides produced by the weathering of an iron meteorite.

shale, jet *sed.* Bituminous shale containing jet. Also named jet rock.

shale, kerogen *sed.* = shale, oil

shale, kerosine *sed.* Bituminous oil shale. Also spelled kerosene shale.

shale, menilite *sed.* Tertiary bituminous shale occurring on the northern slope of the Carpathian range from the boundary of Poland to Rumania; distillation yields useful organic compounds.

shale, micaceous *sed.* Shale with abundant detrital muscovite and considerable matrix sericite; commonly associated with subgraywacke.

shale, mud *sed.* Shale consisting of no more than 10% sand and having a silt to clay ratio between 1:2 and 2:1; a fissile mudstone. (Folk in 1954)

shale, oil *sed.* Term used to describe a wide variety of shales containing a solid hydrocarbonaceous material named kerogen, associated with illite, calcite, dolomite, and other minerals; when heated to about 450-600°C, kerogen undergoes destructive distillation, producing crude oil (up to 10-40 gallons of crude per ton of feed shale); the exact nature of kerogen is uncertain (it may be a highly cycloparaffinic polymer with a molecular weight of more than 3,000, with closely associated other compounds).

shale, paper *sed.* Shale, often highly carbonaceous, that easily separates on weathering into thin layers suggesting sheets of paper.

shale, phosphatic *sed.* Dark gray to black shale consisting of paper-thin layers to beds several inches in thickness and composed of clays and granular quartz in a carbonaceous and phosphatic (collophane) matrix.

shale, potassic *sed.* Shale with 5% or more K_2O in its chemical analysis; a content of 2-5% is normal. (Pettijohn in 1975)

shale, quartzose *sed.* Shale composed essentially of rounded quartz grains of siltsize; often associated with orthoquartzite.

shale, red *sed.* Red silty shale (or siltstone) commonly found in red beds; colored by hematite staining, but the beds are not particularly iron-rich deposits.

shale, siliceous *sed.* Fine-grained, silica-rich (up to 70%) shale containing opal, or cryptocrystalline quartz derived from opal, as well as clay minerals (abundant montmorillonite), glass shards, organic matter, and very small clasts.

shale, silt *sed.* Consolidated sediment consisting of no more than 10% sand and having a silt to clay ratio greater than 2:1.

shale, silty *sed.* Shale composed chiefly of silty (between 1/256 and 1/16 mm diameter) materials; essentially the same as silt shale, although the amount of silt has not been specified.

shale, sodic *sed.* Shale whose Na_2O not only exceeds the 1.3% in average shale but also greatly exceeds the amount of K_2O, which on the average is from 2-5%. (Pettijohn in 1975)

shale, tuffaceous *ig. extru., sed.* IUGS, shaly tuffite whose average clast size is less than 1/256 mm and which, by volume, contains 25-75% pyroclasts, the remaining materials being sedimentary.

shard *ig. extru.* Small fragment of rhyolitic glass, usually with a curved, spiculelike form; generally they are the bubble-wall fragments produced by the disintegration of pumice during or after an eruption. Also spelled sherd. [O.Eng. *sceard,* from *scearan,* to shear]

sharpstone *sed.* 1. General term for any rock fragment larger than sand size (greater than 2 mm diameter) and having angular edges and corners; like rubble. (Shrock in 1948) 2. Fine-grained, nonargillaceous sandstone from Yorkshire, England, that breaks into angular fragments.

shastaite *ig. extru.* Normal vitreous dacite containing andesine and hypersthene phenocrysts in a glassy matrix (63%) that shows microlites of pyroxene, andesine, and iron oxides. [Mt. Shasta, California] (Iddings in 1913)

shastalite *ig. extru.* Unaltered glassy andesite; term used to distinguish this rock from weiselbergite, an altered form of the rock. [Mt. Shasta, California] (Wadsworth in 1884)

shergottite *meteor.* Pyroxene-plagioclase achondrite (eucrite type) in which the plagioclase (usually anorthite) is replaced by maskelynite (glass of plagioclase composition). [Shergotty, India] (Brezina in 1904)

shihlunite *ig. extru.* Anorthoclase trachyte (related to selagite) composed of anorthoclase (61%), augite (16%), olivine (10%), and lesser biotite, iron oxides, apatite, leucite, and glass. [Shihlung lava flow (of 1720), in Wu-ta-lien-chih district, Lung-chiang province, Manchuria] (Ogura and Gen in 1936)

shonkinite *ig. pluton.* 1. Dark-colored syenite composed of essential orthoclase and augite, with or without sodic-plagioclase, olivine, nepheline, and others; there is no quartz. [Am. Ind. Shonkin, name of the Highwood Range, Montana] (Weed and Pirsson in 1895) 2. *IUGS*, plutonic rock in which F is 10-60, P/(A + P) is 10 or less, and M is 60-90.

shonkinite, melanite *ig. pluton.* Melanite-rich variety of normal shonkinite.

shoshonite *ig. extru.* Extrusive rock very closely related to absarokite and banakite but darker and containing some glass. [Shoshone Indians and the Shoshone River, Yellowstone Park, Wyoming] (Iddings in 1895)

shungite *meta.?* Hard, black, amorphous, carbon-rich (98%) substance occurring interbedded with Precambrian schists; probably metamorphosed bitumen. Also spelled schungite. [Schunga, Olonets, Soviet Union] (Inostrantzev in 1884)

sideraerolite *meteor.* = siderolite. [Gk. *sideros,* iron + aerolite]

siderite *meteor.* Meteorite essentially composed of nickel-iron, without important accessory minerals. Also named holosiderite and iron meteorite. The term should not be confused with the ferrous carbonate ($FeCO_3$) mineral siderite. [Gk. *sideros,* iron] (Daubrée in 1867)

siderite rock *sed.* Any one of several rock types containing the mineral siderite as an important constituent; represented by sideritic limestone, cherty iron carbonate rock, sideritic mudstone, iron carbonate concretions, and sphaerosiderite rock.

siderite rock, cherty *sed.* Rock composed of thinly and rhythmically interlaminated chert and siderite in various proportions; the layers range from 1/8 in to about 6 in in thickness and some chert beds are oolitic, in which original siderite ooliths have been replaced by quartz.

sideroferrite *sed.* Grains of native iron occurring in petrified wood; apparently formed by the reduction of iron compounds by the organic matter of the wood. [Gk. *sideros,* iron + L. *ferrum,* iron] (Bahr, date uncertain)

siderolite *meteor.* Stony-iron meteorite, which on the average contains nickel-iron (50%) and silicate minerals (50%). Also named sideraerolite, stony-iron, or stony-iron meteorite. [Gk. *sideros,* iron + *lithos,* stone] (Maskelyne in 1863)

sideromelane *ig. extru.* Black to olive green obsidianlike glass with the composition of basalt. Also named tachylyte or hyalobasalt. [Gk. *sideros,* iron + *melas,* black] (Waltershausen in 1853)

siderophyre *meteor.* Bronzite-tridymite stony-iron meteorite, consisting of a network of nickel-iron metal enclosing granular aggregates of orthopyroxene (bronzite-hypersthene boundary) and minor tridymite. [Gk. *sideros,* iron + porphyry] (Tschermak in 1883)

siegburgite *sed.* Golden yellow to reddish resin, occurring as concretionary masses mixed with quartz sand, in the brown coal near Troisdorf and Siegburg, Lower Rhine. [Siegburg, western Germany] (Lasaulx in 1875)

sienit *ig. pluton.* Old spelling for syenite, esp. for one from area near Dresden, Germany, which is quartz-free and contains hornblende. [Syene (Asswan), Egypt] (Werner in 1788)

sienna *sed.* General term for various brownish yellow earthy limonitic pigments used in oil stains and paints; generally darker in oils than ocher. [Siena, Tuscany, Italy]

sierranite *sed.* Rock containing onyx and chert; from the Sierra Nevada of California. [Sierra Nevada]

sievite *ig. pluton.* Local name for a series of glass-rich andesite rocks of various compositions from Sieva Mountain, Euganea, Italy. [Sieva, Italy] (Marzari-Pencati in 1819)

silcrete *sed.* Conglomerate formed by the cementation of superficial gravels by silica. [siliceous + L. *crescere,* to grow] (Lamplugh in 1902)

silexite *ig. pluton., diaschist.* Rock composed nearly entirely of primary quartz (60-100%); other minor accessory minerals may include feldspar and muscovite. Also named peracidite and quartzolite. [L. *silex,* quartz, flint] (Miller in 1919)

silexite *sed.* General name applied to siliceous rocks like chert, flint, and similar types, esp. in Europe. [L. *silex,* quartz, flint] (Pliny, *Natural History*)

silica rock *sed., meta.* Industrial term for sandstones and quartzites containing at least 95% quartz (silica); used in the manufacture of glass and silicon.

silicalite *sed., meta.* General term for rocks composed of opal or quartz (silica), including such types as diatomite, chert, jasper, quartzite, and others.

silicastone *sed.* General term applied to any sedimentary rock composed of silica minerals, esp. quartz and opal. (Shrock in 1948)

silicified wood *sed.* Petrified wood formed by the mineralization of wood by silica in such a manner that the original form and structure of the wood are preserved; the silica is usually in the form of quartz (chalcedony) or opal and gives rise to the following types: agatized wood, jasperized wood, opalized wood. See petrified wood for other minerals that may replace wood.

silicilith   *sed.* 1. Sedimentary rock composed of quartz. (Grabau in 1924)   2. Sedimentary rock composed essentially of the siliceous remains of organisms, e.g., diatomite, radiolarite, and others. (Pettijohn in 1957)

siliciophite   *meta.* Serpentinite impregnated with opaline silica. [silica + Gk. *ophis,* serpent (for serpentine)] (Schrauf in 1882)

silicocarbonatite   *ig. hypab.* General term for a carbonatite rock containing silicate minerals but in which calcite is more abundant than the silicates. Examples include aegirine sövite and kåsenite. (Brögger in 1921)

silkstone   *sed.* Crystalline quartz containing embedded fibrous goethite; similar to binghamite.

sillar   *ig. extru.* Type of ignimbrite from an ash cloud or nuée ardente, indurated by recrystallization due to escaping gases rather than by welding, as in the case of welded tuff. [Sp. *sillar,* ashlar] (Fenner in 1948)

sillimanite rock   *meta.* Any of several high-grade regional metamorphic rocks in which sillimanite is the principal aluminum silicate. Major examples include the following: s.-biotite schist or gneiss, s.-kyanite schist or gneiss, s.-garnet schist, s.-garnet-quartz-orthoclase granulite (khondalite), quartz sillimanitisé, and massive s. rock (a jadelike rock once used in western Europe for implements by prehistoric man).

sillite   *ig.* Variously defined igneous rock name; considered to be gabbro, diabase, mica syenite, and mica diorite by various authors. [Sillberg, in the Bavarian Alps] (Gümbel in 1861)

silt   *sed.* 1. Fragment of any mineral or rock between 1/256 mm and 1/16 mm in diameter.   2. Loose aggregate of unlithified particles of silt size; varies considerably in composition but commonly has a high content of quartz, feldspar fragments, and clay minerals. [M.Eng. *silte,* sand, from Sw. *sila,* to filter, to strain]

silt, clayey   *sed.* 1. Unconsolidated sediment containing more particles of silt size than of clay size.   2. Unconsolidated sediment containing 40-75% silt, 12.5-50% clay, and 0-20% sand. (Shepard in 1954)

silt, coarse   *sed.* Silt particle size between 1/100 mm and 1/16 mm in diameter.

silt, fine   *sed.* Silt particle having a diameter in the range 1/128 mm to 1/64 mm; in Great Britain the range 1/100 mm to 1/20 mm has been used.

silt, sandy   *sed.* 1. Unconsolidated sediment containing more particles of silt size than of sand size.   2. Unconsolidated sediment containing 10-50% sand and having a ratio of silt to clay greater than 2:1. (Folk in 1954)

siltite   *sed.* = siltstone. [silt] (Kay in 1951)

siltstone   *sed.* Indurated silt having a texture similar to shale but usually lacking its fine lamellar structure; at least two-thirds of the material should be silt size; generally these rocks are intermediate between shale and sandstone in both grain-size and composition. Also named siltite. [silt] (Green in ca. 1870)

siltstone, sandy   *sed.* Consolidated sandy silt.

siltstone, tuffaceous   *ig. extru., sed. IUGS,* tuffite whose average clast size varies between 1/256 mm and 1/16 mm and containing by volume 25-75% pyroclasts, the remaining materials being sedimentary.

Siltstone with fossil *Ophiomorpha* burrow. East of Johnstown, Weld County, Colorado. 10 cm.

silttil   *sed.* Brownish to buff friable silt containing a few siliceous pebbles, representing a chemically decomposed and washed till that may originally have been clayey. (Leighton and MacClintock in 1930)

simetite   *sed.* Deep red variety of amber, having a high content of sulfur and oxygen and a low content of succinic acid; from area near Mt. Etna, Sicily. [Simeto River, south of Catania, Sicily] (Helm and Conwentz in 1881)

sinaite   *ig. pluton.* Obsolete term once proposed for syenite. [Mt. Sinai] (Rosière, pub. by d'Aubuisson in 1819)

sinople   *sed.?* Variety of quartz containing red hematite inclusions; esp. for material from Schemnitz, Hungary. Equal to the German name Eisenkiesel (iron-silica) and also spelled sinopel. [Gk. *sinopis,* from Sinope, a town on the Black Sea where red ocher occurs]

sinter   *sed.* General term for porous to cellular, calcareous or siliceous (opaline) materials deposited by hot or cold mineral waters of springs, lakes, or streams. Important varieties are calcareous s. (tufa, calcareous tufa) and siliceous s. [G. *Sinter,* cinder, iron dross]

sinter, calcareous   *sed.* = tufa. Also named calcareous tufa.

sinter, pearl.   *sed.* Siliceous sinter with a pearly luster well illustrated by material from Santa Fiora, Italy (fiorite).

sinter, siliceous.   *sed.* White or nearly white, lightweight, porous aggregate composed mainly of a delicate network of minute fibers of opal and deposited from hot springs; the same material deposited by geysers is named geyserite.

sismondinite   *meta.* Schist in which the magnesium-rich chloritoid, named sismondine, is the predominant mineral; occurs at St. Marcel, Piemonte, Italy. [Angelo Sismonda, 1807-1878, of Turin, Italy] (Franchi in 1897)

skarn   *meta.* Pyrometasomatic rock composed of various Ca, Mg, Fe, or Al silicates, derived from nearly pure limestone and dolomite rocks with the introduction of large amounts of Si, Al, Fe, and Mg; often the rocks consist chiefly of two or three mineral combinations of the species grossular, diopside, vesuvianite, clinozoisite, anorthite, phlogopite, tremolite, forsterite (altered to serpentine), wollastonite, hedenbergite, and others. Essentially the same as tactite. [Sw. *skarn,* miners' term for silicate gangue minerals]

skedophyre   *ig.* General term for a porphyritic igneous rock characterized by a skedophyric texture, one in which the phenocrysts are more or less uniformly scattered throughout the groundmass. [Gk. *skedao,* from *skedannumi,* to scatter + porphyry]

skleropelite   *meta.* Argillaceous and related rocks indurated by low-grade metamorphism; more dense and massive than shale and without the cleavage of slate. [Gk. *skleros,* hard + *pelos,* mud, clay] (Salomon in 1915)

skomerite   *ig. extru.* Altered andesite that is compact, fine-grained, and dark gray and shows minute feldspar laths (near albite) in a microscopic groundmass containing augite, olivine, and plagioclase (more calcic than the phenocrysts). [Island of Skomer, adjacent to Pembrokeshire, England] (Thomas in 1911)

slabstone   *sed.* Rock easily split into slabs; like flagstone.

slate   *meta.* Fine-grained to aphanitic metamorphic rock with a highly developed foliation, called slaty cleavage, by means of which the rock can be split along closely spaced, parallel, and relatively smooth planar surfaces; the mineralogy is very difficult to determine, but optical and X-ray diffraction studies have shown that the major constituents are quartz, chlorite, muscovite, biotite, magnetite, hematite, pyrite, calcite, dolomite, epidote, albite, illite, and carbonaceous materials. Varieties are often named according to colors: gray s., black s., green s., purple s., red s., white s., and others. [M.Eng. and Scot. *sclate* from O.F. *esclat,* splinter, from O.H.G. *skleizan,* to break]

slate, carbonaceous   *meta.* Black or dark-gray slate containing graphite or black carbon compounds, apparently derived from organic substances in the original sediment. Also named black slate.

slate, chiastolite   *meta.* Highly metamorphosed, alumina-rich slate containing metacrysts of andalusite (chiastolite).

slate, clay   *meta.* 1. Old term once applied to slate in which the mineral matter was supposed to be largely detrital and without the formation of new minerals during metamorphism; it is now recognized that these rocks also are composed mostly of minerals formed by metamorphic processes.   2. Slate derived from an argillaceous (clay) sediment rather than from volcanic ash.

slate, curly   *meta.* Slate crumpled or folded in such a way as to show an undulating or wavy cleavage and surface.

slate, draw   *sed.* Sedimentary shale occurring above a coal seam and collapsing during or shortly after the removal of the coal.

slate, ferruginous   *meta.* Slate containing iron minerals, usually hematite; in some rocks magnetite and iron silicates (stilpnomelane, minnesotaite) may also be important; hematite-rich slates are often named red slate.

slate, flinty   *meta.* Very siliceous slate; can be used as a touchstone (basanite).

slate, graywacke   *sed., meta.* Fine-grained micaceous and sandy slates or shales.

slate, ribbon   *meta.* Slate produced by incomplete metamorphism still showing sedimentary bedding that cuts across the cleavage surfaces; the rock is therefore characterized by a varicolored ribbonlike structure.

slate, roofing   *meta.* Hard slate, with good cleavage and uniform color, used for making roofing shingles.

slate, spotted   *meta.* Shaly, slaty, or schistose argillaceous rock with a spotted appearance due to the incipient growth of metacrysts in response to contact metamorphism of low to medium intensity. Also named fleckschiefer.

slate, talc   *meta.* Impure, hard, slaty variety of talc.

smaragdite   *meta.* Massive to foliated, bright green amphibole-rich rock, composed of actinolite, hornblende, diallage, red corundum, and other minerals. Closely related to goodletite. [Gk. *smaragdos,* emerald, green gem] (Theophrastus, *On Stones*)

snide   *sed.* Local term applied to large pieces of potch (opal) occurring in Queensland, Australia. [Possibly snide, tricky, spurious]

snow   *sed.* Precipitation in the form of minute hexagonal crystals of ice, formed from vapor in the air when the temperature is below 32°F and usually falling in irregular masses or flakes.

soaprock   *meta.* = soapstone

soapstone   *meta.* Massive talc-rich rock in which the talc may be cryptocrystalline, flaky, or bladed (pseudomorphic after tremolite) and in which minor amounts of carbonate minerals (dolomite, calcite, magnesite), chlorite, tremolite, magnetite, chromite, and sulfides may occur. Also named soaprock and steatite. French chalk is a variety. [soap, alluding to its feel + stone]

soda niter rock   *sed.* Calichelike material containing soda niter ($NaNO_3$) with variable amounts of anhydrite, gypsum, halite, glauberite, bloedite, and other rare minerals as well as clastic particles, occurring primarily in the Atacama and Tarapaca deserts of northern Chile. Also named caliche; not to be confused with the more common calcitic caliche.

soda straw   *sed.* Popular term for a tubular stalactite usually composed of aragonite (less commonly, of calcite); the term tubular stalactite is preferred.

sodalithite   *ig. extru.* Violet red, phaneritic igneous rock whose major component is reddish sodalite, although there are small amounts of aegirine, feldspar, and eudialyte; occurs on Nunasarnak, Greenland. Also spelled sodalitite. [sodalite] (Ussing in 1911)

soggendalite   *ig. aschist.* Dark pyroxene-rich diabase with sparse plagioclase. [Soggendal parish, Norway] (Kolderup in 1896)

soil   *sed.* Natural, usually unconsolidated, body consisting of layers or horizons of mineral and/or organic constituents of variable compositions and thicknesses, differing from the parent rock in their morphological, physical, chemical, and mineralogical properties and their biological characteristics. [O.F. *sueil,* from L. *solum,* bottom]

Solenhofen stone   *sed.* Gray to buff, even-grained, compact lithographic limestone of Late Jurassic age, found at Solenhofen (Solnhofen), a village in Bavaria, Germany.

sölvsbergite   *ig. diaschist.* Medium- to fine-grained, rarely porphyritic, dike rock, containing dominant albite with less potassic feldspar, aegirine, or sodium amphibole, and little or no quartz. [Sölvsberget, Gran, Norway] (Brögger in 1894)

sommaite   *ig. pluton.* Light-colored, medium- to coarse-grained essexite, showing crystals of olivine and augite in a groundmass of bytownite, orthoclase, leucite, olivine, augite, biotite, sodalite, and accessories. [Monte Somma, Italy] (Lacroix in 1905)

sondalite   *meta.* Bluish to greenish metamorphic rock, containing cordierite, quartz, garnet, with some tourmaline and kyanite. [Sondalo, Italy] (Stache and von John in 1877)

sordawalite   *ig. aschist.* Glassy rock occurring on the border of diabase porphyry in gneiss and representing diabase glass (essentially the same as hyalodiabase and hyalobasalt). Also spelled sordavalite. [Sordawals, north end of Lake Ladoga, Wiborg, Finland] (Nordenskjöld in 1820)

sörkedalite   *ig. pluton.* Feldspathoid-free monzonite or monzodiorite containing labradorite, sodium orthoclase, olivine, pyroxene, apatite (12%), and iron-titanium oxides (18%). [Sörkedal, Oslo region, Norway] (Brögger in 1933)

sorotiite   *meteor.* Rare meteorite type consisting of approximately equal amounts of nickel-iron and troilite. [Soroti, Uganda] (Henderson and Perry in 1958)

sövite   *ig. hypab.* Nearly pure, coarse-grained, calcitic carbonatite, with less than 10% other species, which can include biotite, apatite, iron oxides, and more. [Söve, Fen district, Telemark, Norway] (Brögger in 1921)

sövite, aegirine   *ig. hypab.* General term for calcitic carbonatite containing 10-50% silicate minerals, esp. aegirine; an example is kåsenite.

sparagmite   *sed.* General term used in Scandinavia for late Precambrian fragmental rocks, including feldspathic sandstone, coarse arkose, subarkose, graywacke, conglomerate. [Gk. *sparagma,* fragment, piece]

sparite   *sed.* Limestone in which sparite cement is more abundant than the micrite matrix; sparite cement consists of crystalline, clean, relatively coarse-grained (usually greater than 10 microns) calcite (or aragonite) that either accumulated during deposition or was introduced later as a cement. (Folk in 1959)

spastolith   *sed.* Oolith markedly irregular in shape, being attenuated, flattened, and indented. [Gk. *spastikos,* drawing, pulling]

spathite   *sed.* Aragonite (or more rarely calcite) speleothem consisting of a vertical succession of small, petal-shaped, thin-walled cones. [Gk. *spathe,* broad flat blade] (White and Stellmack in 1959)

speckstone   *meta.* = soapstone. [G. *Speck,* lard, bacon, alluding to its greasy feel]

spectrolite   *ig. pluton.* Dark variety of labradorite (anorthosite rock) showing a brilliant play of colors, from Finland; used as a gem material. [spectrum, alluding to play of colors]

specularite   *meta.* Black to gray, crystallized metallic hematite with a splendent luster, often showing iridescence, and commonly occurring in micaceous or foliated masses. Also named specular schist or specularite schist. [L. *specularis,* mirror]

speleothem   *sed.* General term for any secondary mineral formation or deposit formed in caves by the action of water; may include carbonates (calcite, aragonite), sulfates, halides, and other mineral classes. [Gk. *spelaion,* cave + *thema,* deposit] (Moore in 1952)

speleothem, dead   *sed.* Speleothem that is dry and flaky; water flow and deposition have ceased.

speleothem, live   *sed.* Speleothem actively being deposited; usually shows a wet and glistening surface.

spergenite   *sed.* Biocalcarenite containing ooliths and fossil debris (such as bryozoan and foraminiferal fragments) and containing less than 10% quartz. [Spergen Hill, near Salem, Indiana] (Pettijohn in 1949)

sperone   *ig. extru.* Light-weight, porous, yellow to brown leucite rock, also containing garnet, yellow pyroxene, nepheline, hauyne, and biotite; from the Albano Hills, Italy. [It. *sperone,* abutment, buttress]

spessartite   *ig. diaschist.* Lamprophyre characterized by abundant mafic phenocrysts and rarer sodic-plagioclase phenocrysts in a holocrystalline groundmass containing plagioclase, orthoclase, and rare quartz. Varieties, depending upon the mafic, include: hornblende s., augite (or diopside) s., and olivine s. Not to be confused with the mineral spessartine, which has also been spelled spessartite. [Spessart Mountains, Germany] (Rosenbusch in 1896)

sphaerosiderite rock   *sed.* Siderite rock, often associated with coal measures, consisting of siderite spherulites in an interstitial and sometimes subordinate clay matrix; spherulites may be somewhat polygonal in shape and often consist of narrow to broad radial fibers of siderite. [Gk. *sphaira,* sphere + siderite] (Hausmann in 1813)

sphenitite   *ig. pluton.* Jacupirangite with more than 50% titanite (sphene); augite may be abundant and there are accessories, including the iron oxides. [sphene, synonym for titanite] (Allen in 1914)

spherite   *ig. extru., sed.* 1. Rock composed of gravel-sized (or larger) aggregates of constructional (nonclastic) origin, simulating in texture a rudite of clastic origin. Examples include rocks formed from an accumulation of volcanic bombs (agglomerate) or formed from an accumulation of concretions.   2. Individual spherical grain in a sedimentary rock, such as an oolith, pisolith, or spherulite. [sphere]

spherophyre   *ig.* General term for an igneous rock in which the phenocrysts are aggregates of crystals in the form of spherulites. [spherulite + porphyry]

spherulite   *ig., sed., meta.* Rounded or spherical mass of crystalline materials, often radiating from a central point and concentric in development. [spherule]

spherulite, igneous   *ig. extru., pluton.* Spherulite varying from microscopic to several feet across and having a radial fibrous structure composed of feldspar (usually potassic) and quartz that occurs in volcanic glass (usually rhyolitic); spherical masses also occur in phaneritic igneous rocks, e.g., spherical orbicules with a concentric development in orbicular granite, orbicular gabbro (corsite), and the like.

spherulite, sedimentary   *sed.* Spherical mineral nodules with a radial structure that have formed in place (in contrast to ooliths and pisoliths). Examples include barite, celestite, siderite, pyrite, marcasite, and the like.

spherulite, metamorphic   *meta.* Spherical to ellipsoidal masses, similar to igneous orbicular forms, occurring in some metamorphic rocks; commonly up to a few inches across and illustrated by quartz-sillimanite orbs in micaceous sillimanite schists and leptites.

spicularite   *sed.* Earthy or consolidated sediment composed chiefly of the siliceous spicules of invertebrates, esp. one composed of sponge spicules (spongolite). Also spelled spiculite. |L. *spiculum* from *spicum,* point, spike|

spicularite, arenaceous   *sed.* Porous, somewhat granular, gray-yellow rock transitional between a glauconitic sandstone cemented by opal and chalcedony, and spicularite derived from sponge spicules.

spiculite   *sed.* = spicularite

spilite   *ig., meta.* Altered basalt in which the feldspar has been albitized and is typically accompanied by chlorite, calcite, epidote, chalcedony, prehnite, and other minerals characteristic of a greenstone; often occurs as submarine lava flows and may be vesicular or amygdaloidal and with pillow structures. [Gk. *spilos,* spot] (Brongniart in 1827)

spilosite   *meta.* Contact metamorphic rock formed when shales or slates are intruded by diabase; adinole is similar but of a higher grade of metamorphism. |Gk. *spilos,* spot] (Zincken in 1841)

spinellite   *ig. pluton.* Igneous rock containing a preponderance of spinel, e.g., picotitfels. [spinel] (Johannsen in 1938)

spongolite   *sed.* Spicularite composed principally of the remains of siliceous sponge spicules. [sponge]

sporadosiderite   *meteor.* Stony meteorite containing disseminated iron. [Gk. *sporados,* scattered + *sideros,* iron] (Daubrée in 1867)

sporbo   *sed.* Local term, used in the San Joaquin Valley, California, for phosphatic ooliths. [smooth + polished + round + black, blue, or brown + object] (Horton, pub. by Galliher in 1931)

staffelite   *sed.* Massive variety of apatite occurring in botryoidal, reniform, or stalactitic masses incrusting the phorphorite of Staffel, Germany. [Staffel, Germany] (Stein in 1866)

stagmalite   *sed.* General term to include both stalactite and stalagmite cave formations (speleothems) produced by dripping water. [Gk. *stagma,* drop] (Farrington in 1901)

stalactiflat   *sed.* Composite stalactite-flowstone speleothem formed when a stalactite stops growing vertically and merges with a flowstone coating over cave fill; cave fill is later removed, leaving the stalactiflat form. [stalactite + flat]

stalactite   *sed.* Vertically hanging speleothem formed by dripping water; generally having a tube or the remnant of a tube at its center; shapes may be tubular or conical. [Gk. *stalaktos, dripping*]

stalactite, volcanic   *ig. extru.* Conical form of lava hanging from the roof or walls of a lava tunnel or similar cavity and developed by the dripping of fluid lava.

stalactostalagmite   *sed.* = column. [stalactite + stalagmite]

stalagmite   *sed.* Conical or cylindrical cave floor or ledge speleothem developed upward by the action of dripping water. [Gk. *stalagmos,* dripping]

stalagmite, volcanic   *ig. extru.* Conical formation of lava built up from the floor of a lava tunnel or similar cavity and with a form corresponding to a cave stalagmite.

stantienite   *sed.* Black, opaque, dull resin occurring with Prussian amber, esp. near Yantarnyy (Palmnicken), and having a very high oxygen content (23%). Also named black amber or black resin (or in German, Schwarzharz). [Whelhelm Stantien, innkeeper and later merchant, who with M. Becker dredged for amber in the Baltic in the last half of the 19th century (located in Königsberg, now Kaliningrad, Soviet Union] (Pieszczek in 1881)

stassfurtite   *sed.* Massive boracite sometimes with a subcolumnar structure and resembling fine-grained, white marble or granular limestone. [Stassfurt, Germany] (Rose in 1856)

statobiolith   *sed.* Biolith, usually limestone, composed mainly of the remains of attached (sessile) reef- or shoal-building organisms in their positions of growth. [Gk. *statos,* standing + biolith]

staurotile   *meta.* Old name for staurolite-rich mica schist that also may contain garnets. [staurolite] (Cordier in 1868)

stavrite   *ig.? meta.?* Coarse-grained dike rock, which may be igneous or perhaps metamorphic in origin, composed of essential tremolite (grammatite) and biotite, with accessory quartz, ilmenite, and apatite. [Stavre, Alnö Sound, Norrland, Sweden] (von Eckermann in 1928)

steargillite   *sed.* White, yellow, or pistachio green, subtranslucent, massive variety of montmorillonite rock; easily cut into cakes that look like soap or wax; from near Virolet and at the tunnel of

Poitiers, France. [L. *steatites,* soapstone + *argilla,* clay] (Meillet in 1862)

steatite   *meta.* = soapstone. [L. *steatites,* soapstone]

stellarite   *sed.* Variety of albertite occurring with bituminous coal at Stellarton, Nova Scotia. Also named stellar coal or oil coal. [Stellarton, Nova Scotia] (How in 1869)

stictolite   *migmat.* Migmatite with a spotted appearance. Also spelled stictolith. [Gk. *stiktos,* dotted]

stigmite   *ig. extru.* Obsolete name for pitchstone. [Gk. *stigma,* mark]

stigmite   *sed.* Variety of chalcedony (carnelian or agate) with decided markings. [Gk. *stigma,* mark]

stillolite   *sed.* = siliceous sinter. [L. *stillare,* to drip; alluding to its mode of formation]

stinkstone   *sed.* Sedimentary rock that emits an odor on being struck, broken, or rubbed, e.g., bituminous limestone or dolomite that gives a fetid odor (from the decomposition of organic matter). Also named stinkstein (from German). (Jameson in 1804)

stipite   *sed.* Variety of lean coal rich in pyrite, from the Lias formation in France. [Gk. *stiptos,* hard, close pressed] (Cordier in 1868)

stomach stone   *sed.* = gastrolith

stone   *ig., meta., sed., meteor.* 1. Loose entity (larger than a sand grain) made up of one or more minerals or rock materials and having its loose aspect as a consequence of natural processes. (Dietrich in 1980) 2. One of the larger fragments (clasts) in a sedimentary rock, as in a conglomerate or breccia. 3. General term for any rock used in construction or decoration, either crushed for use as aggregate or cut into shaped blocks.   4. Synonym for stony meteorite. [O.Eng. *stan*]

stone bubble   *ig. extru.* = lithophysa

stonegall   *sed.* Clay-rich concretion found in certain sandstones. [O.F. *galle,* from L. *galla,* gall nut]

stony meteorite   *meteor.* Meteorite consisting entirely or largely of silicate minerals, e.g., chondrite and achondrite; an abundant type constituting more than 90% of the meteorites seen to fall. Also referred to as a stone.

stony-iron meteorite   *meteor.* Meteorite representing the transition between iron meteorites and stony meteorites, and containing appreciable amounts of both components. Specific examples are represented by pallasite and siderophyre. Also referred to as stony-iron.

stromatite   *migmat.* Migmatite with a layered structure so intricately united that the whole, rather than the individual layers, constitutes the geological field unit. [Gk. *stroma,* bed]

stromatolite   *sed.* Laminated, calcareous sediment owing its origin to blue-green algae, which, because of their mucilaginous and filamentous nature, trap and bind particles together; comprising thick limestone or dolomite sequences in the geological record, particularly in the Precambrian. Also named stromatolith and algal stromatolite. [Gk. *stroma,* bed] (Kalkowsky in 1908)

stromatolite, algal   *sed.* = stromatolite

stronalite   *meta.* Cataclastic biotite gneiss associated with diorite gneiss and kinzigite at the original locality at Strona. [Strona, Ivrea, western Alps, Italy]

strontianite rock   *sed.* Sedimentary rock composed primarily of the mineral strontianite; rocks of this type occur as masses, lenses, and layers, as much as 3 ft thick, in clay near Barstow, California.

strontium rock   *sed.* Sedimentary rock composed of, or at least containing appreciable amounts of, celestite or strontianite. Celestite is widespread in dolomite and limestone, where it forms geode crystals, nodules, veins, layers, and disseminations, and in sands, where it may form sand crystals; strontianite is less widespread but may form rock masses (see strontianite rock).

stubachite   *meta.* Altered pyroxene peridotite containing tremolite, talc, serpentine, magnetite, pyrite, and ferroan magnesite (breunnerite). [Stubachtale, Tyrol] (Weinschenk in 1891)

subaluminous rock   *ig.* Igneous rock whose chemistry meets the following requirement: $Al_2O_3/(K_2O + Na_2O)$ is approximately 1. In these rocks the mafic minerals are all nonaluminous, e.g., olivine, orthopyroxene, diopside. [L. *sub,* under, below + aluminous] (Shand in 1943)

subarkose   *sed.* 1. Feldspathic sandstone that does not have enough feldspar to be classed as an arkose; sandstone intermediate in composition between arkose and pure quartz sandstone. 2. Arkosic sandstone containing 75-95% quartz, less than 15% detrital clay matrix, and 5-25% unstable minerals in which the feldspar grains exceed the rock fragments in amount. (Pettijohn in 1954) [L. *sub,* under, below + arkose]

subarkose, lithic   *sed.* Sandstone composed of subequal amounts of feldspar and rock fragments; specifically, sandstone containing 10-25% feldspar, 10-25% rock fragments, and 50-80% quartz. Also named feldspathic sublitharenite.

subbentonite *meta.* = metabentonite. [L. *sub,* under, below + bentonite]

suberinite *sed.* Kind of provitrain in which the cellular structure is derived from corky material. [L. suber, cork + vitrain]

subgraphite *meta.* = meta-anthracite. [L. *sub,* under, below + graphite]

subgraywacke *sed.* Common type of sandstone, intermediate in composition between graywacke and quartz arenite, containing less than 75% quartz, less than 15% detrital clay matrix, and an abundance (more than 25%) of unstable materials (feldspar grains and rock fragments) in which the rock fragments (at least 15%) exceed the feldspar, and having voids or mineral cement (esp. carbonates) exceeding the amount of clay matrix. [L. *sub,* under, below + graywacke] (Pettijohn in 1957)

subgraywacke, feldspathic *sed.* Sandstone composed of quartz and subequal amounts of rock fragments of igneous and metamorphic derivation; specifically, with 10-25% feldspars and igneous rock fragments, 10-25% micas and metamorphic rock fragments, and 50-80% quartz. (Folk in 1954)

subgraywacke, quartzose *sed.* = protoquartzite

sublitharenite *sed.* Sandstone with too few rock fragments to be classed as a litharenite; sandstone intermediate in composition between litharenite and pure quartz sandstone; specifically, having 5-25% fine-grained rock fragments, 65-95% quartz, and less than 10% feldspar. [L. *sub,* under, below + litharenite] (McBride in 1963)

sublitharenite, feldspathic *sed.* = subarkose, lithic

sublithwacke *sed.* Wacke with 5-25% detrital rock particles; sublitharenite with over 15% matrix. [L. *sub,* under, below + Gk. *lithos,* stone + wacke] (Pettijohn and others in 1973)

subphyllarenite *sed.* Phyllarenite containing 3-25% rock fragments. [L. *sub,* under, below + phyllarenite]

subsoil *sed.* Layer of more or less decomposed and loose fragments of country rock lying between the soil and the bed rock in regions not covered by transported soils. [L. *sub,* under, below + soil]

succinite *sed.* Amber; esp. amber mined in East Prussia or recovered from the Baltic Sea yielding succinic acid when heated; not to be confused with the amber-colored variety of grossular garnet of the same name. [L. *succinum,* amber]

sudburite *ig. extru.* Fine-grained, amygdaloidal, gray basalt, consisting of bytownite, hypersthene, augite, and magnetite; the aphanitic equivalent of hyperite. [Sudbury district, Ontario, Canada] (Coleman in 1914)

suevite *impact.* Gray to yellow, glassy, trachyte-breccia impactite, associated with the Ries crater near Nördlingen, Schwaben, Germany. Also named Ries impactite. [L. *Suevi,* the Suevians, people of medieval Swabia (now Schwaben)] (Sauer in 1920)

sugar stone *ig., altered.* Fine-grained, pink datolite from the Michigan copper district.

sugilite rock *ig.?* Massive variety of manganoan sugilite that varies from opaque to translucent and occurs in gray purple or rich magenta colors; occurs in eastern Africa and is used as a gem and decorative stone.

suldenite *ig. hypab.* Hornblende andesite differing from ortlerite in having an andesitic rather than a micro-dioritic groundmass. [Suldenferner, southern Tyrol, Italy] (Stache and von John in 1879)

sulfur rock *sed.* Rock composed of native sulfur occurring with certain limestone, anhydrite, or gypsum deposits, where it may form layers or lenses of considerable thickness; the sulfur is considered to have been formed by the reduction of sulfates (gypsum, anhydrite) or of sulfides by hydrocarbons.

sumacoite *ig. extru.* Porphyritic trachyandesite with abundant phenocrysts of plagioclase (labradorite to andesine) and augite and rarer olivine, in a groundmass of augite prisms, small plates of andesine, orthoclase, nepheline, hauyne, and abundant magnetite. [Sumaco volcano, eastern Ecuador] (Johannsen in 1938)

sunstone *sed.* Seldom used synonym for amber; not to be confused with sunstone feldspar (a mineral). [alluding to a Gk. myth in which the daughters of the Sun were changed into poplars and wept amber]

superanthracite *meta.* = meta-anthracite. [L. *super,* over, above + anthracite]

sussexite *ig. hypab.* Porphyritic tinguaite with large phenocrysts of nepheline in a groundmass composed of nepheline with abundant aegirine needles, orthoclase, perovskite, and traces of biotite. [Sussex County, New Jersey] (Brögger in 1894)

sussexite, natron *ig. hypab.* Sodium-enriched, sussexite-like rock from Penikkavaara, Kuusamo, Finland, with a $K_2O:Na_2O$ ratio of 1:11.42, whereas normal sussexite has a ratio of 1:2.86. [natrium (sodium)]

sussexite-tinguaite *ig. hypab.* Based on chemical analyses, hypabyssal rock intermediate between sussexite and tinguaite in which the SiO_2 percentage is 50-53. (Brögger in 1894)

sviatonossite   *ig. pluton.* Garnet-bearing syenite composed of orthoclase perthite (45%), albite (25%), aegirine-augite (17%), andradite (9%), and minor accessories. [Sviatoy Noss, Transbaikalia, eastern Siberia, Soviet Union] (Eskola in 1921)

swinestone   *sed.* = anthraconite. A variety of stinkstone. [G. *Schweinstein,* swinestone; for its fetid odor] (Kirwan in 1794)

syenide   *ig. pluton.* General field term for any holocrystalline, medium- to coarse-grained igneous rock containing one or more feldspars, no important quartz, and generally biotite or hornblende; dark minerals comprise less than half of the rock. Term includes syenite and the light-colored diorites. [syenite]

syenite   *ig. pluton.* 1. Phaneritic, holocrystalline igneous rock consisting of alkali feldspar (usually orthoclase, microcline, or perthite), less sodic-plagioclase, and usually some mafic mineral, e.g., biotite or hornblende; quartz is absent or is present only as an accessory. Trachyte is the aphanitic equivalent. [Syene, near modern Aswan, Egypt] (Pliny, *Natural History,* for granite from Syene [see Granite, syene]; name later applied to quartz-free syenite from Dresden by Werner in 1788)   2. *IUGS,* plutonic rock with 0-5 Q, and 10-35 P/(A + P).   3. Ancient name used by Pliny the Elder for granite from Syene (Aswan), Egypt. (See Granite, syene).

syenite, alkali   *ig. pluton.* Syenite typically rich in sodium, in which the potassic feldspars are strongly perthitic or anorthoclase, the plagioclase is albite or sodic-oligoclase, biotite is iron-rich, the amphiboles are hastingsite, arfvedsonite, or riebeckite, the pyroxenes may include aegirine-augite and aegirine, and there may be accessory feldspathoids. (Rosenbusch in 1907)

syenite, alkali-feldspar   *ig. pluton.* IUGS, plutonic rock with 0-5 Q, and P/(A + P) less than 10.

syenite, alkali-feldspar-quartz   *ig. pluton.* IUGS, plutonic rock with 5-20 Q and P/(A + P) less than 10.

syenite, alkali-lime   *ig. pluton.* = syenite, calc-alkali

syenite, analcime   *ig. pluton.* Medium- to coarse-grained, light gray rock, composed of abundant microperthite, rare oligoclase, considerable analcime in small rounded grains (imbedded in feldspar or interstitial to it), subhedral to anhedral dark-green hornblende, and accessories; from Pleasant Mountain, Maine. (Hibsch in 1902)

syenite, augite   *ig. pluton.* Syenite whose chief mafic mineral is augite. (vom Rath in 1875)

syenite, biotite   *ig. pluton.* Syenite whose chief mafic is biotite, usually present only in moderate amounts.

syenite, calc-alkali   *ig. pluton.* General term for the more widespread syenite (normal syenite) composed of perthitic orthoclase or microcline, oligoclase (rarely andesine), biotite, green hornblende, diopside or diopsidic augite, with accessory quartz; used in contrast with alkali syenite, which is more sodic. (Rosenbusch in 1907)

syenite, cancrinite   *ig. pluton.* Foid syenite in which primary cancrinite is an essential constituent (about 12%) and occurs with microperthite (83%) and minor accessories (aegirine, biotite, iron oxides, calcite, apatite); originally from near Särna, Sweden. Also named särnaite. (Törnebohm in 1883)

syenite, catapleiite   *ig. hypab.* Porphyritic rock related to tinguaite with phenocrysts of catapleiite and more rarely eudialyte, in an aphanitic, but holocrystalline, groundmass composed of those minerals with alkali feldspar, nepheline, and aegirine. (Törnebohm in 1906)

syenite, corundum   *ig. pluton.* Medium-grained, light yellow or pale rose-colored syenite, containing crystals of corundum in a groundmass of microperthite with accessory quartz and biotite; originally described from the Urals, Soviet Union. (Morozewicz in 1898)

syenite, elaeolite   *ig. pluton.* = syenite, nepheline. Also spelled eleolite syenite. [Old name for nepheline from Gk. *elaion,* oil; alluding to the mineral's greasy luster]

syenite, eudialyte   *ig. pluton.* Nepheline syenite containing eudialyte as an essential constituent. (Vrba in 1874)

syenite, feldspathoidal   *ig. pluton.* = syenite, foid

syenite, ferro   *ig. pluton.* General term applied to syenites rich in ferrous-iron silicate minerals, e.g., biotite, hedenbergite, pigeonite, fayalite, and others.

syenite, foidal   *ig. pluton.* = syenite, foid. [feldspathoidal]

syenite, foid   *ig. pluton.* 1. Group of phaneritic, holocrystalline feldspathoidal rocks showing a great diversity of mineral compositions, e.g., alkali feldspar varies from about 35% to 80%, feldspathoids vary from 10% to 45%, sodic-plagioclase varies from less than 5% to 45%, and mafics (iron-rich biotite, sodic amphiboles, sodic pyroxenes) vary from 10% to 65%. Examples include nepheline syenite, cancrinite syenite, sodalite syenite, miascite, husebyite, and others.

[feldspathoid] 2. *IUGS*, plutonic rock with 10-60 F and P/(A + P) less than 10.

syenite, hornblende *ig. pluton.* Syenite containing primary hornblende as its principal mafic mineral; quartz or biotite may be accessories.

syenite, leucite *ig. pluton.* Foid syenite in which feldspar and leucite are essential; in most examples the leucite has been altered to a mixture of minerals (pseudoleucite). (Williams in 1891)

syenite, nepheline *ig. pluton.* Foid syenite composed of granular aggregates of alkali feldspar (orthoclase, microcline, microperthite, cryptoperthite, or rarely albite), nepheline, and usually sodium-rich mafics (arfvedsonite, hastingsite, aegirine-augite, aegirine, titanaugite), and accessory feldspathoids (cancrinite, sodalite, hauyne, nosean); phonolite is the aphaitic extrusive equivalent. Also named elaeolite syenite. (Rosenbusch in 1878)

syenite, nosean *ig. pluton.* Foid syenite in which nosean is an essential constituent accompanied by orthoclase (or other alkali feldspar), and lesser aegirine, biotite, apatite, titanite, iron oxides, and possibly analcime; original specimens from Highwood Mountains, Montana. Also named noselite syenite. (Pirsson in 1905)

syenite, pyroxene *ig., meta.?* In the charnockite series of rocks, a quartz-poor member having more microperthite than plagioclase; a mangeritelike syenite. (Tobi in 1971)

syenite, pyroxene-alkali *ig., meta.?* Charnockite with less than 20% quartz and characterized by the presence of microperthite. (Tobi in 1971)

syenite, quartz *ig. pluton.* 1. Calc-alkali syenite with at least 5%, but no more than 10% (some petrologists, 20%) quartz. 2. *IUGS*, plutonic rock with 5-20 Q and 10-35 P/(A + P).

syenite, sodalite *ig. pluton.* Foid syenite in which sodalite is an essential constituent, although it is often accompanied by nepheline; original specimens from Julianehaab, Greenland. (Steenstrup in 1881)

syenite, zircon *ig. pluton.* Zircon-bearing nepheline syenite; originally noted in southern Norway. (Buch in 1810)

syenitite *ig.* Obsolete term for which there were two definitions: syenite aplite (Polenov in 1899) and plagioclase-bearing syenite (Loewinson-Lessing in 1900). [syenite]

syenitoid *ig. pluton.* IUGS, preliminary field term for a plutonic rock with Q less than 20 or F less than 10, and P/(A + P) less than 65.

syenodiorite *ig. pluton.* Rock intermediate between syenite and diorite; one carrying both orthoclase and plagioclase but more of the latter than of the former; nearly a synonym for monzonite, but it also includes rocks intermediate between monzonite and diorite. [syenite + diorite] (Johannsen in 1937)

syenodiorite, sodaclase *ig. pluton.* Syenodiorite in which the plagioclase is albite (sodaclase). (Johannsen in 1937)

syenogabbro *ig. pluton.* Rock differing from normal gabbro in that it contains some orthoclase in addition to normal calcic-plagioclase and mafic minerals. [syenite + gabbro] (Johannsen in 1917)

syenoid *ig. pluton.* Term proposed for foid syenite, which was never widely used. [syenite + feldspathoid] (Shand in 1910)

sylvinhalite *sed.* = halite-sylvite rock.

sylvinite *sed.* = sylvite-halite rock. [sylvite] (pub. by Hintze in 1911)

sylvite-halite rock *sed.* Evaporite rock in which sylvite dominates over halite; sylvite is commonly colored red by hematite, whereas halite is colorless; there are complete gradations to halite rock (rock salt). Also named halitosylvine and sylvinite.

symmictite *ig. extru.* Homogenized eruptive breccia composed of a mixture of country rock and intrusive rock. [Gk. *syn*, together + *miktos*, mixed] (Sederholm in 1924)

symmictite *sed.* = diamictite. [Gk. *syn*, together + *miktos*, mixed]

symplectite *ig.* Intimate secondary intergrowth of two minerals, one a single-crystal host, the other a guest appearing in vermicular, plumose, or micrographic units which are crystallographically continuous. Myrmekite is an example. [Gk. *syn*, together + *plektos*, twisted] (Loewinson-Lessing in 1897)

synnyrite *ig. pluton.* Nepheline syenite containing fine-grained intergrowths of alkali feldspar with kalsilite. [Synnyr, northern Baikal region, Soviet Union] (Zhidkov in 1962)

syntectite *ig., migmat.* Rock resulting from syntexis, the melting of two or more rock types and assimilation of country rock by a magma. [Gk. *syn*, together + *tektos*, molten]

syssiderite *meteor.* = stony-iron meteorite. [Gk. *syn*, together + *sideros*, iron] (Daubrée in 1867)

systyl *meta.* = jasper, basalt. [Gk. *systylos*, from *syn*, together + *stylos*, column; possibly alluding to a contact-metamorphic association] (Zimmerman in 1822)

T

tabbyite   *sed.* Local name for asphaltite occurring in veins in Tabby Canyon, Utah. [Tabby Canyon] (Bradwell in 1913)

tachylyte   *ig. extru.* Black, brown, or green basalt glass resembling obsidian in appearance but chemically poor in silica and close to the gabbro-basalt family; the glass is filled with crystallites; essentially the same as hyalobasalt. [Gk. *tachus,* rapid, swift + *lutos,* dissolving; alluding to the rapidity with which it dissolves in acids] (Breithaupt in 1826)

taconite   *meta.* Fine-grained rock containing quartz (chert, jasper), the hydrous iron silicate minerals (greenalite, minnesotaite, stilpnomelane), magnetite with or without hematite, and usually some siderite; some are banded and very finely laminated, while others have a spotted or mottled appearance. Important varieties include: minnesotaite taconite, stilpnomelane taconite, jaspilite. Also spelled taconyte. [Taconic system (after Taconic range) introduced by Ebenezer Emmons in 1842] (Winchell in 1892)

tactite   *meta.* General term for rocks formed by the contact metamorphism of limestone, dolomite, and other carbonate rocks and into which foreign matter from the magma intrusion has been introduced as hot solutions; usually having complex mineral compositions, including such species as andradite, hedenbergite, epidote, scapolite, vesuvianite, and wollastonite; skarn is nearly synonymous. [L. *tactus,* touch] (Hess in 1919)

tactite, ribboned   *meta.* = wrigglite. (Jahns in 1944)

tadjerite   *meteor.* Black, semiglassy, chondritic stony meteorite composed of clinohypersthene and olivine, with troilite, metal, chromite, ilmenite, and traces of oligoclase. [Tadjera, Algeria]

tahitite   *ig. extru.* Trachyandesite containing phenocrysts of hauyne in a groundmass that is glass (analcime?) with augite, titanomagnetite, hauyne, and occasionally orthoclase and leucite; apparently contains more sodic-plagioclase than orthoclase, but this is uncertain. [Papenoo, Tahiti] (Lacroix in 1917)

taimyrite   *ig. extru.* Light-colored, friable, sandstonelike phonolite with essential anortho-clase and nosean and accessory sanidine, plagioclase, brown hornblende, biotite, melanite, titanite, and more. [Taimyr River, northern Siberia, Soviet Union] (Chrustschoff in 1894)

taiwanite   *ig. extru.* Basaltic rock, composed chiefly of glass (partly altered to palagonite) and minor crystals of olivine and plagioclase. [Eastern Coastal Range, Taiwan] (Juan, Tai, and Chang in 1953)

talc rock, gneissoid   *meta.* = besimaudite

talc rock, massive   *meta.* = soapstone

talcite   *meta.* 1. Massive, scaly talc or soapstone. [talc] (Kirwan in ca. 1784)   2. Old synonym for the variety of hydromica (a mineral) once named damourite. (Thomson in 1836)

talpatate   *sed.* Surficial rock crust formed by the cementing action of calcium carbonate on sand, soil, or volcanic ash; essentially the same as caliche. [Sp., from Aztec *tepetatl,* stone matting]

talus   *sed.* Sloping heap of coarse rock-waste fragments lying at the foot of a cliff or steep slope; nearly equivalent to scree (common British term). [L. *talus,* ankle]

tamaraite   *ig. hypab.* Dark-colored lamprophyric dike rock, composed of augite, hornblende, biotite, nepheline, plagioclase (labradorite ?), orthoclase, and minor accessories; nepheline-rich camptonite. [Island of Tamara, Los Archipelago, Guinea] (Lacroix in 1918)

tangawaite   *meta.* Serpentinite, similar to bowenite, occurring in New Zealand; although softer, it resembles nephrite jade in appearance. Also spelled tangiwaite and tangiwai; also named kawakawa. [New Zealand Maori, *tangiwai,* tear water; native term alluding to the appearance of polished specimens to drops of water] (Koechlin in 1911)

tannbuschite   *ig. extru.* Dark-colored nepheline basalt in which the major minerals are nepheline, pyroxene, and olivine. [Tannbusch, Mittelgebirge, Bohemia] (Johannsen in 1938)

tar   *sed.* Thick, viscous, brown to black asphaltic residue resulting from the partial evaporation or distillation of petroleum or other hydrocarbons of asphaltic base or of wood; composition is variable. [O.Eng. *teru,* tar] tarantulite   *ig. diaschist.* Transitional rock between alaskite and silexite, composed of over 50% quartz and less alkali feldspar (over half of which is orthoclase and the remainder albite) with less than 5% mafics. [Tarantula Spring, Nevada] (Johannsen in 1919)

taraspite   *meta.* Mottled, fine-grained dolomite marble, used for decorative purposes. [Tarasp, Switzerland] (von John in 1891)

tartufite   *sed.* Fetid, fibrous calcite rock from Monte Viale, Venetia, Italy; when struck it emits an odor like that of truffles. An old term used by Venetian mineralogists. [It. *tartufo,* truffle] (pub. by Catullo in 1812)

tasmanite   *ig. pluton.* Zeolite-rich rock that would be melilite ijolite if the zeolites replaced nepheline; contains nepheline, hydronepheline, natrolite, thomsonite, phillipsite, melilite (8%), pyroxene (32%), iron oxides, and others. [Shannon Tier, Tasmania] (Johannsen in 1938)

tasmanite   *sed.* Reddish brown, impure coal transitional between cannel coal and oil shale. [Tasmania] (Milligan in 1852)

taspinite   *ig. pluton.* Local name for granite porphyry that occurs in the Rofna Valley, Upper Rhine, Switzerland. Also named Rofna-porphyry. [Alp Taspin, Canton Graubünden, Switzerland] (Ruetschi in 1903)

taurite   *ig. extru.* Sodium-rich rhyolite, composed of anorthoclase, quartz, orthoclase, aegirine-augite, and sodium hornblende; original rock from Ouraga, Alouchta, Crimea. [L. *Tauri,* from Gk. *Tauroi,* ancient people from southern Crimea] (Lagorio in 1897)

tautirite   *ig. extru.* Nepheline trachyandesite or hawaiite, composed of potassic feldspar, andesine, nepheline, brown hornblende, and accessory titanite. [Tautira, Tahiti] (Iddings in 1918)

tautirite, leucite   *ig. extru.* = columbretite. (Tröger in 1935)

tavite   *ig. pluton.* = tawite

tavolatite   *ig. extru.* Phonolite with large phenocrysts of leucite, hauyne, aegirine-augite, and garnet in a microporphyritic groundmass of the same minerals with orthoclase, labradorite, biotite, and nepheline. [Osteria di Tavolato, Via Appia Nuova, near Rome] (Washington in 1906)

tawite   *ig. pluton.* Similar to ijolite, but sodalite substitutes for nepheline; contains sodalite (63%), aegirine (37%), with very minor nepheline, cancrinite, eudialyte, and secondary natrolite. Also spelled tavite. [Tawajok valley, Kola Peninsula, Soviet Union] (Ramsay in 1894)

tawmawite   *meta.* Massive, chrome-rich variety of epidote, yellow to emerald green in color, from Tawmaw, Upper Burma; jadeite (Tawmaw jade) occurs in the same area. [Tawmaw] (Bleeck in 1907)

taxite   *ig. extru.* Volcanic rock of clastic appearance, formed from the consolidation and aggregation of more than one kind of product from the same flow. Varieties include eutaxite and ataxite. [Gk. *taxis,* arrangement, order] (Loewinson-Lessing in 1891)

taxoite   *meta.* Local name for a green serpentinite from Chester County, Pennsylvania.

taylorite   *sed.* Obsolete early name for bentonite. [William Taylor, Rock Creek, Wyoming, who made the first shipments of bentonite in 1888] (Knight in 1897)

tcheremkhite   *sed.* = cheremchite. Also spelled tscheremkhite.

tear, Apache   *ig. extru.* See Apache tear

tear, Pele's   *ig. extru.* See Pele's tears

tecali   *sed.* Onyx marble (Mexican onyx) occurring at Tecali, Mexico. Also spelled tecalco and tecati. [Tecali, Mexico]

tectite   *tekt.* = tektite

tectonite   *meta.* 1. Mylonitic rock formed from crystalline schists of sedimentary origin that in part have again recrystallized. (Backlund in 1918) 2. Any rock whose fabric reflects the history of its deformation; often displayed are coordinated geometric features that indicate continuous solid flow during formation. Also spelled tektonite. [Gk. *tekton,* carpenter; often used in reference to earth structure]

tectonite, primary   *sed.* Tectonite whose fabric is depositional (usually sedimentary); most tectonites are of secondary origin as a result of deformation.

tektite   *tekt.* Black, greenish, or yellowish, silica-rich (70-80% SiO_2) glass that occurs as small, rounded to elongated, somewhat pitted bodies, in areas that usually preclude a volcanic origin; these were once thought to be of extraterrestrial origin (e.g., lunar) but are now considered to be a product of large hypervelocity meteorite impacts on terrestrial rocks (impactites). Also spelled tectite. [Gk. *tektos,* molten] (Suess in 1900)

telegdite   *sed.* Fossil resin containing sulfur and no succinic acid; occurs in Transylvania. [Karoly Telegdi-Roth, 1886-   ?, of the Geological Survey of Hungary] (Zechmeister and Vrabely in 1927)

tepetate   *sed., ig. extru.* 1. Calcareous crust coating solid rocks on or just beneath the surface in an arid or semiarid region. 2. General term used in Mexico for a volcanic tuff, or a secondary volcanic or chemical nonmarine deposit, that is usually calcareous. [Sp. from Aztec *tepetatl,* stone matting]

tephra *ig. extru.* 1. Term applied to any material, regardless of size, ejected from volcanoes; may be derived from the magma itself or may include fragments from earlier eruptions or may include older country rock; general term for all pyroclastics. 2. *IUGS,* collective term for pyroclastic deposits that are predominantly unconsolidated. [Gk. *tephra,* ashes] (Pliny, *Natural History*)

tephrite *ig. extru.* Extrusive igneous rock characterized by the mineral combination of calcicplagioclase and feldspathoid (usually leucite or nepheline); if olivine is also present the rock is named basanite; the extrusive equivalent of theralite. Variations in feldspathoids have given the following varieties: hauyne t., leucite t., nepheline t., nosean t. [Gk. *tephra,* ashes] (von Fritsch in 1865)

tephrite, phillipsite *ig. extru.* Glassy tephrite altered to phillipsite through later magmatic or hydrothermal activity. (Hibsch in 1927)

tephrite, phonolithoid *ig. extru.* = ventrallite. (Rosenbusch in 1908)

tephritoid *ig. extru.* Seldom used name for a rock intermediate between olivine-free basalt and tephrite. [tephrite] (Bücking in 1880)

teratolite *sed.* Impure clay rock with a varied coloration, showing lavender and other shades of blue, and spots of red, and rarely, pearl gray; composed of kaolinite, iron oxides, quartz, and others, and occurs at Planitz, near Zwickau, Saxony. [Gk. *teratos,* wonder, sign; related to old G. name *Wundererde,* wonderearth] (Glocker in 1839)

terpitzite *sed.* Siliceous sinter grading into chalcedony or hornstone, occurring in crevices in porphyry. [Terpitz, Saxony] (Dürr in 1828)

terra rossa *sed.* Red ferruginous earth formed as a residual product during the subaerial denudation of limestone; typical of the karst areas around the Adriatic Sea, under conditions of Mediterranean-type climate. Also spelled terra rosa. [It. *terra rossa,* red earth]

teschenite *ig. pluton.* Medium- to rather dark-colored, granular rocks, consisting of calcicplagioclase, augite, sometimes hornblende, and a little biotite, and interstitial analcime. Preferred name is analcime theralite. [Teschen, Silesia, Czechoslovakia] (Hohenegger in 1861)

Texas tektite *tekt.* = bediasite

tezontli *ig. extru.* Old term used in southwestern United States for scoriaceous basalt; in modern local usage, term refers to almost any rock other than granite or marble. Also spelled tezontle, tezoncle, tesoncle, and tesontle. [Aztec, *tetl,* rock, *tzontle,* hair, or *zonnectic,* something spongy]

thanite *sed.* Evaporite material consisting of a mixture of kainite and halite, occurring in the Werra district, Prussia. [Károly Than, 1834–1908, chemist of Budapest, Hungary] (Rózsa in 1914)

thelotite *sed.* Undetermined carbonaceous constituent of boghead coal from the Thélots pits at Autun, France. [Thélots pits] (Bertrand and Renault in 1892)

Tektite (bediasite). Texas. 4.3 cm.

theralite   *ig. pluton.* 1. Phaneritic foid gabbro composed of essential calcic-plagioclase and feldspathoid (usually nepheline) with considerable pyroxene and accessory biotite, hornblende, and sodalite group minerals; olivine and magnetite may occur also; tephrite is an extrusive equivalent. [Gk. *theran,* eagerly looked for] (Rosenbusch in 1887) 2. *IUGS,* plutonic rock with 10–60 F, and P/(A + P) greater than 90; foid gabbro is a synonym.

thermite   *sed.* General name for fossil combustible substances, e.g., the various coals. [Gk. *thermos,* warm, hot] (Wadsworth in 1891)

thermuticle   *meta.* = porcellanite. [dim. of Gk. *thermos,* warm, hot]

thinolite   *sed.* = tufa, thinolitic. [Gk. *thinos,* shore, beach] (King in 1878)

thioelaterite   *sed.* Elastic bitumen, containing 3% sulfur, from Bolivia; natural vulcanized elaterite. [Gk. *theion,* sulfur + elaterite] (Dunicz in 1936)

tholeiite   *ig. extru.* Silica-oversaturated (quartz-normative) basalt, composed of calcic-plagioclase, clinopyroxene, and low-calcium pyroxenes (orthopyroxene or pigeonite); olivine and nepheline are absent. Also named subalkaline basalt. [Schaumberg, near Tholei, Rheinland, Germany] (Steininger in 1840)

tholeiite, abyssal   *ig. extru.* = tholeiite, oceanic. [Gk. *a,* without + *byssos,* bottom]

tholeiite, oceanic   *ig. extru.* The principal igneous rock encrusting the deeper parts of the ocean basins; believed to be the parent material for alkaline basalt. Also named abyssal tholeiite.

tholeiite, olivine   *ig. extru.* Basalt that is silica-undersaturated, containing normative olivine, hypersthene, and diopside, with neither quartz nor nepheline. (Yoder and Tilley in 1962)

thomsonite rock   *ig., altered.* Compact and radial masses of thomsonite sometimes occurring in amygdaloidal cavities in altered basaltic rocks. Some of the variety names given to these thomsonite rocks are: comptonite, lintonite, ozarkite.

thrombolite   *sed.* Algal structure like a stromatolite but with an obscurely clotted, rather than laminated, internal structure. [Gk. *thrombos,* clot, lump] (Aitken in 1967)

thucholite   *ig. ?* Brittle, jet-black, carbonaceous material whose ash contains variable amounts of thorium and uranium, from a Canadian pegmatite. [Th + U + C + H + O] (Ellsworth in 1928)

thunder egg   *ig. extru.* Popular term for geodelike, agate-filled bodies that weather out of welded tuffs, esp. in central Oregon; these rounded bodies were originally gas-filled igneous spherulites formed by crystallization in hot plastic glassy ash; pressure exerted by the collection of gases during crystallization caused expansion of the spherulites by rupture along somewhat symmetrically arranged planes, and later the resulting cavities were filled by chalcedony (agate). [thunder, for a myth of Indians in Oregon that the gods on Mt. Hood and Mt. Jefferson hurled the stones at one another when they became angry, and these episodes were accompanied by thunder]

thuresite   *ig. pluton.* Syenite composed of sodium microcline (with albite), and sodium hornblende, with very minor quartz, biotite, apatite, and others. [Thures, on the Thaya above Raabs, Austria] (Waldmann in 1935)

Tibet stone   *meta.?* Term applied to various colors and mixtures of aventurine quartz and quartz porphyry found in Russia; used for ornamental or curio stones; a variety is eosite.

tilaite   *ig. pluton.* Transitional rock between gabbro and peridotite, consisting of green diopside and olivine and a small quantity of calcic-plagioclase (labradorite to bytownite). [Tilai-Kamen, Koswa region, northern Urals, Soviet Union] (Duparc and Pamfil in 1910)

tilestone   *sed.* Term used in England for a flagstone (flaggy sandstone) used for roofing. [tile]

till   *sed.* Unsorted and unstratified glacial drift, consisting of materials deposited by and underneath the ice, with little or no transportation by water; generally an unconsolidated, heterogeneous mixture of clay, sand, gravel, and boulders. [An early English term, related to thill, origin is uncertain] (Used as early as 1842 by Darwin)

till, ablation   *sed.* Loosely consolidated till, formerly in or on a glacier, that accumulated in place as the surface ice was removed by ablation (melting, evaporation, wind erosion). [L. *ablatus,* carried away]

till, basal   *sed.* Firm clay-rich till containing many abraded rocks, dragged along at the base of a moving glacier and deposited upon bedrock or other glacial deposits. [base]

till, berg   *sed.* 1. Lacustrine or marine clay containing boulders and rocks dropped into it by melting icebergs. 2. Glacial till deposited intact by grounded icebergs in fresh or saline water bordering an ice sheet.

till, gumbo   *sed.* 1. Dark-colored (often grayish), leached, deoxidized clay, developed from profoundly weathered clay-rich till under conditions of low relief and poor subsurface drainage;

composed chiefly of illite or beidellite and may also contain decomposed rock fragments. Also spelled gumbotil. [Louisiana F. *gombo,* from Bantu] (Kay in 1916)   2. Fossilized soil beneath a deposit of later till.

tillite   *sed.* Glacial till consolidated or indurated, esp. pre-Pleistocene till. [till] (Penck in 1906)

tilloid   *sed.* Rock resembling tillite in appearance whose origin is in doubt or unknown but is probably nonglacial, e.g., conglomeratic mudstone resulting from mud flows or slides on the margin of a geosyncline. [till] (Blackwelder in 1931)

tillstone   *sed.* Boulder or other individual rock in a glacial till deposit. Also written till stone.

timazite   *ig. extru.* Greenstonelike, altered hornblende-biotite andesite, containing corroded phenocrysts of hornblende and biotite. [Timok Valley, Serbia, Yugoslavia] (Breithaupt in 1861)

tinguaite   *ig. hypab.* Granular to porphyritic, greenish dike rock, characterized by a combination of alkali feldspar and nepheline, with aegirine or aegirine-augite; like a phonolite but occurring as dikes (hypabyssal). [Tinguá Mountains, near Rio de Janeiro, Brazil] (Rosenbusch in 1887)

tinguaite, analcime   *ig. hypab.* Light gray, finely crystalline, dike rock, consisting of microperthite, analcime (feldspar exceeds analcime), accessory biotite and augite but no nepheline; from Essex County, Massachusetts. (Clapp in 1921)

tinguaite, biotite   *ig. hypab.* Tinguaite rich in biotite or altered biotite, where the mineral is replaced by secondary magnetite. (Eakle in 1898)

tinguaite, cancrinite   *ig. hypab.* Tinguaite containing primary cancrinite in addition to minor nepheline, alkali feldspar, and sodium pyroxenes.

tinguaite, hauyne   *ig. hypab.* Similar to tinguaite but with hauyne substituting for nepheline.

tinguaite, leucite   *ig. hypab.* Dike rock of the composition of tinguaite but containing leucite in addition to nepheline. (Wolff in 1902)

tinguaite, nosean   *ig. hypab.* Similar to tinguaite but with nosean substituting for nepheline.

tinguaite, sodalite   *ig. hypab.* Similar to tinguaite but with sodalite substituting for nepheline; consists of sanidine (60%), sodalite (27%), aegirine-augite (12%). (Hibsch in 1910)

tinstone veins   *ig. pluton., autometa.* Vein rock closely associated with greisens in granite and characterized by cassiterite (named tinstone) associated with topaz, white micas, fluorite, quartz, and brown tourmaline. [tinstone, synonym for cassiterite]

tirilite   *ig. pluton.* Alkali granite composed of microcline perthite (50%), andesine (19%), quartz (18%), with hornblende, pyroxene, biotite (lepidomelane), and others; from Simola, near Wiborg, southern Finland. [Possibly Tiri, Finland] (Wahl in 1925)

tiza   *sed.* 1. Mixtures of ulexite, sand, and chlorides and sulfates of sodium and calcium, esp. on the dry plains of Iquique, Tarapaca, Chile.   2. Term also used for ground chalk and for finely divided gypsum. [Sp. *tiza,* chalk, clay]

tjosite   *ig. hypab.* Porphyritic, nepheline-syenite lamprophyre, verging toward jacupirangite; consists of dominant augite with abundant magnetite, ilmenite, and apatite, with some biotite and a little olivine, in a matrix of anorthoclase and nepheline. [Tjose parish, North Larvik, Norway] (Brögger in 1906)

toadstone   *ig. extru.* Amygdaloidal basalt from Derbyshire, England, that resembles the skin of a toad.

toadstone   *sed.* Fossilized object, like a tooth or bone, thought to have formed within a toad and frequently worn as a charm or an antidote to poison.

toellite   *ig. aschist.* = töllite

toensbergite   *ig. pluton.* Reddish, feldspar-rich rock whose composition is nearly identical to larvikite, except that the feldspar contains a little more calcium. Also spelled tönsbergite. [Tönsberg, Norway] (Brögger in 1899)

toienite   *ig. diaschist.* = windsorite. (Brögger in 1931)

Tokay lux sapphire   *ig. extru.* Brownish black obsidian from Hungary. [Tokay (Tokaj) on Tisza River, northeastern Hungary + lux (F. *luxe,* superfine quality) + sapphire]

tokeite   *ig. extru.* Very dark olivine basalt with phenocrysts of augite, olivine, and magnetite, in a groundmass of magnetite, augite, labradorite, biotite, and no glass. [Arête de Toké, Gouder Valley, Ethiopia] (Duparc and Molly in 1928)

töllite   *ig. aschist.* Garnet-bearing quartz diorite porphyry, with a greenish gray groundmass of quartz and feldspar and phenocrysts of hornblende, andesine, and some biotite, white mica, quartz, orthoclase, and garnet. Also spelled toellite. [Töll, near Meran, Tyrol] (Pichler in 1875)

tolypite   *meta.* Chlorite ball made up of irregularly arranged fibers, from Saxon Vogtland. [Gk. *tolupe,* ball of thread] (Uhlemann in 1909)

tomite   *sed.* = coal, boghead. [Tom River, Tomsk, Siberia, Soviet Union] (Zalessky in 1915) tonalite *ig. pluton.* 1. Variously defined term usually synonymous with quartz diorite or hornblende-biotite quartz diorite; usually consists of essential quartz and andesine with accessory orthoclase and mafics. [Monte Tonale, Tyrol, Italy] (vom Rath in 1864) 2. *IUGS,* plutonic rock with 20-60 Q and P/(A + P) greater than 90.

tonalite, orbicular *ig. pluton.* Tonalite in which there are abundant orbicles; one from Hankasalmi in central Finland contains small spheroids (6-8 cm across) consisting of feldspar and biotite, which make up 66% of the volume of the rock.

tonalite, quartz *ig. pluton.* Quartz-rich quartz diorite in which over 50% of the light-colored components are quartz. (Johannsen in 1919)

tonalite, sodaclase *ig. pluton.* Rare tonalite (quartz diorite) composed of quartz, albite (sodaclase), and 5-50% mafics; resembles granite but contains only a trace of potassic feldspar. (Johannsen in 1932)

tönsbergite *ig. pluton.* = toensbergite

tonstein *sed.* Rock composed of kaolinite with occasional detrital and carbonaceous material, commonly occurring as thin bands in coal seams. [G. *Ton,* clay + *Stein,* stone]

topasfels *ig. diaschist., autometa.* = topazite. [G. *Topas,* topaz]

topaz rock *ig. diaschist., autometa.* 1. = topazite. 2. High-temperature, metasomatic rock in which topaz is an important component; some are composed almost exclusively of cryptocrystalline to microcrystalline topaz (Jefferson, South Carolina), while others have abundant topaz with muscovite, corundum, rutile, magnetite, kyanite, and fluorite (India); essentially the same as topazite.

topazite *ig. diaschist., autometa.* Dike rock composed nearly exclusively of quartz and topaz; mode of formation is similar to that of greisen. (Johannsen in 1919)

topazogene *ig. diaschist., autometa.* = topazite. (von Charpentier in 1778)

topazoseme *ig. diaschist., autometa.* = topazite. (Haüy in 1822)

tophus *sed.* = tufa. [L. *tophus,* tufa]

topsailite *ig. hypab.* Lamprophyre intermediate between camptonite and kersantite, containing phenocrysts of plagioclase (andesine-labradorite), augite, apatite, and ilmenite, in a groundmass composed of andesine, biotite, barkevikite, augite, and titanite. [Topsail Point, Tamara Island, the Los Islands, Guinea] (Lacroix in 1911)

torbanite *sed.* Substance closely related to cannel coal, but with a very uniform composition, occurring with bathvillite at Torbane Hill, on the grounds of Bathville, Scotland. [Torbane Hill, Scotland] (Greg and Lettsom in 1858)

tordrillite *ig. extru.* Light-colored, albite-bearing rhyolite; generally equivalent to phaneritic alaskite. [Tordrillo Mountains, Alaska] (Spurr in 1900)

toryhillite *ig. pluton.* Albite-rich urtite, composed of nepheline (51%), albite (19%), pyroxene, garnet, iron oxides, apatite, and calcite, with no potassic feldspar. [Toryhill, Monmouth Township, Ontario] (Johannsen in 1919)

tosca *sed.* 1. White deposit of calcium carbonate, occurring in the loess on the open treeless plains (pampas) of southern Chile and Argentina. 2. Soft coral limestone, esp. in Puerto Rico. 3. In Mexico, term applied to several materials including clayey vein matter, talc seams, and soft, decomposed porphyry. [Sp. feminine of *tosco,* rough, coarse, unpolished]

toscanite *ig. extru.* Dellenite in which the plagioclase, which accompanies orthoclase, is calcic; the original rock also contained hypersthene, biotite, and much glass (67%). [Vivo, Tuscany (Toscana), Italy] (Washington in 1897)

touchstone *sed.* = basanite. [touch, for use in streak test for precious metals + stone]

tourmaline rock *ig. diaschist., autometa.* = tourmalite

tourmaline-corundum rock *meta.* Very hard, fine-grained, blue black rock composed of tourmaline and corundum; the oolitic structure (seen under the microscope) indicates it was probably formed by metamorphism of an oolitic chert by granite; found in Kinta, Malay States. (Scrivenor in 1910)

tourmalinite *ig. diaschist., autometa.* = tourmalite. [tourmaline]

tourmalite *ig. diaschist., autometa.* Mottled, black and white, coarse- to fine-grained, rarely schistose, rock, consisting of white to gray quartz and black or brown tourmaline, and sometimes minor dark mica, feldspar, cassiterite, and topaz; mode of formation is similar to that of topazite and greisen. [tourmaline] (Johannsen in 1919)

trachite *ig. extru.* = trachyte

trachorheite *ig. extru.* Collective name for the four rocks trachyte, rhyolite, andesite, and propylite. [Possibly trachyte + rhyolite] (Endlich in 1873)

trachyandesite *ig. extru.* General term for a rock intermediate between trachyte and andesite, usually containing potassic feldspar and sodic-plagioclase in about equal amounts; nearly synonymous with latite as well as rocks between latite and andesite. [trachyte + andesite] (Michel-Lévy in 1894)

trachybasalt   *ig. extru.* Variously defined term, usually applied to an extrusive rock intermediate in composition between trachyte and basalt, characterized by the presence of both calcic-plagioclase and potassic feldspar, along with clinopyroxene, olivine, and possibly minor analcime or leucite; generally synonymous with latite as well as rocks between latite and basalt, including trachyandesite. [trachyte + basalt] (Bořický in 1873)

trachydolerite   *ig. extru.* = trachybasalt. [trachyte + dolerite] (Rosenbusch in 1908)

trachyte   *ig. extru.* The extrusive, aphanitic, often porphyritic, equivalent of syenite, composed mainly of alkali feldspar and minor mafic minerals (biotite, hornblende, or pyroxene). As there are two principal groups of syenite, there are likewise two groups of trachyte: alkali trachyte and calc-alkali trachyte. See alkali syenite and calc-alkali syenite for the mineralogical differences. [Gk. *trachys,* rough; from its rough appearance] (Haüy, pub. by Brongniart in 1813)

trachyte, acmite   *ig. extru.* Aegirine-bearing alkali trachyte in which acmite, the sharp-pointed variety of aegirine, is an important accessory. (Mügge in 1883)

trachyte, aegirine   *ig. extru.* Alkali trachyte consisting of sodium sanidine, anorthoclase, aegirine, and aegirine-augite. (Rosenbusch in 1896)

trachyte, alkali   *ig. extru.* Sodium-rich trachyte whose composition is like alkali syenite. See alkali syenite for a discussion of the mineralogy.

trachyte, analcime   *ig. extru.* = shackanite

trachyte, anorthoclase   *ig. extru.* Alkali trachyte in which the major feldspar is anorthoclase. (Skeats and Summers in 1912)

trachyte, arfvedsonite   *ig. extru.* Alkali trachyte in which arfvedsonite (amphibole) is the major mafic.

trachyte, Arso-type   *ig. extru.* = arsoite. (Rosenbusch in 1923)

trachyte, calc-alkali   *ig. extru.* General term for the more widespread trachyte (normal trachyte) whose composition is like calc-alkali syenite. See calc-alkali syenite for a discussion of the mineralogy.

trachyte, calcite   *ig. extru.* Variety of trachyte

containing over 10% calcite, which is apparently primary. (Washington in 1917)

trachyte, Drachenfels   *ig. extru.* = drakonite

trachyte, feldspathoidal   *ig. extru.* = trachyte, foid

trachyte, foid   *ig. extru.* Group of extrusive feldspathoid trachytes resembling phonolite but containing, instead of nepheline (essential to phonolite), a different feldspathoid associated with alkali feldspar (usually sodium sanidine) and mafics. Important types include the following: analcime t., hauyne t., leucite t., nosean t., sodalite t. Also named feldspathoidal trachyte and foidal trachyte. [feldspathoidal]

trachyte, foidal   *ig. extru.* = trachyte, foid

trachyte, katophorite   *ig. extru.* Alkali trachyte consisting of sodium sanidine and anorthoclase (83%), katophorite (an amphibole) (17%), with minor aegirine, apatite, glass, and others; from Naivasha Lake, Kenya, eastern Africa.

trachyte, leuco-sodaclase   *ig. extru.* Light-colored trachyte containing, in addition to the normal potassic feldspar, some albite (sodaclase). (Johannsen in 1937)

trachyte, phonolitic   *ig. extru.* Extrusive rock closely related to trachyte, characterized by the presence of small amounts of the typical minerals of phonolite (alkali feldspars, nepheline, other feldspathoids, sodium pyroxenes, sodium amphiboles) and the total absence of quartz; differs from phonolite in containing less nepheline; the extrusive equivalent of alkali syenite or pulaskite. (Rosenbusch in 1908)

trachyte, Ponza   *ig. extru.* Group of rocks similar to drakonite but differing in the more or less complete absence of biotite and amphibole among the phenocrysts and with augite mantled by aegirine-augite or aegirine, and augite-magnetite pseudomorphs after biotite; there are both feldspathoid-free and feldspathoid-bearing members of this group; ponzite is a feldspathoid-free member of this group. Also named ponzaite. [Ponza Islands, west of Naples, Italy] (Rosenbusch in 1896)

trachyte, quartz   *ig. extru.* Obsolete name for rhyolite. (Zirkel in 1866)

trachyte, riebeckite   *ig. extru.* Alkali trachyte containing the sodium-rich amphibole riebeckite. (McRobert in 1914)

trachyte, tridymite   *ig. extru.* Normal-type trachyte carrying tridymite as cavity fillings or saturating the groundmass. (Kolenko in 1884)

tractionite   *sed.* Deposit of well-bedded, winnowed, clastic sediments of sand size or larger, made by moving wind or water. [L. *tractus,* drawn] (Natland in 1976)

trainite　*sed.* Originally described as impure banded variscite but later shown to be a mixture of vashegyite with a mineral close to natrolite (laubanite); from Manhattan, Nevada. [Percy Train, fl. 1917, Manhattan, Nevada] (pub. by Schaller in 1918)

trap　*ig., meta.* General field name for any dark, fine-grained, crystalline igneous rock, whether fresh or altered, such as basalt, diabase, fine-grained gabbro, peridotite, andesite, lamprophyre, greenstone, epidiabase, and others; also applied to any such rock used as crushed stone. Also spelled trapp; synonyms include trap rock and trappide. [Sw. *trappa,* stair, step; alluding to stairstep appearance of some of the rocks in the field]

trap, mica　*ig. dike.* English field name for dark, dike rocks rich in mica, esp. with biotite.

trap, white　*meta.* Intrusive igneous rock, usually of mafic composition, that has been "bleached" at the contact with coal or other carbonaceous rock, by the conversion of its ferromagnesian and feldspar minerals into a mixture of light-colored carbonates and clay minerals; term esp. used in the Midland Valley of Scotland.

trass　*ig. extru.* General term for trachytic tuffs, often nepheline-bearing, widely distributed in central Italy and in the Eifel, Germany, and used in the preparation of a hydraulic cement. Numerous spellings include tarrace, tarras, terrace, terras, tarasse, and tiras. [G. dial., *Trass*]

traversoite　*altered ore?* Mixture of chrysocolla and gibbsite found as bright blue, amorphous masses at Arenas, Sardinia. [Giovanni Battista Traverso, 1843-1914, Italian mining engineer of Genoa] (D'Ambrosio in 1924)

travertine　*sed.* Porous, fairly dense, white, tan or cream-colored and banded deposit of $CaCO_3$, esp. common in caverns and also formed by evaporation of spring and river waters; often used as a decorative stone. [L. *lapis Tiburtinus,* stone formed by the water of Anio at Tibur (now Tivoli), Italy, where there are extensive deposits] (pub. by von Buch in 1809)

tremenheerite　*meta.?* Impure graphite rock from India. [George Borlase Tremenheere, 1809-1896, English military figure who traveled in India] (Piddington in 1847)

trevalganite　*ig. pluton.* Variety of granite in which large phenocrysts of pink feldspar or quartz occur in a groundmass of schorl rock. [Possibly Trevalga, England]

trinacrite　*ig. extru., weathered.* Dull brown, somewhat cleavable or micaceous, variety of palagonite tuff. [L. *Trinacria,* Sicily, the locality of the material] (Waltershausen in 1846)

trinkerite　*sed.* Compact, greasy, hyacinth-red to chestnut-brown, oxygenated hydrocarbon containing sulfur; occurs as large, compact masses in brown coal at Carpano near Albona in Istria, Italy. [Joseph Trinker, fl. 1853, Austrian geologist of Tyrol, from whom it was received] (Tschermak in 1870)

tripestone　*sed.* 1. Concretionary variety of anhydrite composed of contorted plates suggesting pieces of tripe.　2. Stalactite resembling intestines. [tripe]

tripoli　*sed.* 1. Light-colored, porous, friable, siliceous earthy rock resulting from the weathering of chert or of siliceous limestone; composed largely of chalcedony, with a harsh, rough feel and used for polishing. Rottenstone is nearly the same material.　2. Term originally, but now incorrectly, applied to other siliceous earths such as diatomaceous earth and radiolarian earth. [Tripoli, Libya, where it was found] (Wallerius in 1747)

tripolite　*sed.* General term for a siliceous earthy rock, having been applied as a synonym for diatomite and less commonly as a synonym for tripoli.

tristanite　*ig. extru.* Silica-saturated to undersaturated rock intermediate between trachyandesite and trachyte, with a differentiation index of 65-75 and K_2O:Na_2O greater than 1:2. [Tristan da Cunha, island in the southern Atlantic] (Tilley and Muir in 1964)

troctolite　*ig. pluton.* 1. Gabbro composed of calcic-plagioclase and olivine and in which pyroxene is practically absent. Also named troutstone and forellenstein (German for troutstone). [Gk. *troktes,* sea fish (taken as meaning trout); alluding to speckled trout-skin appearance] (von Lasaulx in 1875)　2. *IUGS,* plutonic rock satisfying the definition of gabbro, in which pl/(pl + px + ol) is 10-90, and px/(pl + px + ol) is less than 5.

trona rock　*sed.* Massive trona-rich rock occurring in gray to yellowish-white fibrous or columnar layers, and thick beds, in saline evaporite deposits, esp. in northern Africa, Venezuela, and in Nevada and California.

trondhjemite　*ig. pluton.* Quartz diorite whose essential constituents are sodic-plagioclase (oligoclase or andesine) and quartz; there are small quantities of biotite, and potassic feldspar is absent or subordinate. Also spelled trondjemite and trondheimite. [Trondhjem, Norway] (Gold-schmidt in 1916)

trowlesworthite   *ig. pluton., autometa.* Rather coarsely crystalline rock consisting of red orthoclase, tourmaline, and fluorite, with occasional quartz; apparently formed by the auto-metamorphic alteration of granite along fissures by gases. [Trowlesworthy, Devonshire, England] (Worth in 1887)

truffite   *sed.* Large, nodular, woody masses of lignite, occurring within Cretaceous lignite at Pont-Saint-Esprit, department of Gard, France. [F. *truffe,* truffle; alluding to the odor when struck] (Dumas in 1876)

tscheremkhite   *sed.* = cheremchite

tsingtauite   *ig. aschist.* Granite porphyry containing only phenocrysts of feldspar (microperthite and lesser plagioclase), with quartz confined to the groundmass. [Tsingtau, Shantung, China] (Rinne in 1904)

tufa   *sed.* Spongy, porous rock, composed of calcium carbonate (calcite or aragonite), that forms a thin surficial deposit about springs and rivers; has a reticulated structure and is weak, semi-friable, and of limited extent; travertine is related but is more massive and less porous. Also named calcareous tufa, calctufa, tophus, calcareous sinter, and calc-sinter. [It. *tufo,* soft, sandy stone]

tufa, calcareous   *sed.* = tufa

tufa, dendritic   *sed.* Gray tufa occurring in spherical to hemispherical bolsterlike bodies with a coarse radial structure and occurring as sizable mounds or domes in many places along ancient lake shores (e.g., extinct Lake Lahontan, Nevada). [Gk. *dendron,* tree; for treelike structures]

tufa, dendroid   *sed.* = tufa, dendritic

tufa, lithoid   *sed.* Comblike coating of tufa on various other materials, esp. at extinct Lake Lahontan, Nevada. [Gk. *lithos,* stone]

tufa, reef   *sed.* Drusy, prismatic, fibrous calcite, deposited directly from supersaturated water upon the void-filling internal sediment of a calcite mudstone or a reef knoll.

tufa, thinolitic   *sed.* Tufa consisting in part of layers of interlaced delicate prismatic skeletal crystals of thinolite (calcite, perhaps pseudomorphic after gaylussite), up to 20 cm long and 1 cm thick; occurs esp. at extinct Lake Lahontan, Nevada. Also named thinolite rock.

tuff   *ig. extru.* 1. General term for consolidated pyroclastic rocks, esp. those whose fragments are less than 2 mm across. Several factors may be considered in naming tuffs: the source of the materials (recent or former eruption); the physical nature of the fragments (glass, rock fragments, crystals pieces); the igneous rock type (rhyolite, trachyte, andesite, basalt, and others); the nature of any sedimentary fraction contaminating the rock (argillaceous, arenaceous, calcareous). Numerous varieties are listed below. [French *tuf,* from It. *tufo,* soft, sandy stone] 2. *IUGS,* pyroclastic rock whose average pyroclast size is less than 2 mm. Two varieties include coarse tuff, 1/16 mm to 2 mm, and fine tuff (dust tuff), less than 1/16 mm. Also named ash tuff.

tuff, accessory   *ig. extru.* Tuff containing earlier

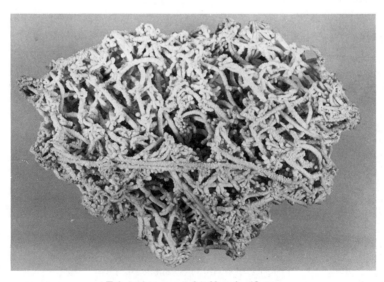

Tufa (calcareous tufa). Nevada. 18 cm.

clasts and lava from the same source as the newer materials.

tuff, accidental *ig. extru.* Tuff containing rock clasts unrelated to the source of the newer materials, e.g., strata or wall-rock through which the volcanic conduit extended.

tuff, air-fall *ig. extru.* Tuff consisting of tephra blown into the atmosphere and deposited by gravitational settling of the individual particles; often shows the effects of eolian differentiation, the largest and densest particles accumulating closer to the parent volcano.

tuff, andesite *ig. extru.* Tuff with andesite composition; although less common than rhyolite tuff, glassy, stony, and crystal tuffs of this composition are known.

tuff, aquagene *ig. extru.* = hyaloclastite

tuff, ash *ig. extru.* = tuff

tuff, ash-fall *ig. extru.* General term for tuffs simply formed through the fallout of volcanic fragments ejected from a vent or fissure; usually subaerial.

tuff, ash-flow *ig. extru.* Tuff formed by pyroclastic flow mechanisms, e.g., ignimbrites, base-surge tuffs, and lahar deposits.

tuff, basalt *ig. extru.* Lapilli, ash, and other fine fragmental materials of basalt composition, cemented to form a rock; fragments may be glass, crystals, or rock pieces.

tuff, base-surge *ig. extru.* Tuff formed where subaerial flows come into contact with water; there is a fast-moving, turbulent mixture of pyroclastic particles, water, and gas (steam); stratification, both planar and cross bedding, is a distinguishing feature.

tuff, cindery lapilli *ig. extru.* Tuff composed of volcanic cinders (vesicular) whose grain size is 4-32 mm.

tuff, coarse *ig. extru.* Tuff composed of coarse ash, defined as having a grain size 0.25-4 mm in diameter.

tuff, crystal *ig. extru.* Tuff in which crystals and crystal fragments are predominant; in a rhyolite these could be crystals of quartz, potassic feldspar, biotite, and hornblende; some limit the term to rocks containing more than 75% crystals by volume. (Cohen in 1871)

tuff, crystal-lithic *ig. extru.* Tuff intermediate between crystal tuff and lithic tuff or one that is predominantly the latter.

tuff, crystal-vitric *ig. extru.* Tuff consisting of crystals and crystal fragments with clasts of volcanic glass; some limit the term to rocks containing 50-75% crystals by volume.

tuff, devitrified *ig. extru.* Tuff that has lost its glassy characteristics; after a long span of time, apparently aided by the influence of heat and pressure, vitric tuffs, like other glassy rocks, may become devitrified and pass into a crystalline condition; in the widespread rhyolitic types, the volcanic glass yields microcrystalline quartz and feldspars.

tuff, dust *ig. extru.* 1. Tuff composed of indurated volcanic dust; essentially equivalent to fine tuff. 2. *IUGS,* pyroclastic rock whose average pyroclast size is less than 1/16 mm. Also named fine ash tuff.

tuff, explosion *ig. extru.* Tuff in which the components have been dropped directly into place after being ejected from a volcanic vent; in contrast to the more ordinary types, which are washed into place. (Green in 1919)

tuff, fine *ig. extru.* Tuff composed of pyroclastic ash or dust having a diameter less than 1/4 mm.

tuff, flood *ig. extru.* = ignimbrite

tuff, glassy *ig. extru.* = tuff, vitric

tuff, hybrid *ig. extru., sed.* Tuff composed of pyroclasts that have been eroded, retransported, and redeposited by running water and are mixed with sedimentary materials like detrital quartz, other minerals, and clays; may grade into tuffaceous sandstones and tuffaceous clays. Varieties include arenaceous tuff and argillaceous tuff. Tuffite is nearly synonymous.

tuff, kalitrachyte *ig. extru.* Tuff composed of kalitrachyte clasts; very rare, but some types of trass are tuffs that approach this composition.

tuff, keratophyre *ig. extru.* Tuff whose composition is like that of keratophyre; relatively common type.

tuff, lapilli *ig. extru.* 1. Tuff composed of fragments of lapilli size, considered to be 4-32 mm by some petrologists. 2. *IUGS,* pyroclastic rock whose average pyroclast size is 2-64 mm.

tuff, lithic *ig. extru.* Tuff in which the most conspicuous elements are fragments of rocks; the rock fragments may be frothy, glassy, fine-grained, porphyritic, or phaneritic and may even consist of fragments of sedimentary and metamorphic rocks from the conduit of the volcano. (Pirsson in 1915)

tuff, lithic-crystal *ig. extru.* Tuff intermediate between lithic tuff and crystal tuff but predominantly the latter.

tuff, palagonite   *ig. extru.* Tuff consisting of fragments of hydrothermally altered or weathered basaltic glass (palagonite).

tuff, rhyolite   *ig. extru.* General term for tuff that has the composition of a rhyolite and can be vitric, crystalline, or composed of rock fragments (lithic).

tuff, sand   *ig. extru., sed.*   1. = sandstone, tuffaceous.   2. Obsolete name for tuff whose component fragments are in the size range of sand.

tuff, scoria   *ig. extru.* Rock composed of fragmented scoria in a fine-grained tuff matrix.

tuff, sedimentary   *ig. extru., sed.* = tuffite

tuff, sideromelane   *ig. extru.* Vitric tuff composed of solid pieces of the glassy basalt named sideromelane (the same as hyalobasalt); the most abundant of these have been altered to palagonite tuff while they were still hot and in contact with water vapor.

tuff, silicified   *ig. extru., altered.* Alteration or diagenetic changes may produce a tuff high in silica (opal, chalcedony, microgranular quartz) and resembling sedimentary chert or porcellanite; relict glass shards are often visible in these rocks.

tuff, stony   *ig. extru.* = tuff, lithic

tuff, tephrite   *ig. extru.* Tuff consisting of the feldspathoid-bearing rock tephrite.

tuff, vitric   *ig. extru.* Tuff composed mainly of glass particles, often consisting of shards of glass and very fine-grained clasts; the most abundant are those derived from rhyolitic and dacitic magmas, less frequently those from trachytic or andesitic magmas, and still less frequently those from basaltic magma (sideromelane tuff).

tuff, volcanic   *ig. extru.* = tuff

tuff, welded   *ig. extru.* Glassy, usually banded or streaked, siliceous tuff, indurated by the welding together of its glass shards under the combined action of the heat retained by the particles, the weight of the overlying material, and hot gases.

tuff, zeolite   *ig. extru., altered.* Tuff altered to one or more zeolite minerals (e.g., chabazite, clinoptilolite, erionite, mordenite, or phillipsite) as well as to clays, calcite, quartz, and others. Included among the several varieties are clinoptilolite tuff and mordenite tuff.

tuffisite   *ig.* Term applied to fragmented country rock occurring in volcanic pipes located in Swabia, southwestern Germany. [tuff] (Cloos in 1941)

tuffite   *ig. extru., sed.* 1. Hybrid tuff consisting of both pyroclastic and detrital materials, usually with predominant pyroclasts. [tuff] (Mügge in 1893)   2. *IUGS,* hybrid rock consisting of mixtures of pyroclasts (less than 75% by volume) and epiclasts (greater than 25% by volume); here, epiclasts are crystals, crystal fragments, glass fragments, and rock fragments that have been liberated from any type of pre-existing rock, volcanic or nonvolcanic, by erosion.

tufflava   *ig. extru.* Rock containing both pyroclastic and lava-flow characteristics, so that it is considered to be an intermediate form between lava flow and a welded-tuff type of ignimbrite. Also spelled tuffolava, tuff lava, and tuflava. [tuff + lava]

tuffsite   *ig.* = tuffisite

tuffstone   *sed.* Sandstone containing pyroclasts of sand-grain size.

turbidite   *sed.* Sediment or sedimentary rock deposited by decelerating turbidity currents; a common type is deepwater sandstone, but also included are some evaporite rocks and other types. [L. *turbidus,* disturbance]

turjaite   *ig. pluton.* Gray, medium- to coarse-grained, feldspar-free rock, composed of about equal amounts of melilite and biotite, nepheline (17%), apatite, and small amounts of perovskite, ilmenite, magnetite, and calcite; nearly equivalent to extrusive melilite nephelinite. [Turja, Kola Peninsula, Soviet Union] (Ramsay in 1921)

turjite   *ig. hypab.* Rather variable hybrid rock, formed by the mixing of carbonate magma with monchiquite or alnoite, and composed of bright yellow mica (40%), analcime (20%), primary calcite (20%), melanite (18%), and minor aegirine, apatite, perovskite, and others. [Turja, Kola Peninsula, Soviet Union] (Beljankin in 1924)

Turkey stone   *sed.* 1. Very fine-grained siliceous novaculite, containing up to 25% calcite, quarried in Turkey and used as a whetstone.   2. Synonym for the mineral turquoise.

turmalinschiefer   *ig. diaschist., autometa.* Tourmalite with a schistose structure, resulting from the parallel arrangement of the tourmaline crystals. [G. *Turmalin,* tourmaline + *Schiefer,* schist, slate]

turrelite   *sed.* Tough, gray, bituminous coquina, found in Uvalde County, Texas. Also spelled turrellitè. [Possibly fossil turreted Turritella shells in rocks at the locality]

turtle stone   *sed.* Large, flattened, oval septarium whose system of cracks, exposed by weathering or sawing, resembles a turtle's shell.

tusculite   *ig. extru.* Variety of melilite-bearing leucitite containing only small amounts of pyroxene, ilmenite, and feldspar. [Tusculum, Italy] (Cordier in 1868) tutvetite   *ig. hypab.* Altered, light-colored, reddish trachyte, consisting predominantly of albite, microcline, and a much decomposed mafic that may have been aegirine; accessories include pyrite and possibly anatase and nordenskiöldine. [Tutvet, Hedrum, Norway] (Johannsen in 1938)

tuvinite   *ig. pluton.* Urtite consisting of calcite and often as much as 96% nepheline. [Balyktya-Khim River, Tuva Region, Soviet Union]

tuxtlite   *meta.* Jadelike rock composed of pyroxene in which the diopside and jadeite components are about equal in amount; may contain albite and form a series with rocks commonly named mayaite, the "jade" of the Mayas. [Tuxtla, Mexico] (Washington in 1922)

tveitåsite   *ig. pluton.* Dark, medium- to fine-grained, contact rock, composed chiefly of aegirine-diopside and alkali feldspar (orthoclase, cryptoperthite, microperthite, and albite), with titanite, apatite, pyroxene, and possibly with nepheline and calcite as accessories. [Tveitåsen, Fen region, Norway] (Brögger in 1921)

U

ugandite *ig. extru.* Dark-colored olivine leucitite, containing leucite, augite, and abundant olivine, in a sodium-rich glassy groundmass. [Muganza crater, southwestern Uganda, eastern Africa] (Holmes and Harwood in 1937)

uigite *meta.* White to yellow, banded rock, having a pearly sheen and composed primarily of prehnite. [Uig, Island of Skye, Scotland] (Heddle in 1856)

uintaite *sed.* Bright, lustrous, black variety of asphalt, having a good conchoidal fracture and fusing easily; from Uintah Valley, Utah. Also spelled uintahite and also named gilsonite. [Uintah County, Utah] (Blake in 1885)

ukrainite *ig. pluton.* Quartz-poor (about 11%) quartz monzonite composed of andesine (40%), orthoclase perthite (26%), and minor diallage, hornblende, apatite, and others. [Chermalyk, near Mariupol, Ukraine] (Bezborodko in 1935)

ulmite *sed.* Form of humus coating the grains of black, friable sandstone from New South Wales, Australia. [Possibly ulmin, a constituent of humus] (Steel in 1921)

ulrichite *ig. hypab.* Olivine-bearing tinguaite porphyry, containing phenocrysts of nepheline, alkali feldspar, sodic feldspar (anorthoclase), sodic pyroxene, and amphibole, and smaller accessory olivine phenocrysts in a groundmass of feldspar, analcime, pyroxene, and amphibole. [George Henry Ulrich, 1830-1900, German-Australian-New Zealand petrologist who first discovered alkaline rocks at Dunedin, New Zealand] (Marshall in 1906)

ultrabasic rock *ig.* Collective term applied to igneous rocks containing little or no feldspar but characterized by one or more of the common mafic minerals, such as olivine, pyroxene, hornblende, and others; chemically defined as having a percentage of silica less than that of anorthite, the limiting figure being about 45%; nearly synonymous with ultramafic rock. [L. *ultra,* beyond + basic, from concept that it is low in silicic acid] (Judd in 1881)

ultrafenite, alkaline *ig., meta.* In situ metasomatite that has undergone the most extensive series of mineralogical changes, e.g., the transformation to rocks consisting of combinations of alkali feldspar, sodic pyroxene, and nepheline, in which the original contrast between light and dark components of the ancestral migmatite can still be discerned. [L. *ultra,* beyond + fenite] (von Eckermann in 1948)

ultramafic rock *ig.* 1. Igneous rock composed chiefly of mafic minerals, e.g., rocks composed of olivine, pyroxene, or hornblende, without the presence of feldspar; essentially the same as ultrabasic rock. 2. *IUGS,* general name for plutonic rocks with color index M greater than or equal to 90, illustrated by dunite, peridotite, and pyroxenite. [L. *ultra,* beyond + mafic (magnesium + ferrum)]

ultramylonite *meta.* Variety of mylonite in which all primary structures have been entirely obliterated, so that the rock becomes homogeneous, aphanitic, and often vitreous, with little sign, if any, of parallel structure. Flinty crushrock is an example. [L. *ultra,* beyond + mylonite] (Quensel in 1916)

ultra-urtite *ig. pluton.* Urtite containing a very large amount of nepheline (as much as 96%), e.g., tuvinite, from the Tuva Region, Soviet Union. [L. *ultra,* beyond + urtite]

umber *sed.* Earth mainly composed of the hydrated oxides of iron and manganese and used in its natural state as a brown pigment. [F. *terre d'ombre,* earth from Umbria, Italy]

umptekite *ig. pluton.* Alkali syenite composed of potassic feldspar (orthoclase-microperthite, cryptoperthite, or microcline-microperthite) and sodic amphibole (arfvedsonite), with accessory aegirine, titanite, apatite, iron oxides, and small amounts of interstitial nepheline; similar to pulaskite but containing more sodium than potassium. [Umptek (Khibina), Kola Peninsula, Soviet Union] (Ramsay in 1894)

unakite *ig. pluton., altered.* Altered igneous rock consisting of pink orthoclase (including some microcline), white to smoky quartz, and much yellow-green epidote (secondary after an earlier feldspar or mafic mineral); minor constituents include zircon, apatite, magnetite, biotite, chlorite, and pumpellyite; common in the Blue Ridge Mountains of Virginia, the Great Smoky Mountains of North Carolina and Tennessee, and elsewhere. [Unaka Range, Great Smoky Mountains] (Bradley in 1874)

uncompahgrite   *ig. pluton.* Coarse-grained, feldspar-free, igneous rock, composed of melilite (72%), diopside (14%), iron oxides (10%), and small amounts of cancrinite, perovskite, phlogopite, melanite, calcite, and others. [Mt. Uncompahgre, Gunnison County, Colorado] (Larsen and Hunter in 1914)

undaturbidite   *sed.* Sediment formed from a suspension produced by violent storms; the deposit is intermediate between an ordinary wave deposit and a turbidite. [L. *unda,* wave + turbidite] (Rizzini and Passega in 1964)

underclay   *sed.* = seatearth

underearth   *sed.* 1. = seatearth.   2. The soil beneath the earth's surface.

ungaite   *ig. extru.* Glassy oligoclase-bearing dacite. [Unga Island, Aleutian Islands, Alaska] (Iddings in 1913)

unghwarite   *sed.* = chloropal. Also spelled unghvarite and ungvarite. [Ungvár, Hungary (now in Czechoslovakia)] (Glocker in 1837)

uralite   *ig., altered.* Amphibole formed from the alteration of pyroxene; usually pseudomorphic and consisting of an aggregation or bundle of slender amphibole prisms retaining the shape of the original pyroxene crystal, sometimes with the parent pyroxene remaining near the core. [Maisk, Ural Mountains, where it was first observed] (Rose in 1831)

uralitite   *ig. dike, altered.* Diabase whose augite

has completely altered to uralite. Also named uralite diabase. [uralite] (Kloss in 1885)

urbainite   *ig. pluton.* Rutile-rich ilmenitite, composed of rutile (23%), ilmenite (58%), hematite, sapphirine, ferroan spinel (pleonaste), andesine, and biotite. [St. Urbain, parish in Quebec, Canada] (Warren in 1912)

ureilite   *meteor.* = achondrite, olivine-pigeonite. [Novo Urei, Soviet Union] (Jerofejeff and Latschinoff in 1888)

urpethite   *sed.* Soft, waxy, brown hydrocarbon, found in Urpeth Colliery, County Durham, England. [Urpeth Colliery] (Johnston in 1832)

urtite   *ig. pluton.* 1. Light-colored member of the ijolite series, composed of more than 70% nepheline; other minerals may include aegirine, apatite, and calcite, and minor feldspars may occur but only as accessories. [Lujavr-Urt (now Lovozero), Kola Peninsula, Soviet Union] (Ramsay in 1896)   2. *IUGS,* plutonic rock in which F is 60-100, M is 30 or less, and sodium exceeds potassium.

usamerite   *sed.* Rock comparable to graywacke, characterized by size grades ranging from gravel to sand, by poor sorting with a substantial quantity of matrix, and by variable rock and mineral fragments that are predominantly angular to subangular. [United States of America] (Boswell in 1960)

utahlite   *sed.* Synonym for the mineral variscite, esp. for the large, compact, nodular masses found in Cedar Valley, Tooele County, Utah. [Utah] (Kunz in 1895)

V

vabanite   *sed.* Local name for brown red jasper with yellow flecks, from California.

vake   *sed.* Soft, compact, mixed claylike material that has a flat, even fracture and is commonly associated with basaltic rocks. [Fr. *vake,* wacke]

vakite   *sed.* Rock composed primarily of vake. [vake]

valaite   *sed.* Massive to crystalline, pitch black, aromatic resin occurring as thin crusts on dolomite and calcite in the Rossitz-Oslawaner coal formation, Moravia. Also spelled walaite. [Josef Vála, 1820-1881, Bohemian mineralogist and personal friend of W. Helmhacker, who described the resin] (Helmhacker in 1867)

valamite   *ig. extru.* Quartz and orthoclase-bearing hypersthene diabase, containing labradorite (46%), hypersthene (26%), quartz (9%), orthoclase (8%), with apatite and oxide minerals. [Valamo Island in the Ladoga Sea, Finland] (Hackman in 1931)

valbellite   *ig. pluton.* Black, fine-grained rock composed of bronzite, olivine, hornblende, and magnetite; weigelith in which bronzite proxies for enstatite and that contains a greater amount of iron-oxide minerals. [Valbella valley, Piemonte, Italy] (Schaefer in 1898)

valchovite   *sed.* = walchowite. [Valchov (Walchow), Czechoslovakia]

vallevarite   *ig. pluton.* Light-colored, potassium-rich andesinite, composed of andesine, microcline, and antiperthite, with small amounts of clinopyroxene (diopside), biotite, apatite, and iron oxides. [Vallevara, Sweden] (Gavelin in 1915)

valverdite   *ig. extru., weathered.* Rounded or lenticular, weathered, obsidian glass, containing crystalline inclusions; occurs near Del Rio, Texas. [Val Verde County, Texas]

variolite   *ig. extru.* Aphanitic, fine-grained, grayish-green basaltic igneous rock, containing varioles, small spherulites composed of radiating crystals of feldspar (plagioclase), with augite, and various alteration products; the groundmass apparently is devitrified glass. Also named pearl diabase. [L.L. *variola,* smallpox; from the appearance upon weathering] (Aldrovande in 1648)

varnish, desert   *sed.* Thin, dark, shiny film, composed of iron oxides accompanied by traces of manganese oxides and silica, formed on the surfaces of rock fragments and on rock outcrops in desert regions after long exposure; apparently caused by exudation of solutions from within and deposition by evaporation on the surface.

värnsingite   *ig. hypab.* Coarse-grained, light-colored, dike rock, consisting of albite (66%), augite (27%), and minor titanite, magnetite, apatite, and secondary epidote, prehnite, chlorite, amphibole, and muscovite. [West Värnsingen, Ulfo Island, Nordingrå district, Sweden] (Sobral in 1913)

varve   *sed.* 1. Sedimentary bed (lamina or sequence of laminae) deposited in a body of still water within one year's time, e.g., a unit of layers seasonally deposited, usually by meltwater streams, in a glacial lake or other body of still water, in front of a glacier (usually a coarse-grained summer layer and a very fine-grained winter layer). 2. Any cyclic sedimentary laminae, as in certain shales and evaporites. [Sw. *varv,* layer]

varvite   *sed.* Lithified equivalent of varve. Nearly synonymous with pelodite. [varve]

vase   *sed.* Freshwater silt, consisting of a mixture of sandy and pulverulent grains of quartz, calcite, clay minerals, and diatom shells, with a binder of algon, and occurring along the Atlantic coast of Europe and Africa. [Fr. *vase,* mud, slime, ooze]

vaughanite   *sed.* Pure, dense, homogeneous, fine-grained, dove-colored limestone that breaks with a smooth and more or less conchoidal fracture and that contains relatively few fossils; on weathering it becomes white and chalky. [Thomas Wayland Vaughan, 1870-1952, American paleontologist and geologist, Scripps Institution, California] (Kindle in 1923)

vaugnerite   *ig. diaschist.* Rough, coarse-textured, dark quartz diorite, consisting of biotite, hornblende, plagioclase, and quartz, with accessory orthoclase, apatite, magnetite, pyrite, and titanite. [Vaugneray, near Lyons, France] (Fournet in 1861) vein rock   *hydrotherm.* Mineral aggregates with tabular or lenticular shapes, formed by the filling of pre-existing fractures or fissures; deposits may have sharply defined walls or indefinite boundaries caused by metamorphic changes upon the intruded rock; veins vary in composition from nearly pure milky quartz to quartz with carbonate minerals (dolomite) to very complex assemblages including sulfide ores, native metals, and gangue minerals.

veinite   *migmat.* = venite

velikhovite   *sed.* Variety of pyrobitumen having a bright conchoidal fracture and occurring in veins; partly soluble in organic solvents and apparently a weathering product of albertite. [Velikhovka, Guberlin Mountains, southern Urals, Soviet Union] (Lopukhov in 1931)

venanzite   *ig. extru.* Porphyritic leucitite or melilitite containing olivine and phlogopite phenocrysts in a fine-grained, holocrystalline, groundmass of olivine, phlogopite, melilite, leucite, and magnetite. Also named euktolite and euktolith. [Venanzo, Umbria, Italy] (Sabatini in 1898)

vendeenite   *sed.* Fossil resin from the coal measures of Vendée, France. Also spelled vendeennite. [Vendée, France] (Rivière in 1840)

venite   *migmat.* Veined migmatite in which some of the fluid was formed by partial anatexis of the parent rock. Also spelled veinite. [L. *vena*, vein] (Holmquist in 1921)

ventifact   *sed.* General term for any rock shaped by the abrasive action of wind-blown sand. Varieties include dreikanter, einkanter, windkanter, and zweikanter. [L. *ventus*, wind + *facere*, to make] (Evans in 1911)

ventrallite   *ig. extru.* Phonolite carrying more potassic feldspar than calcic-plagioclase and having nepheline as the only essential feldspathoid; other minerals present include orthoclase, labradorite, augite, and biotite phenocrysts in a groundmass of the same minerals, except biotite, with additional nepheline and magnetite. Incorrectly spelled vetrallite. Also named phonolithoid tephrite. [Ventralla, Monte Vico, Ciminian district, Italy] (Johannsen in 1938)

venturaite   *sed.* Broad term for a great variety of gaseous, liquid, and solid bitumens containing hydrobenzoles, esters of pyridin bases and similar compounds; characterized as nitrogen bitumens. Also named California oils. [Ventura County, California] (Peckham in 1895)

verd antique   *meta.* Dark-green serpentinite, usually crisscrossed with white veinlets of calcite and magnesium carbonates; at times the rock is mottled with red due to unusual amounts of hematite (rosso di Levanto marble); well-known ornamental stone, often classed as marble in commerce. Also named verde antico (Italian). [F. *verd*, green + *antique*, ancient]

verde di Corsica duro   *ig. pluton.* Igneous rock composed of green diallage and labradorite, having a changing greenish color; used for vases, inlaying, and other ornamental purposes; from the Island of Corsica. [It. *verde*, green + *duro*, hard + Corsica]

verdite   *meta.?* Green, massive rock, composed of fine-grained, chromian muscovite (fuchsite) and other minor minerals. Transvaal, South Africa, is a major source. [F. *verd*, green] (pub. by Kunz in 1908)

verdolite   *meta.* Talcose dolomitic breccia, from New Jersey; used as a decorative stone. [F. *verd*, green]

verite   *ig. extru.* Black, pitchlike trachyte, containing many small phenocrysts of phlogopite and smaller ones of olivine, in a glassy groundmass that contains a second generation of phlogopite and nearly colorless diopside microlites; differs from fortunite, with which it is associated, in containing more glass and having olivine instead of orthopyroxene. [Vera, Cabo de Gata, Spain] (Osann in 1899)

vermilite   *sed.* Opaline rock colored red by cinnabar; variety of opalite. [vermilion]

veronite   *hydrotherm.* = earth, Verona. [Verona, Italy]

verrankohle   *sed.* Rolled fragments of brown coal occurring on the coast of Norway. [Verran, near Trondheim, Norway + Nor. *kohle*, coal] (Kjerulf in 1871)

vesbite   *ig. extru.* Transitional rock between melilitite and leucitite, composed essentially of leucite and pyroxene (augite) in a groundmass of melilite and alteration products. [Vesbius, L. variant name for Vesuvius, Italy] (Washington in 1920)

vesecite   *ig. hypab.* Lamprophyre composed of olivine, monticellite, melilite, phlogopite, and nepheline; type of polzenite with monticellite. [Vesec Svetla, Bohemia, Czechoslovakia] (Scheumann in 1912)

vesuvite   *ig. extru.* Variety of leucite tephrite with nearly 40% leucite accompanied by augite, calcic-plagioclase, and minor olivine, nepheline, and sodalite. [Vesuvius, Italy] (Lacroix in 1917)

vetrallite   *ig. extru.* = ventrallite

viandite   *sed.* = geyserite. [F. *viande*, meat, flesh; alluding to its appearance] (Goldsmith in 1883) vibetoite *ig. pluton.* Coarse-grained, biotite-hornblende pyroxenite, containing much calcite (14%) and apatite (9%); either an altered ultramafic rock or a hybrid product of carbonatite with ultramafic silicate rock affinities. [Vibeto farm, Fen complex, Norway] (Brögger in 1921)

vicoite *ig. extru.* Leucite tephrite rich in leucite (40%) and containing sodium sanidine (20%) and labradorite (20%), with augite and some olivine. [Vico volcano, near Viterbo, Italy] (Washington in 1906)

vierzonite *sed.* 1. = grossouvreite. (de Grossouvre in 1901) 2. Yellow ochreous clay from Vierzon, Cher, France. [Vierzon] (pub. by Bristow in 1861)

vintlite *ig. aschist.* Fine-grained quartz diorite porphyry, containing a few phenocrysts of quartz, much black hornblende, and oligoclase, in a dark, greenish-black groundmass, composed of platy feldspar, hornblende, diopside, and rare bronzite and biotite. [Vintl, near Klausen, Tyrol] (Pichler in 1871)

violite *sed.* Compact variety of purple chalcedony from San Diego County, California. Not to be confused with the trade name for purple synthetic sapphire or for a variety of copiapite, each of which have been given the same name. [violet] (Trenchard, fl. 1905)

viridite *altered.* General name for obscure green rock alteration products (including the chlorite and serpentine minerals) that cannot be, or have not yet been, specifically determined. [L. *viridis,* green] (Vogelsang in 1872)

vistaite *sed.* Term for varieties of jasper and agate that, when cut properly, show patterns resembling landscapes and other scenes, e.g., jasper rocks from Ahalt Creek area, Prineville, Crook County, Oregon. [vista, alluding to the scenes presented by the rock]

viterbite *ig. extru.* Leucite trachyte containing labradorite in addition to potassic feldspar; the abundant leucite phenocrysts in an ash-gray, aphanitic groundmass give the rock a spotted appearance, whence the popular Italian name *occhio di pesce,* fish eye. [Viterbo, Ciminian district, Italy] (Washington in 1906)

viterbite *sed.* Compact, chocolate-colored or white powdery material determined to be a mixture of allophane clay and wavellite. [Santa Rosa de Viterbo, Boyacá, Colombia] (Codazzi in 1925)

vitrain *sed.* In banded coal, the thin bands of bright, glassy-looking, jetlike material with a conchoidal fracture. Also named vitrite. [L. *vitreus, vitrum,* glass] (Stopes in 1919)

vitrophyre *ig. extru.* Volcanic glass containing phenocrysts; these rocks represent rhyolite glasses whose groundmasses did not crystallize and contain phenocrysts of quartz, glassy sanidine, minor clear plagioclase, and lesser biotite, hornblende, or augite. The most common variety is pitchstone vitrophyre, but obsidian vitrophyre, perlite vitrophyre, and pumice vitrophyre occur. If the composition differs from rhyolite, appropriate modifying terms must be used, e.g., basalt (or hyalobasalt) vitrophyre and dacite vitrophyre. [L. *vitreus, vitrum,* glass + porphyry] (Vogelsang in 1872)

vitrophyre, andesite *ig. extru.* Andesitic glass with phenocrysts; relatively rare.

vitrophyre, basalt *ig. extru.* 1. Basalt with dominant vitreous matrix (hyalobasalt) and some subordinate phenocrysts. 2. Basalt glass without phenocrysts; a misnomer, better named hyalobasalt.

vitrophyre, dacite *ig. extru.* Dacitic glass with phenocrysts, commonly plagioclase, quartz, amphibole, and biotite.

vitrophyre, diabase *ig. aschist.* Glassy phase of diabase containing phenocrysts and occurring at the borders of diabase intrusions.

vitrophyre, nepheline *ig. extru.* Glassy nephelinite showing phenocrysts.

vitrophyre, rhyolite *ig. extru.* = vitrophyre

vitrophyride *ig. extru.* General field term for any volcanic glass containing phenocrysts. [vitrophyre] (Johannsen in 1931)

vltavine *tekt.* Name proposed for the tektites traditionally called moldavite; it was suggested that the derivation of the name be from the Czech term Vltava River, rather than from the German Moldau River; because the earlier term has been in use for over 100 years, a change in terminology would only lead to confusion. Also spelled vltavite. [Vltava River (Moldau River) of central Bohemia (Czechoslovakia)]

vogesite *ig. diaschist.* Lamprophyre related to syenite, consisting of phenocrysts of hornblende or diopside in a groundmass of the same minerals with orthoclase and minor plagioclase, quartz, apatite, and rare olivine. [Vosges Mountains, France] (Rosenbusch in 1887)

vogesite, pilite *ig. diaschist., altered.* Vogesite in which former olivine has been altered to actinolite (pilite). [L. *pilus,* hair]

volcaniclastic rock *ig. extru., sed.* Clastic rock containing various amounts of volcanic material, without regard to its environment or origin.

volcanite *ig. extru.* 1. Porphyritic rock, occurring as volcanic bombs and containing abundant phenocrysts of anorthoclase, andesine, and augite, in a glassy groundmass containing micro-lites of feldspar and augite. [Volcano, Lipari Island, Italy] (Hobbs in 1893)   2. = vulcanite

volhynite   *ig. hypab.* Quartz kersantite containing phenocrysts of plagioclase and hornblende, with or without biotite, in a groundmass of quartz and feldspar (plagioclase and orthoclase) with abundant chlorite. [Volhynia, in the Soviet Union] (Ossovski in 1871)

vorhauserite   *meta.* Variety of retinalite (serpen-tinite), brown to greenish-black, occurs at Fleimsthal, Tyrol. [Johann Vorhauser, 1784-1865, of Innsbruck, Austria, expert on the minerals of Tyrol] (Kenngott in 1856/57)

vredefortite   *ig. pluton.* Granogabbro porphyry, consisting of phenocrysts of labradorite, hypersthene, and minor biotite, in a micrographic groundmass (46% of rock) composed of quartz and potassic feldspar. [Vredefort, south of Pretoria, Union of South Africa] (Niggli in 1936)

vulcanite   *ig. extru.* General name for any extru-sive or volcanic rock. Also spelled vulkanite and volcanite. [Vulcan, Roman god of fire] (Scheerer in 1864)

vullinite   *meta.* Schistose rock consisting of orthoclase, oligoclase, diopside, hornblende, biotite, epidote, titanite, magnetite, and apatite, apparently formed by the contact metamorphism of a calcareous sediment by aplite dikes. [Aultivullin, brook in northwestern Scotland] (Shand in 1910)

vulpinite   *sed.* Scaly, granular anhydrite rock, from Vulpino in Lombardy, Italy; cut and polished for ornamental purposes. [Vulpino, Italy] (Ludwig in 1804)

vulsinite   *ig. extru.* Alkali trachyte; the extrusive equivalent of a nepheline-bearing syenite containing a calcic-plagioclase. [Vulsinii, name of the Etruscan tribe that inhabited the region of Italy where the rock occurs] (Washington in 1896)

vulsinite-vicoite   *ig. extru.* Leucite-poor trachytic vicoite, composed of sanidine (40%), plagioclase (34%), mafics (aegirine-augite, hornblende, titanaugite, biotite), and leucite (9%); from Urujon Island, Korea. (Tsuboi in 1920)

W

wabanite   *meta.* Banded, cream to black, gray to purple, or chocolate-colored slate, from Massachusetts, sometimes used as a decorative stone. [Possibly Waban, part of Newton, Massachusetts]

wachenrodite   *sed.* Variety of wad that contains up to 12% lead, from Baden, Germany. [Heinrich Wilhelm Ferdinand Wachenroder, 1798-1854, German chemist and pharmacist of Jena, who first analyzed it] (Adam in 1869)

wacke   *sed.* 1. Impure sandstone consisting of a mixture of angular and unsorted or poorly sorted mineral (quartz and feldspar) and rock fragments and of an abundant matrix of clay and fine silt (often specified as 15-75%); major category of sandstone. Varieties include arkosic w., feldspathic w., lithic w., quartz w., and others.   2. Often a shortened form of graywacke but not considered a synonym.   3. Originally applied to a soft, earthy basalt, or to the dark claylike residue resulting from the decomposition of basalt, basaltic tuff, and related igneous rocks. [G. *Wacke,* variously defined provincial miners' term for a rock]

wacke, arkosic   *sed.* Sandstone composed of abundant quartz, more than 10% argillaceous matrix, and more than 25% feldspar and characterized by an abundance of unstable materials in which the feldspar grains exceed the fine-grained rock fragments.

wacke, feldspathic   *sed.* Sandstone containing abundant quartz, more than 10% argillaceous matrix, and 10-25% feldspar (usually sodic-plagioclase) and characterized by an abundance of unstable materials in which the feldspar grains exceed the fine-grained rock fragments; less feldspathic than arkosic wacke.

wacke, lithic   *sed.* Sandstone containing abundant quartz, more than 10% argillaceous matrix, and more than 10% feldspar (usually sodic-plagioclase) and characterized by an abundance of unstable materials in which the fine-grained rock fragments exceed feldspar grains.

wacke, lithic-arkosic   *sed.* Graywacke in which feldspar exceeds rock particles. Feldspathic graywacke is a synonym.

wacke, lithic-subarkosic   *sed.* Wacke with nearly equal proportions of feldspar and rock fragments but not more than 25% of either.

wacke, quartz   *sed.* Sandstone containing up to 90% quartz, and with more than 10% argillaceous matrix, less than 10% feldspar, and less than 10% rock fragments; similar to subgraywacke.

wacke, quartz-free   *sed.* Wacke with more than 90% unstable materials.

wacke, subarkosic   *sed.* Wacke with over 15% matrix and with 5-25% feldspar. Essentially equal to feldspathic wacke and subfeldspathic-lithic wacke.

wacke, subfeldspathic-lithic   *sed.* Lithic wacke containing less than 10% feldspar; essentially the same as subarkosic wacke.

wacke, volcanic   *ig. extru., sed.* Lithic wacke composed chiefly of pyroclastics derived from andesitic and basaltic volcanic rocks and having a low quartz content.

wackestone   *sed.* Mud-supported carbonate sedimentary rock containing more than 10% particles with diameters greater than 20 microns, e.g., calcarenite. (Dunham in 1962)

wad   *sed.* 1. Dull, bluish- or brownish black, soft, earthy to compact masses, containing mixtures of manganese oxides with oxides of other metals, including iron, cobalt, and copper. Also spelled wadd. Also named black ocher and bog manganese.   2. Obsolete name applied to graphite. [English term of obscure origin] (used as early as 1614)

walchowite   *sed.* Waxy, honey-yellow, variety of retinite, occurring as nodules in coal and sandstone of the Cretaceous rocks at Walchow (Valchov), Moravia, Czechoslovakia. Also spelled valchovite. (Haidinger in 1843)

wallongite   *sed.* = wollongongite

warrenite   *sed.* General name for a large group of gaseous and liquid bitumens consisting mainly of a mixture of paraffins and isoparaffins; variety of petroleum rich in paraffins. Also known as Pennsylvania oils. [Cyrus More Warren, 1824-1891, American chemist who did research on volatile hydrocarbons and asphalt for roofing and paving] (Peckham in 1895) wascoite   *sed.* Brown, jasper-rich, rock showing a large variety of patterns closely resembling petrified wood; apparently derived from a silica-rich sediment from ancient Lake Wasco, Oregon; contains fossil fish. [Wasco County, Oregon]

Washita stone   *sed.* Porous, uniformly textured, novaculite, found in the Ouachita (Washita) Mountains, Arkansas, and used for sharpening tools. Also named Ouachita stone.

waterstone   *sed.* 1. Term used in England for stratum whose surface has the appearance of watered silk, which is generally understood to express the water-bearing quality of the rock; specifically, micaceous sandstones and marls in the Keuper of the English Midlands, from which some water is available.   2. Stone used in grinding whose cutting fragments break away rapidly from their bond; forms a gritty paste with the addition of water.

websterite   *ig. pluton.* 1. Pyroxenite composed of reddish brown hypersthene and greenish black diallage in approximately equal amounts. [Webster, North Carolina] (Williams in 1890)   2. *IUGS,* plutonic rock with M equal to or greater than 90, ol/(ol + opx + cpx) less than 5, and both opx/(ol + opx + cpx) and cpx/(ol + opx + cpx) less than 90.

websterite, olivine   *ig. pluton. IUGS,* plutonic rock with M equal to or greater than 90, 5-40 ol/(ol + opx + cpx), opx/(ol + opx + cpx) greater than 5, and cpx/(ol + opx + cpx) greater than 5.

wehrlite   *ig. pluton.* 1. Peridotite consisting of olivine and clinopyroxene (diallage in original specimens). [Alois Wehrle, 1791-1835, Austrian mineralogist and chemist, who first analyzed it] (von Kobell in 1838)   2. *IUGS,* plutonic rock with M equal to or greater than 90, ol/(ol + opx + cpx) is 40-90, and opx/(ol + opx + cpx) less than 5.

weidgerite   *sed.* = wiedgerite

weigelith   *ig. pluton.* Peridotite composed of olivine, hornblende, and enstatite; in the original rock, much of the enstatite was altered to actinolite. [Weigelsberg, north of Ebersdorf, Moravia (Czechoslovakia)] (Kretschmer in 1917)

weiselbergite   *ig. extru.* Augite andesite showing phenocrysts of plagioclase and augite, rarely of enstatite, in a groundmass of augite needles, plagioclase microlites, and brownish glass. [Weiselberg, near St. Wendel, Germany] (Rosenbusch in 1887)

wennebergite   *ig. extru.* Quartz-bearing porphyry consisting of phenocrysts of sanidine, biotite, and quartz, in a groundmass of oligoclase, quartz,

apatite, and titanite. [Wenneberg, Ries, Bavaria, Germany] (Schowalter in 1904)

wesselite   *ig. hypab.* Nephelinite composed of nepheline, biotite (anomite), barkevikite, titanaugite, hauyne, and analcime. [Wesseln, Bohemia (Czechoslovakia)] (Scheumann in 1922)

westerwaldite   *ig. extru.* Trachybasalt consisting of phenocrysts of serpentinized olivine, often surrounded by a zone of augite, in a groundmass of labradorite, sanidine, augite, and biotite; nepheline occurs in the spaces between the minerals. [Westerwald, Germany] (Johannsen in 1938)

wetherilite   *sed.* Bitumen that fuses with difficulty; from Canada. [Charles Mayer Wetherill, 1825-1871, American chemist] (Danby in 1913)

wheelerite   *sed.* Yellowish resin, filling fissures or forming interstratified layers in lignite in the Cretaceous beds of northern New Mexico, [George Montague Wheeler, 1842-1905, American soldier and engineer] (Loew in 1874)

whetstone   *sed.* Any hard, fine-grained, naturally occurring rock, usually siliceous, used for whetting the edges of knives, axes, and other instruments. [O.Eng. *hwettan,* to sharpen]

whinstone   *ig.* Popular British name for any dark-colored igneous rock, such as dolerite, diabase, basalt, and andesite, that is comparatively unaltered and fine-grained. In some instances, term also includes altered types (greenstone, spilite) and then is nearly equivalent to trap. [Whin sill, northern England]

whitleyite   *meteor.* Enstatite achondrite meteorite containing some fragments of black chondrite. [Whitley County, Kentucky]

wiborgite   *ig. pluton.* = rapakivi. [Wiborg, southern Finland] (Wahl in 1925)

wichtisite   *ig. aschist.* Glassy diabase (hyalodiabase) occurring as the selvage of a diabase dike in granite. Also named wichtyn. [Kukkaron, parish of Wichtis, Finland] (Laurent in 1835)

wichtyn   *ig. aschist.* = wichtisite

wiedgerite   *sed.* Trade name for a soft, liver brown bitumen resembling elaterite but containing much sulfur and water. [Origin unknown] (pub. by Day in 1909)

wildmigmatite   *migmat.* Migmatite with a schlieren structure indicating that rotations and torsions have occurred; as it formed, it approached magmatic conditions with a high degree of mechanical mobility. [wild + migmatite] (Berthelsen, Bondesen, and Jensen in 1962)

wilkinite   *sed.* Trade name for a highly colloidal bentonitic clay, occurring in certain parts of the intermountain region of the United States. Also spelled wilkonite. Also named jelly rock. [Possibly Wilkins, Elko County, Nevada] (pub. by Wells in 1920)

wilkite   sed. Local name for a delicately colored yellow, purple, pink, and green variety of jasper, found near Willow Creek, north of Eagle, Ada County, Idaho. A preferred name is Willow Creek jasper. [Willow Creek]

williamsite   meta. Lamellar, yellow to apple-green serpentinite resembling nephrite in appearance; used for decorative purposes; originally from Texas, Pennsylvania. Incorrectly spelled williamsonite. [Lewis White Williams, 1804?-1873, of West Chester, Pennsylvania, who discovered it] (Shepard in 1848).

wilsonite   ig. extru. = owharoite. The name owharoite was substituted because of a possible conflict with the mineral named wilsonite (a decomposed scapolite). [Possibly Wilsons Bay, Coromandel Peninsula, Auckland, New Zealand] (Sollas in 1905)

winchellite   ig., altered. = lintonite. [Newton Horace Winchell, 1839-1914, American geologist, University of Minnesota, who made an optical study of the material]

windkanter   sed. Ventifact, usually highly polished, and bounded by one or more smooth faces that intersect in one or more sharp edges. Depending upon the number of sharp edges, varieties include einkanter (one edge), zweikanter (two edges), and dreikanter (three edges). [G. Wind, wind + Kante, edge, corner]

windsorite   ig. diaschist. Normal granite aplite, composed of orthoclase, microperthite, calcic-oligoclase, quartz, and biotite. [Windsor, Vermont] (Daly in 1903)

winkworthite   sed. Colorless to white nodules, crystalline on fracture and consisting of a mixture of howlite and gypsum; found in gypsum at Winkworth, Nova Scotia. [Winkworth] (How in 1871)

wocheinite   sed. = bauxite. [Wochein, Carniola (now in Yugoslavia)] (Flechner in 1866)

wolgidite   ig. extru. Leucitite containing leucite, titanian potassian richterite (magnophorite), diopside, and minor amounts of olivine and phlogopite. [Wolgidee Hills, Western Australia] (Wade and Prider in 1940)

wollastonite rock   meta. Wollastonite-rich rock, formed from thermally metamorphosed impure limestones; various amounts of calcite, garnet, vesuvianite, diopside, titanite, feldspars, and other minerals, may be present. Also named wollastonitfels. A variety includes wollastonite gneiss.

wollongongite   sed. Carbonaceous shale (kerosine shale) occurring in blocks, without lamination, but with a conchoidal fracture, in New South Wales, Australia. Also spelled wollongite. [Wollongong, New South Wales, Australia] (Silliman in 1869)

wonderstone   ig. extru., altered. Porous, fine-grained, rhyolite showing bands of strongly contrasting colors due to successive waves of infiltration of water solutions containing oxides of iron and manganese; esp. common in the American West, in Montana, Nevada, Utah, New Mexico (elixirite), and California.

woodendite   ig. extru. Dull, black, largely glassy rock related to macedonite, showing, under the microscope, augite, enstatite, magnetite, ilmenite, and altered olivine, in a brown glassy matrix. [Woodend, Macedon district, Victoria, Australia] (Skeats and Summers in 1912)

woodstone   sed. Fossilized wood. Terms like wood agate, wood jasper, and wood opal can also be used. See discussions under petrified wood and silicified wood.

wrigglite   meta. Dark, fine-grained skarn, showing contorted rhythmic laminar patterns of alternating light and dark bands (0.5 mm wide); composed of magnetite (dark bands) and vesuvianite and fluorite $\pm$ adularia, pyroxene, garnet) (light bands); applied esp. to rocks from Mt. Garnet region, North Queensland, Australia. Also named ribbon-rock tactite, ribboned tactite, rhythmically banded rock, or apo-carbonate greisen. [wriggle] (Askins in 1975)

wulstdiabase   ig., meta. Variety of spilite with a pillow structure; derived from pillowy lava. [G. Wulst, roll, bulge, hump + diabase] (Brauns in ca. 1906)

wurtzilite   sed. Massive, black or light brown, elastic asphaltite, used for waterproof and insulating paints; from veins in Utah. [Henry Wurtz, 1828-1910, American mineral chemist of New York, employed by Thomas A. Edison in 1888] (Blake in 1889)

wyomingite   ig. extru. Lamproite containing phlogopite phenocrysts in a fine-grained matrix composed of leucite, diopside, and glass; phlogopite-leucite trachyte. [Boar's Tusk, Leucite Hills, Wyoming] (Cross in 1897) xenogenite *ig., hydrotherm., sed.* General term for any rock or mineral deposit of later origin than its enclosing wall rock. [Gk. *xenos,* guest, foreign + *genes,* born] (Posepny in 1893)

xenolith *ig., sed., meta.* Fragment of a foreign rock or of an earlier crystallized portion of the same magma, enclosed in igneous rock. [Gk. *xenos,* guest, foreign + *lithos,* stone] (Sollas in 1894)

xerasite *ig., meta.* Old name for a porphyritic greenstone or an aphanite. [Gk. *xeros,* dry] (Haüy in 1822)

xylanthite *sed.* Fossil resin. [Gk. *xylon,* wood] (Groth and Mieleitner in 1921)

xylith *sed.* Variety of lignite composed almost entirely of anthraxylon, vitreous coal materials derived from woody tissues of plants. [Gk. *xylon,* wood]

xylopal *sed.* Wood that has been replaced by opal. Also spelled xilopal. See silicified wood.

xyloretine *sed.* Fossil resin occurring with the remains of coniferous stems in peat and brown coal deposits. Also xyloretinite. [Gk. *xylon,* wood + *rhetine,* resin] (Forchhammer in 1839)

xylovitrain *sed.* Variety of vitrain. [Gk. *xylon,* wood + vitrain]

Xenolith of hornblendite in pink granite. Locality unknown. 7.3 cm.

Y

yakatagite *sed.* Conglomeratic sandy mudstone poorly indurated and containing angular gravel-sized fragments that show its glaciomarine origin. [Yakataga, southeastern Alaska] (Miller in 1953)

yalakomite *meta.* Local name for a decorative stone composed of magnesite and quartz, with an attractive coloration imparted by pink jasper and green chromian phengite (mariposite); related to the serpentinization of ultrabasic rocks in the same region of British Columbia; similar to listvenite. [Yalakom River, northwest of Lillooet, British Columbia, Canada]

yamaskite *ig. pluton.* Pyroxenite related to jacupirangite; medium to fine-grained, dark, and containing pyroxene (titanaugite), hornblende, iron oxides, and small and varying amounts of anorthite. [Mt. Yamaska, Quebec, Canada] (Young in 1906)

yatalite *ig. diaschist.* Coarse-grained, pegmatitic rock, composed of uralite (actinolite after diopside), titaniferous magnetite, and albite, with lesser amounts of microcline, titanite, quartz, and apatite. [Hundred of Yatala, Houghton district, South Australia] (Benson in 1909)

yellow ground *ig. pipe.* Weathered kimberlite of yellowish color, found at the surface of diamond pipes above the fresher kimberlite (blue ground), esp. in South Africa. Miners' term.

yentnite *ig. pluton.* Rock originally considered to be diorite containing sodic-plagioclase, biotite, and scapolite; later shown to contain quartz, rather than scapolite, so it is actually quartz-bearing diorite. [Yentna River, Alaska] (Spurr in 1900)

yogoite *ig. pluton.* Obsolete term originally proposed for syenite with about equal amounts of orthoclase and augite; in addition to these minerals (each about 20%), also contains andesine, biotite, hornblende, and other minor minerals. [Yogo Peak, Little Belt Mountains, Montana] (Weed and Pirsson in 1895)

yosemitite *ig. pluton.* The normal granite forming El Capitan, Yosemite Park, California. [Yosemite Park] (Niggli in 1923)

youngite *sed.* Brecciated chalcedony-jasper rock, consisting of fragments of pastel red and yellow jasper cemented by white to gray chalcedony, and occurring northwest of Guernsey, Platte County, Wyoming. [Rex Emmett Young, 1891-1950, American businessman and mineral collector of Torrington, Wyoming, who discovered the deposit in 1939]

youstone *meta.* Early British name for jade. [Probably Chin. *yù,* jade]

Yowah nuts *sed.* Ironstone concretions, up to several inches in diameter, that may contain precious opal, either in the central portion, or as concentric rings, or as veins radiating toward the periphery. [Yowah station, West Queensland, Australia]

yù *meta.* Chinese word for jade or for any other very precious stone. Also spelled yuh.

yù yen stone *meta.* Massive, greenish gray serpentinite resembling jade.

yukonite *ig. diaschist.* Quartz diorite aplite, composed of oligoclase and quartz, with minor biotite, calcite, apatite, and iron oxides. [Fort Hamlin, Yukon River, Alaska] (Spurr in 1904)

Z

zaccab *sed.* Calcium-rich, white earth used by the natives of Yucatan, Mexico, for plaster and stucco.

zeasite *sed.* Term originally applied to a variety of fire opal but now applied to opalized wood. Also spelled zeazite. [Francisco Antonio Zea, 1770-1822, Colombian statesman and naturalist, envoy to France and England] (Larivière in 1826)

zebra rock *sed.* 1. Variety of dolomite showing narrow banding, consisting of alternating black layers (rich in organic matter) and white, coarse-grained, layers; occurs in the Colville district in northeastern Washington. 2. Fine-grained argillite (siliceous siltstone or claystone) that, when sliced properly, shows rhythmic patterns of red bands or spots contrasting with a white background; from Kununurra in the East Kimberley region of northwestern Australia. 3. General term for various rocks whose striped appearance resembles the markings on a zebra, e.g., zebra jasper. Also named zebra stone.

zeolite rock *sed., meta.* General term for rocks containing one or more of the zeolite group minerals. These include the following: sedimentary rocks (e.g., some sediments of saline lakes, marine sediments, and alkaline soils of semiarid regions); hydrothermally altered rocks (e.g., florinite, phillipsite tephrite, and tasmanite); very low-grade metamorphic rocks derived from pyroclastic and other sediments (e.g., zeolite tuff and zeolite metagraywacke).

zeyringite *sed.* Finely fibrous, greenish-white or sky-blue, calcareous (aragonite?) tufa, colored by aurichalcite (originally believed to be colored by nickel compounds); from Austria. [Zeyring, Styria, Austria] (Pantz in 1811)

zeyssatite *sed.* = randanite. More correctly spelled ceyssatite. [Ceyssat, Auvergne, France]

zietrisikite *sed.* = pietricikite. [Zietrisika (Pietricica), Moldavia] (Dana in 1868)

zillerite *meta.* Matted, fibrous asbestos whose composition is related to tremolite-actinolite; variety of mountain cork. Also named zillerthite. [Zillertal, Austria] (Fersmann in 1912)

zillerthite *meta.* = zillerite. (Delamétherie in 1795)

zirkelite *ig. extru., altered.* Altered basaltic glass; to be used in contrast with unaltered hyalobasalt; essentially equal to palagonite. [Ferdinand Zirkel, 1838-1912, German petrographer, University of Leipzig] (Wadsworth in 1887)

zittavite *sed.* Lustrous black lignite, harder and more brittle than dopplerite, which it resembles. [Zittau, Saxony] (Glöckner in 1912)

zobtenfels *ig., meta.* = zobtenite. [Zobtenberg, Silesia] (von Buch in 1797)

zobtenite *ig., meta.* Rock closely related to flaser gabbro, containing knots of diallage surrounded by uralite and embedded in a granular mass of epidote and plagioclase (saussuritized). Also named zobtenfels. [Zobtenberg, Silesia] (Roth in 1887)

zoisite rock *meta.* General term for any rock containing abundant zoisite; includes such rocks as biotite-zoisite schist, zoisite amphibolite, zoisite granulite, zoisitite, as well as some skarns and marbles. A bright green zoisite-rich rock, with red corundum (ruby), occurs near Longido, northeastern Tanzania.

zoisitite *meta.* Variety of prasinite rich in zoisite and amphibole and containing some muscovite and minor albite. [zoisite] (Riva in 1897)

zonite *sed.* Local term for jasper and chert specimens of various colors from Arizona. [Possibly Arizona]

zonochlorite *ig., altered.* Variety of pumpellyite occurring as rounded masses, filling amygdaloidal cavities (up to 2 in across) at Neepigon Bay, Lake Superior, Canada; composed of alternate layers of light-green and of dark green color; often cut as a gem; essentially equal to chlorastrolite. [Gk. *zone*, girdle, belt + *chloros*, green; alluding to the green, banded structure] (Foote in 1873)

zoogenic rock *sed.* Biogenic rock directly formed by the presence or activities of animals, e.g., shell limestone, coral reef, guano. Also named zoogenous rock. [Gk. *zoon*, animal + *genes*, born]

zoolith *sed.* Biolith formed by the presence or

activities of animals; zoogenic rock. [Gk. *zoon,* animal + *lithos,* stone]

zweikanter   *sed.* Wind-worn pebble (ventifact) having faces intersecting to form two sharp edges; a variety of windkanter. [G. *Zwei,* two + *Kante,* edge, corner]

zwitter   *ig. pluton., autometa.* Aphanitic, quartz-rich, dark-gray, fine-grained to dense rock usually containing chlorite, cassiterite, arsenopyrite, biotite, and quartz, although the composition is quite variable; related to greisen in origin. Zwittergestein is a synonym. Old Saxon miners' term. [G. *Zwitter,* hybrid, mongrel]

GLOSSARY

accessory mineral   Mineral whose presence does not define the root name of a rock; a minor accessory mineral has no bearing whatsoever on the name; a characterizing accessory may modify the root name, e.g., corundum in corundum syenite.

algae   Group of primitive, chiefly aquatic, chlorophyll-bearing plants that lack true stems, roots, and leaves; included are diatoms, seaweeds, kelps, pond scums, and stoneworts.

algon   Viscous, organic binding material (as in vase) that consists of finely divided remains of algae (or of land vegetation in some circumstances) and iron, usually in the form of FeS.

alkali   Said of silicate minerals and rocks containing alkali metals (sodium, potassium, and more rarely, lithium, rubidium, and cesium) but little calcium; in chemistry, any strongly basic substance, such as a hydroxide or carbonate of the alkali metals.

alkali amphibole   Division of the amphibole group in which the structural sites that can be occupied by Ca, Na, and K are predominantly occupied by Na, e.g., glaucophane, riebeckite, richterite, katophorite, eckermannite, and arfvedsonite.

alkali feldspar   Any one of several feldspars rich in potassium and/or sodium and in which the proportion of the anorthite molecule is less than 20%, e.g., orthoclase, sanidine, adularia, microcline, anorthoclase, and albite.

alkali pyroxene   Division of the pyroxene group in which the structural sites that can be occupied by Ca, Na, and Mg are predominantly occupied by Na, e.g., aegirine (acmite), aegirine-augite, and jadeite.

alkalic rock   Igneous rock containing more alkali metals (potassium and/or sodium) than is considered average for the group of rocks to which it belongs.

alkaline   See alkali.

allochem   One of several varieties of discrete carbonate aggregates that serve as the coarser framework grains in mechanically deposited limestones, as distinguished from the cement and mud matrix; included are silt, sand, gravel, ooliths, lumps, fossils.

alteration   Any change in the composition of a rock or mineral brought about by physical or chemical means, including weathering and hydrothermal processes.

alteration mineral   Secondary mineral derived from a primary mineral by alteration.

alteration product   The rock or mineral formed as a result of alteration; often referred to as secondary.

amygdaloidal   Said of a structural feature of volcanic rocks, where cavities, often almond-shaped, are filled entirely or in part with secondary minerals.

anaerobic   Said of conditions that can exist only in the absence of free oxygen; also said of an organism (e.g., bacterium) that can live in the absence of free oxygen.

anatexis   The process, deep within the crust, where masses of rock can be melted as a result of changes in the environment, esp. from elevated temperatures and pressures.

anhedral   Said of a mineral grain that has no crystal outline; not bounded by its own crystal faces.

aphanitic   Said of a rock texture (esp. igneous) in which the components are not distinguishable by the unaided eye; included are microcrystalline and cryptocrystalline rocks.

arenaceous Containing sand or sand-size fragments or having a sandy appearance or texture.

argillaceous Composed of or containing clay-size particles or clay minerals; of any other reference to clay, e.g., example, odor.

aschistic Said of rocks with the same composition as their parent mass, e.g., some dikes and extrusive rocks; they are undifferentiated.

assimilation The incorporation and digestion of fluid or solid foreign materials by magma.

augen Lenticular or rounded aggregates of minerals in gneisses and other metamorphic rocks that suggest eyes.

authigenic Said of a mineral, usually in sedimentary rocks, that crystallized from solutions within that environment; it was not brought in from outside as a particle.

autometamorphism The alteration of newly formed igneous rock by its own volatiles or residual liquids.

automorphic Said of a holocrystalline rock texture characterized by crystals showing their own faces; often used as a synonym for euhedral.

biogenic Said of any mineral or rock deposit resulting from the physiological activities of organisms.

blast-, -blast Prefix in metamorphic rock names to signify a relict texture, e.g., blastoporphyritic; used as a suffix in metamorphic rock names to signify a texture formed entirely by metamorphism, e.g., porphyroblast.

blastoporphyritic Said of a texture in a metamorphic rock derived from a porphyritic igneous rock, in which the porphyritic character remains as a relict feature.

botryoidal Said of a mineral or rock struc-

ture consisting of closely united spherical masses resembling a bunch of grapes.

brecciation The formation of a breccia by the crushing of an older rock into angular fragments.

calc-alkaline Said of igneous rocks in which the weight percentage of silica is 56-61 when the weight percentages of CaO and $K_2O + Na_2O$ are equal; illustrated by a variety of rocks including basalt, andesite, dacite, and rhyolite.

calcareous General term, of various rocks containing calcium carbonate.

calcic-plagioclase Collective term, for those members of the plagioclase feldspar series that fall between $Ab_{50}An_{50}$ and Ab_0An_{100}, including labradorite, bytownite, and anorthite.

cataclastic Said of certain metamorphic rocks having a fragmental texture caused by crushing during dynamic metamorphism.

characterizing accessory Accessory mineral abundant enough to modify the root name of the rock, e.g., corundum in corundum syenite and biotite in biotite granite.

chatoyant Having a changeable, undulating, or wavy color or luster; resembling the changing luster of the eye of the cat at night.

chondrule In meteorites, spheroidal granule or aggregate, usually about 1 m in diameter, consisting mainly of olivine or orthopyroxene, and occurring embedded more or less abundantly in a fragmental matrix; sometimes occurring singly in marine sediments.

clinopyroxene General name for any one of the monoclinic pyroxenes, e.g., augite and diopside.

color index The ratio of dark- to light-colored constituents of a rock; estimated by microscopic measurements or computed from the norm. For example, a color index greater than 40 is dark-colored, less than 40 is light-colored.

conchoidal fracture   Fracture that is curved and resembles a seashell; exhibited by obsidian.

concordant   Said of an igneous rock body that parallels the bedding or foliation into which it has intruded; also applied to sedimentary beds that are parallel and without angular junctions.

contact   The place or surface where two different kinds of rocks come together.

contact metamorphism   Metamorphism genetically related to the contact of an older rock with the intrusion (or extrusion) of an igneous magma.

crinoid   Any of various marine invertebrates, including sea lilies and feather stars, characterized by feathery, radiating arms and a stalk by which they are attached to a surface.

cryptocrystalline   Said of a rock consisting of grains or crystals that are too small to be recognized and separately distinguished even under the ordinary light microscope (although an electron microscope may reveal them).

crystallite   Minute body that does not react to polarized light, occurring in glassy igneous rocks; represents incipient crystallization and cannot be referred to definite mineral species.

crystalloblast   Crystal of a mineral produced entirely by metamorphic processes.

cumulate layering   Layering of igneous rock formed by the accumulation of crystals that settle out from a magma by the action of gravity.

detrital   Said of a rock made up of the debris (detritus) resulting from the mechanical disintegration of older rock.

devitrification   The conversion of glass to crystalline material by solid diffusion.

diagenesis   The chemical, physical, or biological changes that occur in a sediment after its initial deposition and during and after its lithification, exclusive of weathering and metamorphism.

diaschistic   Said of an intruded rock consisting of a differentiate with a composition differing from that of the parent magma.

diatom   Any of numerous minute unicellular or colonial algae, having siliceous cell walls consisting of two overlapping symmetrical parts.

differentiate   Igneous rocks formed as a result of the magmatic differentiation of a parent magma.

differentiation index   In igneous rocks, the number that represents the sum of the weight percentages of normative quartz, orthoclase, albite, nepheline, leucite, and kalsilite; a numerical expression of the extent of the differentiation of a magma.

dike   Tabular intrusion of igneous rock that cuts across the bedding or other layered structure of the country rock.

discordant   Said of an igneous intrusion that is not parallel to the foliation or bedding planes of the country rock, e.g., dike; in sedimentary rocks, said of strata lacking conformity or parallelism of bedding or structure.

drusy   Said of a rock or mineral surface that is rough due to a large number of small, closely crowded crystals.

essential mineral   Mineral whose presence is necessary before the root name of a rock can be applied, e.g., the essential minerals of a granite are quartz and potassic feldspar.

euhedral   Said of a mineral grain that is completely bounded by its own crystal faces.

exsolution   The process whereby an initially homogeneous solid crystalline phase separates into two (or possibly more) distinct crystalline phases without the addition or removal of material (without change in the bulk composition).

extrusive   Said of those igneous rocks that have cooled after their magma has reached the surface, e.g., in lava flows; also named effusive.

fabric   That part of a rock texture that depends on the shapes and mutual arrangements of the mineral constituents of a rock, e.g., ophitic and trachytic.

feldspathic   Said of a rock containing feldspar.

felsic   Said of the light-colored minerals in a rock, esp. igneous rocks; also applied to an igneous rock containing abundant light-colored minerals. [feldspar + lenad (feldspathoid) + silica]

felted   Said of a matted, felty rock fabric.

fenitization   The widespread alkali (sodium) metasomatism of quartzo-feldspathic country rocks in the environs of carbonatite complexes.

ferruginous   General term, of various rocks containing iron, e.g., hematite-bearing sandstone.

fetid odor   Odor of rotten eggs caused by the liberation of hydrogen sulfide, e.g., in some bituminous limestones.

fissile   Said of a rock capable of being split into thin, more or less papery, sheets.

foliated   Said of the planar arrangement of the textural or structural features of any type of rock; esp. of the planar structures exhibited in metamorphic rocks, such as gneiss, schist, phyllite, and slate.

foraminiferan   Any of several unicellular microorganisms characteristically having a calcareous shell with perforations through which numerous pseudopodia protrude.

friable   Said of a rock or mineral that is readily crumbled.

fritted   Said of a rock that is decomposed and partly fused or vitrified by heating.

frustule   The hard, siliceous (usually opaline) shell of a diatom.

geosynclinal   Pertaining to a geosyncline, a mobile downwarping of the Earth's crust, measured in scores of kilometers, in which sedimentary and volcanic rocks accumulate to thicknesses of thousands of meters.

granitization   The metasomatic process whereby a solid rock is converted or transformed into a granitic rock by the entry and exit of material, without passing through a magmatic stage.

granoblastic   Said of the texture of nonfoliated metamorphic rocks consisting essentially of equidimensional crystals with normally well-sutured boundaries; illustrated by most quartzite and marble.

granulation   In metamorphism, the fragmentation of minerals strained beyond their elastic limit; usually no visible openings result.

graphic intergrowth   Intergrowth of two crystals (commonly quartz and feldspar) that produces a fairly regular geometric outline and orientation resembling cuneiform writing.

groundmass   Usually, the finer matrix of a rock containing scattered coarser units, such as phenocrysts, metacrysts, pebbles, fossils.

hemipelagic   Said of a deep-sea sediment in which more than 25% of the fraction coarser than 5 microns is of terrigenous, volcanogenic, or neritic origin.

holocrystalline   Rock made up entirely of crystals (no glass).

holohyaline   Rock made up entirely of glass.

hydrothermal   Said of rock and mineral deposits produced by the action of hot water or precipitated from hot aqueous solutions; some writers restrict the water to that of magmatic origin.

hypabyssal   Said of shallow igneous intrusions, such as dikes and sills, as distinguished from surface volcanic rocks, on the one hand, and deep-seated plutonic rocks (like batholiths), on the other.

hypocrystalline   Rock made up partly of crystals and partly of glass; intermediate between holocrystalline and holohyaline and sometimes named hypohyaline.

igneous rock   Any rock formed from the solidification (usually crystallization) of a magma, a high-temperature, silicate-rich, fluid rock material.

impact crater   Crater formed on the surface of the earth, moon, and so forth, by the impact of an unspecified projectile, e.g., meteorite, asteroid, comet.

indurated   Said of a sediment or similar rock material hardened or consolidated by pressure, cementation, or heat.

interstitial material   The finer materials that fill the pores or voids between larger materials of the host rock.

intraformational   Formed within a geological formation, more or less contemporaneously with the enclosing sediments.

intrusive   Said of igneous rock bodies that have solidified below the surface of the earth; they usually have penetrated into or between older rocks while in the magma stage.

isotropic   In crystal optics, said of those substances whose indices of refraction do not vary with crystal direction; said of cubic and amorphous substances.

karst   Type of topography formed on limestone, gypsum rock, and other soluble rocks by dissolution and characterized by sink holes, caverns, and underground streams, interspersed with abrupt ridges and irregular protuberant rocks.

kelyphitic rim   Rim composed of pyroxene, hornblende, or other minerals that sometimes surrounds the garnets of peridotites and other igneous rocks; less commonly, similar rims appearing around olivine.

lacustrine   Said of deposits that form in the bottom of lakes.

lamellar   Composed of thin layers, scales, or plates, like the leaves of a book.

lath   Crystal habit that is long and thin and of moderate to narrow width; in rock thin sections, lath-shaped minerals are usually cross sections of tabular or platy crystals, e.g., plagioclase plates.

lithic   Of or pertaining to stone.

lithified   Said of the consolidation of sediments or other materials into stone.

macrocrystalline   Rock in which the constituents are distinctly visible to the unaided eye.

mafic   Said of the dark-colored minerals in a rock, esp. igneous rocks; also applied to an igneous rock containing abundant dark-colored minerals. [magnesium + ferrum]

matrix   The surrounding substance within which a mineral, fossil, or other feature is contained.

mechanically deposited   Said of a sediment deposited through the agency of physical processes (e.g., wind and water) without involving chemical processes or changes.

megascopic   Said of observations made on minerals and rocks, and their characteristics, by means of the unaided eye or pocket lens but not with a microscope.

metacryst   Well-developed crystal, like garnet or staurolite, resembling a phenocryst, embedded in the groundmass of a comparatively fine-grained metamorphic rock.

metamorphic rock   Rock derived from a preexisting rock by changes in texture or mineralogy in an essentially solid state, usually through processes involving increased temperatures, increased pressures, or chemically active fluids.

metasomatism   Process of practically simultaneous capillary solution and deposition by which a new mineral of partly or wholly different chemical composition may grow in the body of an old mineral or mineral aggregate.

miarolitic cavity   Opening in an igneous rock lined with euhedral crystals coarser than the grain of the rock around them but, for the most part, of similar minerals.

micrographic   Said of a graphic intergrowth of an igneous rock that is distinguishable only with the aid of a microscope; also applied to a rock having such a texture.

microlite   Microscopic crystal that polarizes light and has some determinable optical properties; these are a bit larger than crystallites.

micron   Unit of length equal to one-millionth of a meter; one-thousandth part of a millimeter.

mineral   Substance formed in nature without the direct influence of organisms that has a characteristic chemical composition and usually an ordered atomic arrangement, which is often expressed in external geometrical forms (crystals).

mineraloid   Naturally occurring substance that has distinctive properties but does not fully meet the requirement for a mineral; some are formed directly by organisms (amber, pearls, coal), while others are amorphous (obsidian) or are mixtures (bauxite).

mode, modal analysis   The actual mineral composition of a rock, usually expressed in weight or volume percentages.

monomineralic   Said of a rock composed of only one essential mineral, e.g., dunite (olivine), anorthosite (plagioclase), limestone (calcite), and serpentinite (a serpentine mineral).

neritic   Said of an ocean depth environment between low-tide level and 100 fathoms (or approximately the edge of the continental shelf).

nonpolarizing   Said of those isotropic substances (cubic crystals and amorphous) that do not polarize light that passes through them.

norm, normal analysis   The theoretical mineral composition of a rock expressed in terms of mineral molecules that have been determined by chemical analyses; the composition that might be expected had all the components crystallized under equilibrium conditions.

normative mineral   Mineral whose presence in a rock is theoretically possible on the basis of certain chemical analyses; it may or may not actually be present in the rock.

nuées ardentes   Swiftly flowing, turbulent gaseous cloud erupted from a volcano and containing ash and other fragments in its lower part.

opalescent   Said of a milky to pearly reflection exhibited by certain minerals in strong light, such as that shown by some feldspars (moonstone), some opals, and some agates.

ophitic   Said of a rock fabric consisting of interlacing thin plates (lathlike in cross section) of feldspar (usually plagioclase) whose interspaces are chiefly filled by pyroxene of later growth.

orbicular   Containing spheroidal aggregates of megascopic crystals, generally in concentric shells composed of two or more of the constituent minerals of the rock matrix; usually said of igneous rocks, e.g., corsite.

orthopyroxene   General name for any one of the orthorhombic pyroxenes, e.g., enstatite, bronzite, and hypersthene.

pelitic   Composed of fine, argillaceous sediment or clay.

penecontemporaneous   Existing or formed at almost the same time.

petrography   The systematic description of rocks based on observations in the field, on hand specimens, and on thin sections; two major divisions are megascopic (or macroscopic) petrography and microscopic petrography.

petrographic   The adjectival form of petrography but usually restricted to microscopic work.

petrologist One who is engaged in the study of rocks (petrology).

petrology The study or science of rocks, dealing with the origin, occurrence, present conditions, alterations, classification, and many other aspects of rocks.

phaneritic Said of a rock texture (usually igneous) in which the components are distinguishable by the unaided eye; the rock is macrocrystalline.

phanerocrystalline Said of an igneous rock in which all the crystals of the essential minerals can be distinguished individually by the unaided eye; essentially equal to macrocrystalline and phaneritic.

phenoclast Large fragment in a nonuniformly sized sediment, e.g., a pebble in a pebbly sandstone.

phenocryst Large crystal in a nonuniformly sized igneous rock, such as a crystal of plagioclase in an andesite porphyry; these are the conspicuous crystals of the earliest generation in a porphyritic igneous rock.

phyllosilicate General term for those several silicate minerals that have sheetlike structures; their SiO_4 tetrahedrons are linked together to form continuous sheets, and they have a Si:O ratio of 2:5. Examples include the micas, the chlorites, talc, and most of the clay minerals.

plagioclase feldspar Series of six triclinic feldspars whose compositions vary isomorphously from albite, $NaAlSi_3O_8$ ($Ab_{100}An_0$) to anorthite, $CaAl_2Si_2O_8$ (Ab_0An_{100}). See calcic-plagioclase and sodic-plagioclase.

pleochroism Variation in the color of a single crystal of a mineral, resulting from the differential absorption of white light (usually polarized) in different crystallographic directions.

plumose Said of a feathery structure exhibited by some mineral masses.

pluton Igneous rock body that has formed far below the surface of the earth by the crystallization of a magma.

plutonic Said of rocks of igneous origin; usually applied to those igneous rocks that have crystallized at great depth and have granular textures.

poikilitic Said of a texture feature found in igneous rocks in which small grains of one mineral are irregularly scattered, without common orientation, in a typically larger crystal of another mineral, e.g., grains of plagioclase in a large single crystal of pyroxene.

poikiloblastic In nonigneous rocks, said of a texture formed by the crystallization of a large new mineral around numerous grains or relics of older minerals, thus simulating the poikilitic texture of igneous rocks; the term usually applies to metamorphic rocks (e.g., quartz and mica grains in staurolite) but has also been applied to sedimentary sand crystals where a calcite crystal may be filled with quartz sand grains.

porphyritic Igneous texture term, said of those rocks in which larger crystals (phenocrysts) are set in a finer groundmass, which may be crystalline, glassy, or both.

porphyroblast Metacryst

potassic feldspar Alkali feldspar containing a significant amount of potassium, e.g., orthoclase and microcline.

primary mineral Mineral that was deposited in the original rock-forming episode; to be contrasted with secondary mineral.

protozoan Any of the single-celled, usually microscopic organisms, including the most primitive forms of animal life.

proxied by Said of those situations, esp. in igneous rocks, where one mineral may substitute for a more common mineral of similar nature in a given rock, e.g., where one mafic may proxy (substitute) for another mafic, or where one feldspathoid may proxy for another feldspathoid. Mica malchite is an example.

proximal Next to or nearby; near the central part or point of origin.

pseudomorph Mineral whose outward form (usually crystal shape) is that of another mineral species; these develop by alteration, substitution, incrustation, paramorphism, and metamictization.

pyribole Term indicating pyroxene and/or amphibole; used esp. in reference to the constituents of igneous rocks.

quartzo-feldspathic General term, said of rocks containing quartz and feldspars, such as granite, rhyolite, arkose, gneiss, and others.

radiolarian Any of various marine protozoans, having rigid siliceous (opaline) skeletons and spicules.

residual rock The unconsolidated or partly weathered parent material of a soil, presumed to have developed in place, by weathering, from the consolidated rock on which it lies.

resorption The re-fusion or solution, by and in a magma, of previously formed crystals or minerals with which the magma is not now in equilibrium or, owing to changes of temperature, pressure, or chemical composition, with which it has ceased to be in equilibrium.

retrograde metamorphism Metamorphic adjustment in which metamorphic minerals of a lower grade are formed at the expense of minerals characteristic of a higher grade of metamorphism, necessitated by a change in physical conditions, e.g., a lowering of temperatures.

rhombic General term, said of any mineral crystal having the shape of a rhombus; shape is often exhibited in rocks by feldspars and by the carbonates calcite and dolomite.

rock Naturally formed aggregate composed of one or more minerals or mineraloids, including loose incoherent masses as well as firm solid masses, and constituting an essential or appreciable part of the Earth or other celestial body.

saccharoidal Said of a granular texture resembling that of crystalline sugar.

saturated See silica-saturated.

schiller Bronzelike luster or iridescence due to internal reflections on minute inclusions in crystals, e.g., in the orthopyroxenes enstatite, bronzite, hypersthene.

schistose Said of a foliated metamorphic rock texture in which the individual folia are mineralogically similar and whose major minerals are phaneritic.

scoriaceous Said of the texture of a coarsely cellular volcanic rock that resembles scoria.

secondary mineral Mineral resulting from the alteration of a primary mineral, as a result of weathering, solutions, or metamorphism.

sedimentary rock Any rock consisting of chemical and mineralogical materials that are derived from the destruction of previous rocks (and more rarely, organisms); usually the materials are transported from their source and are deposited by the action of water, wind, or glacial ice in surface environments.

sessile Said of an animal or plant that is permanently attached to a substrate and cannot move about.

silica-oversaturated Said of an igneous rock formed from a magma that contains an excess of silica so that free quartz (or equivalent SiO_2 mineral) is in the rock.

silica-saturated Said of an igneous rock formed from a magma that contains just enough silica (SiO_2) so that there is neither free quartz (or equivalent SiO_2 mineral) nor silica-deficient minerals (e.g., feldspathoids).

silica-undersaturated Said of an igneous rock formed from a magma that contains a deficiency of silica, so the rock contains silica-deficient minerals that cannot exist in the presence of quartz (feldspathoids, olivine, and corundum).

siliceous   Said of a rock that contains abundant silica (SiO_2), usually as quartz or opal.

silicification   The introduction of, or replacement by, silica, often resulting in the formation of fine-grained quartz, chalcedony, or opal.

sill   Intrusive concordant sheet of igneous rock of approximately uniform thickness, which is relatively thin compared to its lateral extent.

sodic-plagioclase   Collective term for those members of the plagioclase feldspar series that fall between $Ab_{100}An_0$ and $Ab_{50}An_{50}$, including albite, oligoclase, and andesine.

sodium amphibole   See alkali amphibole.

sodium pyroxene   See alkali pyroxene.

sorted   Said of a sediment, either unconsolidated or cemented, consisting of particles of essentially uniform size or within the limits of a single grade.

spicule   Small, needlelike structure or part, such as one of the opaline processes supporting the soft tissue of certain invertebrates, esp. sponges.

spilitization   The albitization of a basalt to form a spilite.

spinifex   Said of a texture in igneous rocks (e.g., komatiite) consisting of an array of crisscrossing sheaves of closely spaced subparallel blades or plates (often skeletal) of olivine or other mafic.

stratified   Said of a rock displaying stratification where there has been an accumulation or deposition of material in horizontal layers; nearly synonymous with sedimentary.

stratiform   Having the form of a bed, layer, or stratum; consisting of roughly parallel bands of sheets.

stylolite   Highly irregular surface within a sedimentary rock that is typically marked by a thin seam of insoluble residue; in cross section it resembles a suture or the tracing of a stylus.

subaerial   Formed, produced, or deposited in the open air.

subaqueous   Formed, produced, or deposited beneath a body of water.

subhedral   Said of a mineral grain that is only partly bounded by its own crystal faces; intermediate between euhedral and anhedral.

tectonic   Said of rock structures and external forms resulting from the deformation of the earth's crust.

Tertiary period   The oldest of the two periods of the Cenozoic era, thought to have covered a span between 65 and 3-2 million years ago.

test   The shell or other hard supporting structure of many invertebrate animals; some are calcareous, and others are siliceous.

titanaugite   Variety of augite rich in titanium and occurring in silica-poor alkaline rocks, e.g., teschenite, essexite, nepheline dolerite, and others; more accurately named titanian augite.

trachytic   Said of an igneous-rock fabric feature in which neighboring feldspar plates (laths in cross section) of a microlitic groundmass have a sub-parallel disposition, corresponding to the stream lines of the nearly consolidated magma.

unconformity   Erosional break in the continuity of sedimentation; a substantial break or gap in the geologic record may result.

undersaturated   See silica-undersaturated.

vermicular   Wormlike; having the shape of a worm or with wormlike markings.

vesicular   Composed of or containing vesicles, small, bladderlike cells or cavities.

vitreous   Resembling or having the nature of glass; glassy.

vitrified   Changed or made into glass or a

similar substance, esp. through heat or fusion.

vitrophyric   Said of an igneous rock having phenocrysts in a glassy groundmass.

volcanic   Pertaining to, situated in or upon, formed in, or derived from a volcano.

vug   Unfilled cavity in a vein or rock, usually lined with a crystalline incrustation; often the lining has a different composition from the immediately surrounding rock.

vuggy   Said of a rock containing many small drusy cavities.